Mathematical Engineering

Today, the development of high-tech systems is unthinkable without mathematical modeling and analysis of system behavior. As such, many fields in the modern engineering sciences (e.g. control engineering, communications engineering, mechanical engineering, and robotics) call for sophisticated mathematical methods in order to solve the tasks at hand.

The series Mathematical Engineering presents new or heretofore little-known methods to support engineers in finding suitable answers to their questions, presenting those methods in such manner as to make them ideally comprehensible and applicable in practice.

Therefore, the primary focus is—without neglecting mathematical accuracy—on comprehensibility and real-world applicability.

To submit a proposal or request further information, please use the PDF Proposal Form or contact directly: Dr. Thomas Ditzinger (thomas.ditzinger@springer.com)

Indexed by SCOPUS, zbMATH, SCImago.

Tobias Köppl · Rainer Helmig

Dimension Reduced Modeling of Blood Flow in Large Arteries

An Introduction for Master Students and First Year Doctoral Students

Springer

Tobias Köppl
Computational Mathematics,
Faculty of Sciences
Hasselt University
Diepenbeek, Belgium

Rainer Helmig
Institute for Modelling Hydraulic
and Environmental Systems (IWS)
University of Stuttgart
Stuttgart, Germany

ISSN 2192-4732 ISSN 2192-4740 (electronic)
Mathematical Engineering
ISBN 978-3-031-33089-6 ISBN 978-3-031-33087-2 (eBook)
https://doi.org/10.1007/978-3-031-33087-2

This Springer imprint is published by the registered company Springer Nature Switzerland AG
The registered company address is: Gewerbestrasse 11, 6330 Cham, Switzerland

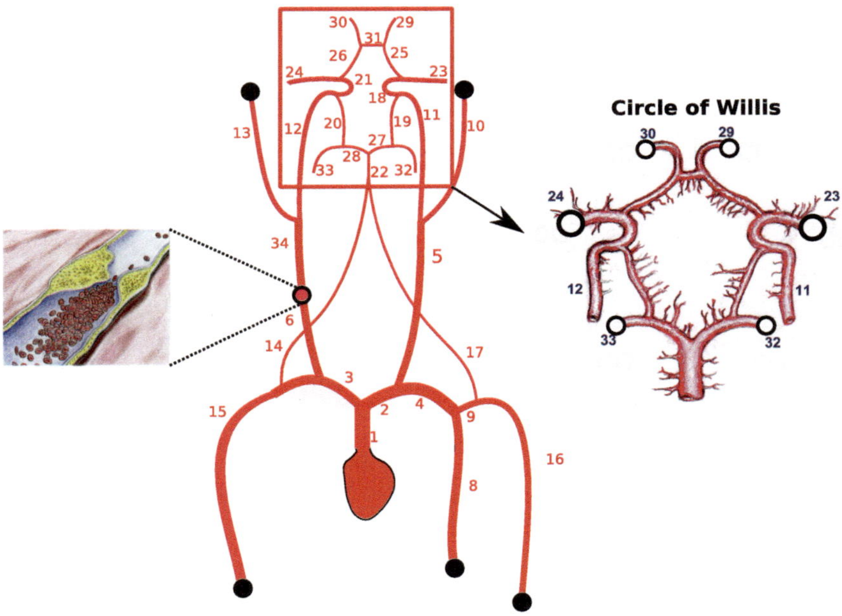

30
29
31
26
25
24
23
21
18
12
20
19
11
13
10
27
28
33
22
32
34
5
6
14
17
3
15
2
4
9
1
16
8
Circle of Willis
30
29
24
23
12
11
33
32

Nature is written in the language of mathematics.

—Galileo Galilei

Preface

Cardiovascular diseases are among the most common causes of death in Western industrialized countries. Thanks to the continuous improvement of computers and software, mathematical models and numerical simulations contribute significantly to the analysis and treatment of cardiovascular diseases. Since these methods are non-invasive, physiological processes can be studied and understood with less cost and risks for a patient. In addition, computer simulations enable the determination of medical data, which cannot be measured in a living human being for ethical and technical reasons. Knowledge that is obtained by means of these techniques can form a basis for the planning of operations and the production of drugs.

However, modeling the human cardiovascular system is very challenging, due to the high number of vessels, the complex composition of the vessels and the different types of blood flows. While convection-dominated flow is present in large vessels, diffusion-dominated flow can be observed in small vessels and tissue. Another difficulty is the interaction of flows with the vessel walls and surrounding tissue. The mathematical equations obtained by modeling these phenomena are coupled with each other and non-linear. For these reasons, numerical methods are required to compute a solution of these equations.

The submitted book has been developed based on a lecture, we designed for students of Simulation Technology at the University of Stuttgart. In this lecture, we started with some basics from physiology and developed a mathematical model for simulating blood flow in large arteries. Furthermore, the properties of suitable numerical solution methods were investigated. Thus, this book is ideal for master's students writing a thesis in bioengineering or applied mathematics as well as for Ph.D. students starting to work on a thesis in these fields. Reading this book, they get to know the core elements of Scientific Computing. These read as follows: Formulating an issue from an application field such as medicine, casting this issue into a mathematical model, analysis of this model, the choice of an appropriate numerical solution method and finally the numerical simulation.

We wish the reader much joy reading this book!

Hasselt, Belgium
Stuttgart, Germany
April 2023

Tobias Köppl
Rainer Helmig

Acknowledgements After finishing the work on this book, we would like to thank some persons who have supported us in terms of this project. First of all, we would like to thank Dr. Martin Schneider, who supported us in terms of the lecture SimTech A. Further, we thank StD a.D. Roland Köppl for the drawings of blood vessels and cardiovascular systems.

Finally, it has to be mentioned that the work on this book has been supported by the Research Foundation – Flanders (FWO) through the Odysseus programme (Project G0G1316N).

Contents

Acronyms

0D	Zero dimensional, ordinary systems of equations
1D	One dimensional
3D	Three dimensional
AVM	Arteriovenous malformation
CoW	Circle of Willis
FGF	Fibroblast Growth Factor
FSI	Fluid-Structure Interaction
i.e.	Abbreviation of id est, that means
mmHg	Medical unit for pressure, abbreviating the expression millimeter of mercury. 1 mmHg is equal to 133.322 Pa
NIDUS	Vessel system forming between the arteries and venes due to an AVM
ODE	Ordinary Differential Equation
PDE	Partial Differential Equation
RKM	Runge-Kutta method
TVD	Total variation diminishing
VEGF	Vascular Endothelial Growth Factor
WHO	World Health Organization
wlog.	Without loss of generality
w.r.t.	With respect to

Symbols

Density of blood: $\rho\left[\frac{kg}{m^3}\right]$
Dynamic viscosity: $\eta\ [Pa \cdot s]$
Kinematic viscosity: $\mu = \frac{\eta}{\rho}\left[\frac{m^2}{s}\right]$
Real numbers: $\mathbb{R}$
Positive real numbers: $\mathbb{R}_+$
Negative real numbers: $\mathbb{R}_-$
Real vector space of dimension n: $\mathbb{R}^n$
Laplace operator: Δ
Nabla operator: ∇
Shear stress: τ
Shear rate: $\dot{\gamma}$
Divergence of a vector field: div
Divergence of a tensor: **div**
Deformation tensor: **D**
Dot product of two vectors: $\langle v,\ w\rangle,\ v,\ w\ \in\ \mathbb{R}^n$
Normal derivative: $\frac{\partial}{\partial n} = \langle \nabla,\ n\rangle$
I-th component of a vector **v**: $\mathbf{v}_i$

Space of real functions that are k times differentiable: $C^k\ (\mathbf{\Omega})$, $\mathbf{\Omega}$ is the definition domain of the respective functions
Space of real functions that are infinitely often differentiable: $C^\infty\ (\mathbf{\Omega})$, $\mathbf{\Omega}$ is the definition domain of the respective functions
Space of real functions with compact support and infinite differentiability: $C_0^\infty\ (\mathbf{\Omega})$, $\mathbf{\Omega}$ is the definition domain of the respective functions
Space of real functions that are bounded almost everywhere: $L^\infty\ (\mathbf{\Omega})$, $\mathbf{\Omega}$ is the definition domain of the respective functions
Space of real L^p-functions, $p \in [1,\ \infty)$: $L^p\ (\mathbf{\Omega})$, $\mathbf{\Omega}$ is the definition domain of the respective functions: $f \in L^p\ (\mathbf{\Omega})$, if there is a constant $C > 0$ such that:

$$\left(\int_{\Omega} |f(x)|^p \, dx \right)^{\frac{1}{p}} < C.$$

Supremums norm of continuous functions: $f \in C(\mathbf{\Omega})$:

$$\|f\|_{C(\Omega)} = \sup_{x \in \Omega} |f(x)|.$$

Supremums norm of continuous differentiable functions: $f \in C^1(\mathbf{\Omega})$:

$$\|f\|_{C^1(\Omega)} = \sup_{x \in \Omega} |f(x)| + \sup_{x \in \Omega} \left| f'(x) \right|.$$

Partial derivative of a function w.r.t. z: $\frac{\partial u}{\partial z}$ or u_z
Jacobian matrix: Let **F** be a vector valued function:

$$\mathbf{F} : \mathbb{R}^n \to \mathbb{R}^n, \; \mathbf{U} \mapsto \mathbf{F}(\mathbf{U})$$

and let each component of **F** be smooth enough. Then the Jacobian matrix of **F** is denoted by $\frac{\partial \mathbf{F}}{\partial \mathbf{U}} \in \mathbb{R}^{n \times n}$:

$$\frac{\partial \mathbf{F}}{\partial \mathbf{U}} = \begin{pmatrix} \frac{\partial \mathbf{F}_1}{\partial \mathbf{U}_1} & \cdots & \frac{\partial \mathbf{F}_1}{\partial \mathbf{U}_n} \\ \vdots & \ddots & \vdots \\ \frac{\partial \mathbf{F}_n}{\partial \mathbf{U}_1} & \cdots & \frac{\partial \mathbf{F}_n}{\partial \mathbf{U}_n} \end{pmatrix}.$$

For all elements z in a set M: $\forall z \in M$
Kronecker symbol: $\delta_{jk} = \begin{cases} 1 \text{ if } j = k \\ 0 \text{ if } j \neq k \end{cases}$
Symbols for an oxygen molecule and a carbon dioxide molecule: O_2, CO_2
Sign function: $\text{sign}(\xi) = \begin{cases} +1, & \text{if } \xi \geq 0, \\ -1, & \text{if } \xi < 0. \end{cases}$
Reynolds number: *Re*
Péclet number: *Pe*
Equivalence sign: $\Leftrightarrow$
Logical Or: $\vee$
Logical And: $\wedge$

Chapter 1
Properties of the Cardiovascular System

Abstract In this chapter we explain basic terms from physiology. Moreover, the basic structure of the human vascular system and the mechanical properties of the human cardiovascular system are discussed. This knowledge forms the basis for the following chapters, which deal with the mathematical modeling of realistic blood flows and their simulation. At the end of this chapter, there is a brief overview on the most important cardiovascular diseases, whose diagnosis and curing can be improved by computer-assisted treatment methods.

1.1 The Cardiovascular System

At the center of the human circulatory system is the heart, which pumps oxygen-rich blood into the vascular network. Oxygen-rich blood flows through blood vessels to the organs, from which oxygen-poor blood is transported back to the heart. In order to add oxygen (O_2) to the deoxygenated and carbon dioxide (CO_2) enriched blood, the heart pushes blood back to the lungs (see Fig. 1.1). The lungs enrich the blood with oxygen from respiration and discharge CO_2 by the respiratory tract.

Thus, blood circulation can be divided into two parts: First, into a *pulmonary circulation* between the heart and the lungs and a *systemic circulation* between the heart and other organs (see Fig. 1.2). The vessels of both circulations, in turn, can be divided into three major groups: The first group is called *veins*, the second group *arteries* and the third group *capillaries*.

In the systemic circulation, veins transport deoxygenated blood from organs to the heart, and in the pulmonary circulation, they transport oxygenated blood from the lungs to the heart. The role of the arteries is vice versa: In the systemic circulation, blood enriched with oxygen is transported from the heart to the organs, while in the pulmonary circulation, deoxygenated blood flows from the heart through the arteries to the lungs. Capillaries are tiny blood vessels that connect arteries and veins. The exchange of substances between blood and tissue takes place in the capillaries. Thereby, oxygen is removed from the red blood cells and transported through the capillary walls to the cells, while CO_2 and other waste products are removed by the capillaries and veins. In the pulmonary circulation, the roles are again switched, i.e.:

T. Köppl and R. Helmig, *Dimension Reduced Modeling of Blood Flow in Large Arteries*, Mathematical Engineering, https://doi.org/10.1007/978-3-031-33087-2_1

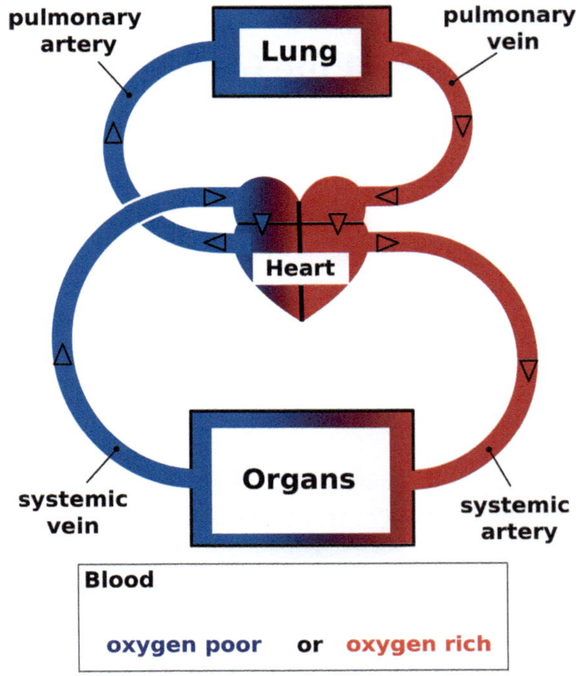

Fig. 1.1 Components of the blood circulation

In the capillaries, blood takes up oxygen and removes CO_2. The following sections are concerned with the most important components of the *systemic circulation* and their interaction, while the *pulmonary circulation* will not be considered in more detail.

1.1.1 Structure and Function of the Heart

The heart is a fist-sized muscle that has the shape of a rounded cone and is located behind the sternum, offset to the left. The tip of this cone points forward to the left. Unlike other muscles, this muscle contains cavities. In total there are four cavities. Two of these cavities form the left and right halves of the heart, respectively. The two halves are separated by a muscular wall, also called the cardiac septum. On both sides, the cavities are separated by heart valves, which can be opened and closed on demand (see Figs. 1.3 and 1.4).

These cavities are denoted as right and left atrium as well as right and left ventricle, respectively. The right and left cavities act as two synchronized pumps. While the right side of the heart is accumulating deoxygenated blood from the systemic circulation and pumping it to the lungs, the left side of the heart receives oxygenated blood from the lungs and pushes it with high pressure into the arterial system of the

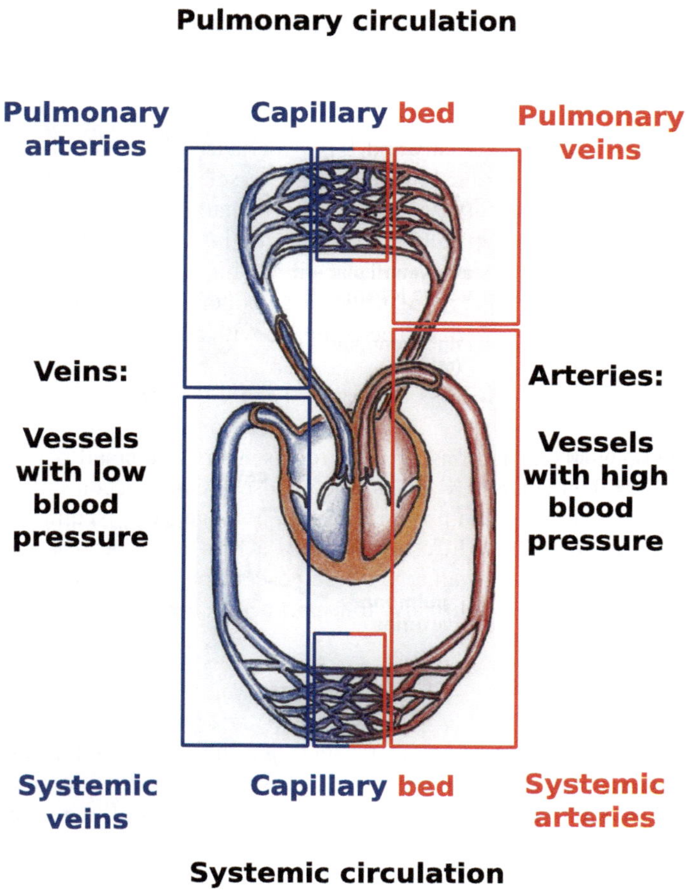

Fig. 1.2 Role of arteries and veins in blood circulation

systemic circulation. The right atrium is connected to the systemic circulation via the inferior and superior vena cava, while the left atrium is connected to the pulmonary circulation via pulmonary veins. On the other hand the two ventricles of the heart are connected to the sub-circulations via the aorta and pulmonary arteries. Both the aorta and the largest pulmonary artery are equipped with valves (semilunar valves) at their inlets, which open and close during the cardiac cycle.

The cardiac cycle can be divided into two phases: *diastole* and *systole*. During diastole, the ventricles are filled with blood. This is achieved by contraction of the atria and relaxation of the ventricles. During this phase, the heart valves (atrioventricular valves) between the atria and the ventricles are open to enable blood flow from the atria into the ventricles. As more and more blood accumulates in the ventricles, internal pressure is gradually built up. This increase in pressure is only possible since

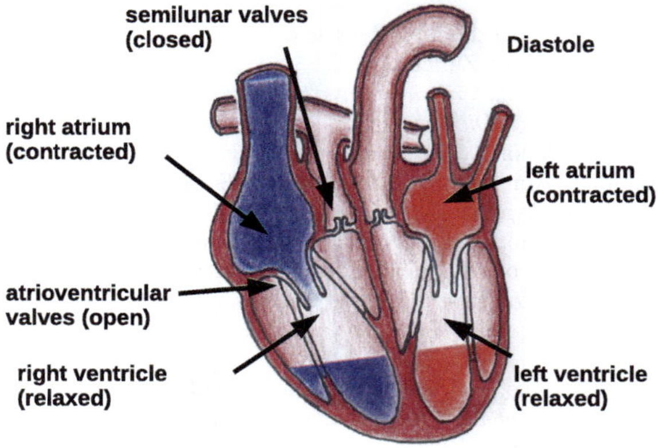

Fig. 1.3 Position of the heart valves during diastole

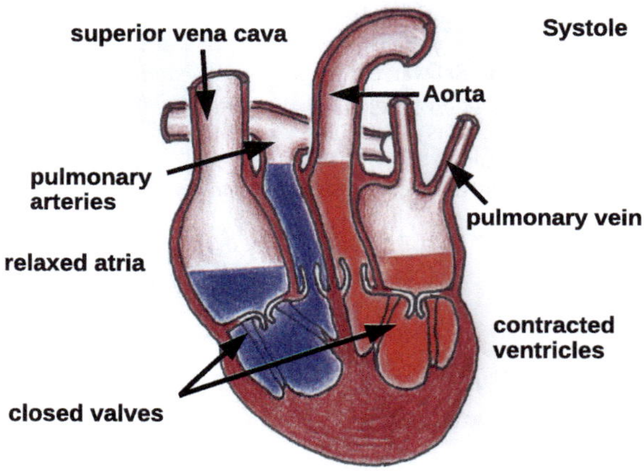

Fig. 1.4 Position of the heart valves during systole

the closing atrioventricular valves prevent blood from flowing back into the atria. During this phase, the other pair of heart valves, the semilunar valves are closed to prevent backflow of blood from the arteries and consequently a decrease of arterial blood pressure (diastolic blood pressure) occurs.

When the ventricular muscle contracts, the internal pressure rises very fast, such that the semilunar valves open and blood is pressed into the pulmonary and systemic arteries. The contraction of the heart muscle is also referred to as heartbeat. It can be recognized as a pulse wave in the larger arteries. After that the pressure in the heart chambers drops again and the heart valves at the outlets of the atria open again. Meanwhile, the atria have been filled again, due to a suction effect. The heart is therefore a suction-pressure pump. Thus it generates a pulsating flow that is propagating into large parts of the vascular system. The duration of diastole is about 0.7 s, while the duration of systole comprises 0.15–0.30 s. In total, there is 5.7–6.0 l of blood in the body, which is transported through the entire circulation in

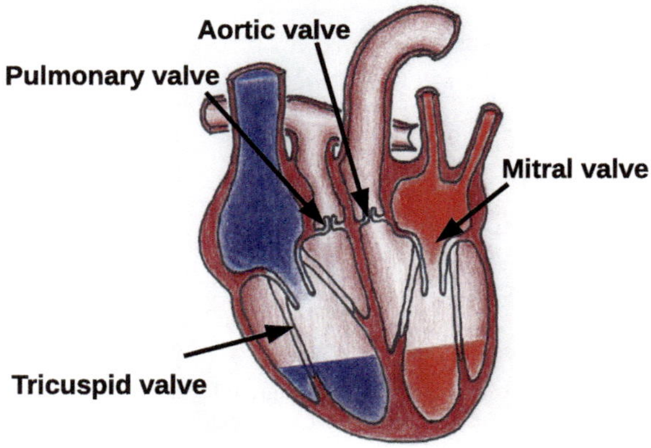

Fig. 1.5 Medical terms for the individual heart valves

one minute. In addition, a healthy adult exhibits about 60–70 heartbeats per minute at rest, i.e. the heart chambers have to contract about 60–70 times. Figure 1.6 shows simulated pressure curves in the different heart chambers, as well as a volume-pressure diagram for the two ventricles. The data presented in these figures are very close to physiological relevant curves. Note that the pressure is not expressed in Pa as usual, but in mmHg. mmHg is a widely used unit in medicine and abbreviates the expression millimeters of mercury:

$$1 \text{ mmHg} \approx 133.322 \text{ Pa}.$$

We observe that the pressure in the *contraction phase* increases rapidly in the right and left ventricle, whereas the volumes in the left (approximately 122 cm^3) and right ventricle (approximately 125 cm^3) remain constant because all the heart valves are closed. When the pressure in the left ventricle reaches approximately 79 mmHg, the aortic pressure is exceeded and the *aortic valve* (see Fig. 1.5) at the inlet of the aorta opens. Similarly, if the pressure in the right ventricle exceeds the pressure in the pulmonary artery, the *pulmonary valve* opens and deoxygenated blood can flow into the pulmonary circulation. As a result (*expulsion phase*), blood volume in the two ventricles is decreasing. During the *relaxation phase*, the pressure decreases, whereas the blood volume in the heart chambers remains constant, due to the fact that the aortic valve and the pulmonary valve are closed again. At a certain point, the atrial pressures (approximately 5.0 mmHg (right atrium) and 10 mmHg (left atrium)) exceed the ventricular pressures again. As a result, the valves between the atria and the ventricles open (*tricuspid valve* and *mitral valve*). Now the *filling phase* of the two ventricles begins. In this phase, the pressures in the atria are always slightly larger as the pressures of the ventricles, so that there is a continuous flow of blood from the atria into the ventricles. In the last third of the filling phase, blood flow into the ventricles is further enhanced by *atrial contraction* before the contraction phase begins again.

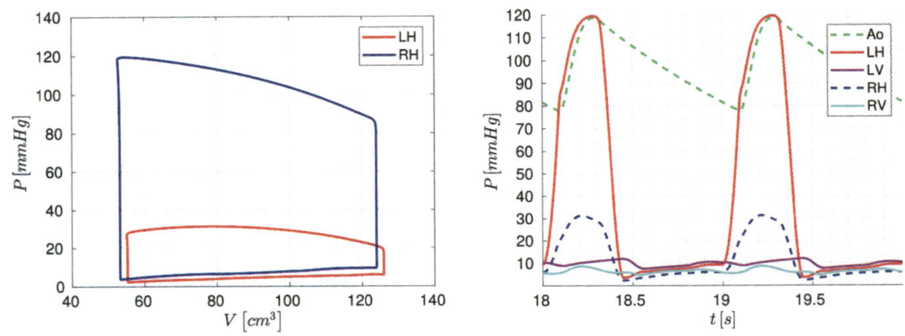

Fig. 1.6 Left: Volume-pressure diagram for the two ventricles. Right: Simulated pressure curves in the different cavities (LH: left ventricle, RH: right ventricle), atria (LV: left atrium, RV: right atrium) and the aorta (Ao)

In order to be able to perform the pumping, the heart muscle must be supplied with oxygen and nutrients. Furthermore, waste products must be removed. This is achieved by means of the coronary vessels, which branch off from the aorta and the pulmonary arteries.

1.1.2 Structure of the Vascular System

In the human body, blood is pumped by the heart (see previous section) in discrete pulses with high pressure into the main arteries. Then blood flows through a network of branching arteries (see Fig. 1.7). After several levels of branching, the arteries decrease in length and diameter and become arterioles.

The capillary bed is connected to the arterioles. Due to branching and narrowing of the vessels blood pressure and flow velocity are reduced, such that exchange processes between tissue and the vascular system can take place at the capillary level. In addition, blood pressure between the arteries and the capillary bed can be adapted by small muscles surrounding the vessel walls of the arterioles. Attached to the capillary bed is the venous system, which is the connection to the heart. It is estimated that there are approximately 20 levels of branching between the heart and the most distant capillary bed, in both the systemic and pulmonary circulation. Therefore, each of the two sub-circulations consist of three parts: Arteries, capillaries, and veins, where each part has a specific task to fulfill.

The main purpose of the arteries is to bring blood to the organs, while the veins return blood from the organs to the heart. The capillaries are responsible for the blood supply of the organs. The arterial and venous systems are almost symmetrical, where the main arteries and veins are often located side by side, separated by connective tissue. A major anatomical difference between veins and arteries is that, unlike arteries, veins have valves preventing backflow of blood into the capillary bed of the

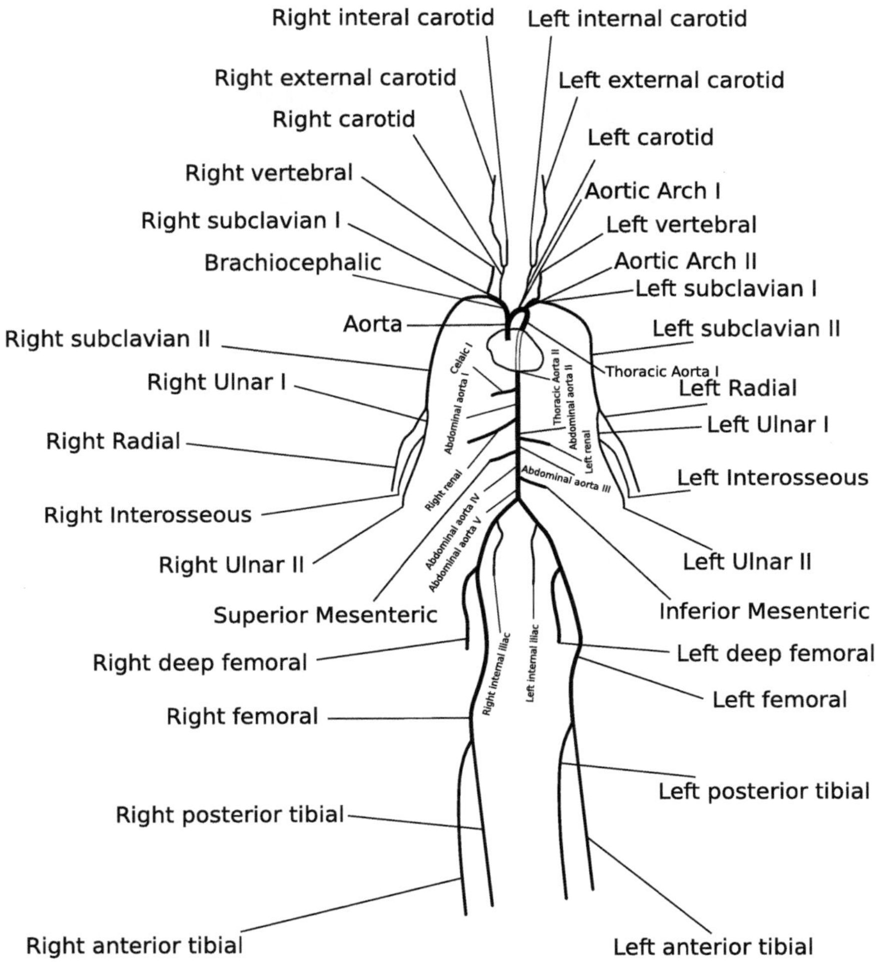

Fig. 1.7 Network of the largest arteries branching out of the heart [1, 2]

organs. In addition, veins usually have thinner walls than arteries. For this reason, most of the blood volume is contained in the veins (see Table 1.1 and Fig. 1.8). The vessel wall of arteries (and also veins) has three layers:

- **Inner layer (intima)**: The intima is lining the inside of an artery with endothelial cells and connective tissue.

- **Thick middle layer (media)**: In vessels close to the heart, this layer exhibits predominantly elastic fibers, but also muscles (elastic arteries). In the media of arteries distant from the heart and small arteries, there are predominantly smooth muscle cells (muscular arteries).

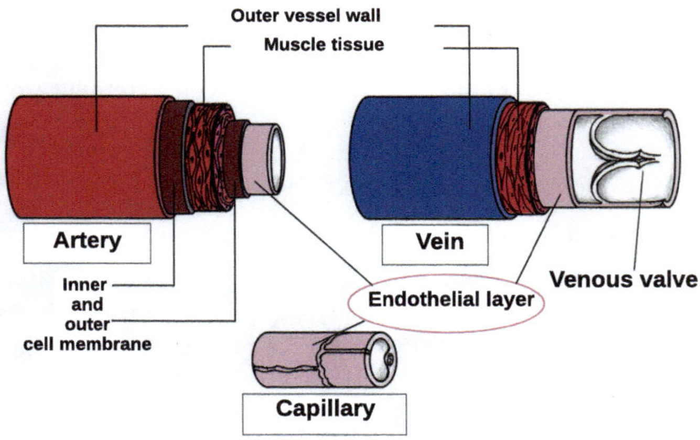

Fig. 1.8 Structure of vessel walls [3, Chap. 1]

- **Outer layer (adventitia)**: The outer layer of an artery consists of elastic fibers and connective tissue.

The elastic vessel walls of the larger arteries in the systemic circulation are responsible for the so-called Windkessel effect. The Windkessel effect is a mechanism that mitigates the large difference between systolic and diastolic pressure. This is accomplished by the elasticity of the arterial walls near the heart (in particular the aorta). In addition, due to the elastic vessel walls, blood volume can be stored in the respective vessels. To transport blood towards the organs, the muscles in the vessel walls contract so that the stored blood volume is pushed further. The muscles are supported by elastic fibers exerting counterpressure on the stored blood volume. This results in a continuous blood supply of tissue in the body.

The name of this mechanism originates from the early fire brigades, in which a large chamber filled with air (Windkessel) was installed (see Fig. 1.9). The heart corresponds to the pump of the fire brigade and the Windkessel to the arteries with elastic vessel walls. The Windkessel, which is located between the pumping device and its outlet, provides the opportunity to store water during the pumping process. As the Windkessel is filled with water, air in the Windkessel is compressed, building up a counterpressure on the inflowing water. Overall, this results in a steady or weakly pulsating flow of water through the pump's outlet. The deformable arteries of the systemic circulation are therefore comparable to the Windkessel since, such as the Windkessel, they have the ability to store fluid volume. The contraction of the muscles and the tension in the elastic arterial walls corresponds to the counterpressure that the compressed air exerts on the water in the Windkessel.

Figure 1.10 summarizes the main features of the systemic circulation. We observe that at the level of the large arteries there is a pulsatile flow and a high flow velocity. In a healthy human, the pressure varies between 60 and 80 mmHg during the diastole and between 100 and 130 mmHg during the systole (see Fig. 1.10). In Fig. 1.11, it can

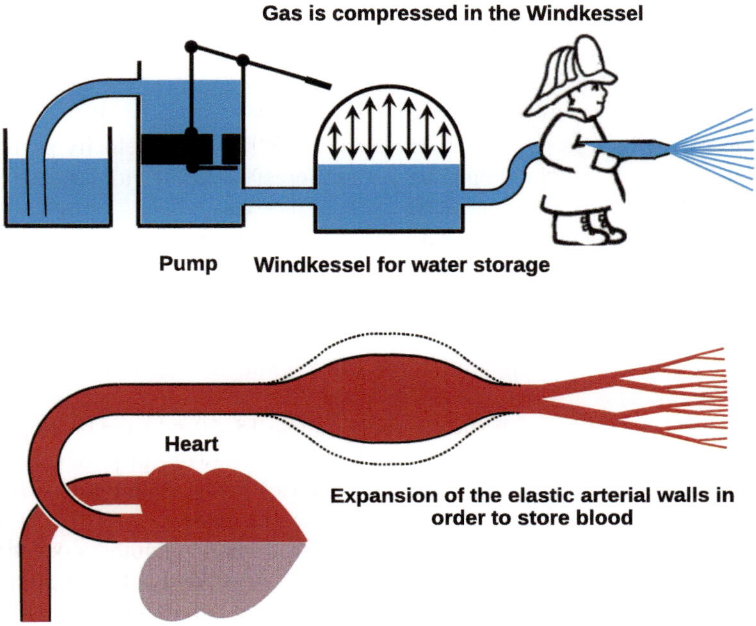

Fig. 1.9 Comparison of a water pump with the cardiovascular system (Windkessel effect)

be seen that the diastolic pressure decreases slightly towards the peripheral or smaller vessels. However, the systolic pressure increases slightly. A pulsating flow can be characterized by the Womersley number. The Womersley number is a dimensionless number that compares the frequency ω of the pulse and the viscosity of a fluid:

$$Wo = D\sqrt{\frac{\omega\rho}{\eta}}, \tag{1.1}$$

where we use ρ to denote the density of blood, D is the diameter of the vessel, and η represents the dynamic viscosity of blood. According to Table 1.1, the Womersley number is very large in the large vessels compared to the smaller vessels. This indicates that there is a pulsatile flow in the larger vessels. The small arteries and arterioles form a complex network with many branching levels. In this network, pulsating flow with high Reynolds and Womersley numbers is transformed into uniform flow with low Reynolds and Womersley numbers (see Fig. 1.11 and Table 1.1). The dimensionless Reynolds number Re, which is given by the ratio between inertia and viscous forces, is computed for blood vessels with a diameter D as follows:

$$Re = \frac{\rho D U}{\eta}. \tag{1.2}$$

Table 1.1 Further properties of the vessels in the systemic circulation

Vessel	Thickness vessel wall [mm]	Volume [%]	Re	Wo
Aorta	2.0	2.0	4000.0	10.5
Large arteries	1.0	5.0	2500.0	7.7
Small arteries	1.0	5.0	100.0	1.4
Arterioles	0.03	5.0	0.5	0.014
Capillaries	0.001	5.0	0.003	0.007
Venules	0.003	25.0	0.5	0.014
Large veins	0.5	50.0	1600.0	5.6

U is the mean flow velocity. The flow velocity (about 15–20 $\mathrm{cm/s}$) and blood pressure decrease towards to the capillary bed. In the capillary network, the velocity is very small compared to the velocity in the larger arteries and pressure ranges from 20 mmHg to 40 mmHg. In the veins, velocity increases again to 7–10 $\mathrm{cm/s}$ at a very low blood pressure. According to Table 1.1, high Reynolds numbers occur in the larger arteries. In the smaller vessels, one observes smaller Reynolds numbers, which increase again in the larger veins. However, they do not attain the level of the large arteries. Thus, turbulent flow can be expected in the larger vessels, while turbulence-free (laminar) flow can be observed in the remainder of the vascular system.

Overall, the human vascular system can be characterized by two attributes: Complexity and diversity. Complexity results from the structure of the network and the number of vessels (see Table 1.1 and Fig. 1.10). Diversity results from the different types of flows and the different properties of the vessels. In addition, physiological data can vary significantly from person to person. These features are causing several challenges for mathematical modeling. On the one hand, one would like to develop models that are generally applicable, allowing us to predict the effect of surgical interventions or diseases. On the other hand, one is interested in developing models for patient-specific issues. To achieve these goals, large amounts of data may have to be processed. Furthermore, the physical properties and network structure of the vascular system must be approximated as good as possible.

1.1.3 Properties of Blood

In medicine, blood is often referred to as liquid tissue; it consists of a blood plasma and special cells. The cells can be divided into three major groups:

- red blood cells (Erythrocytes)
- white blood cells (Leukocytes)
- platelets (Thrombocyths).

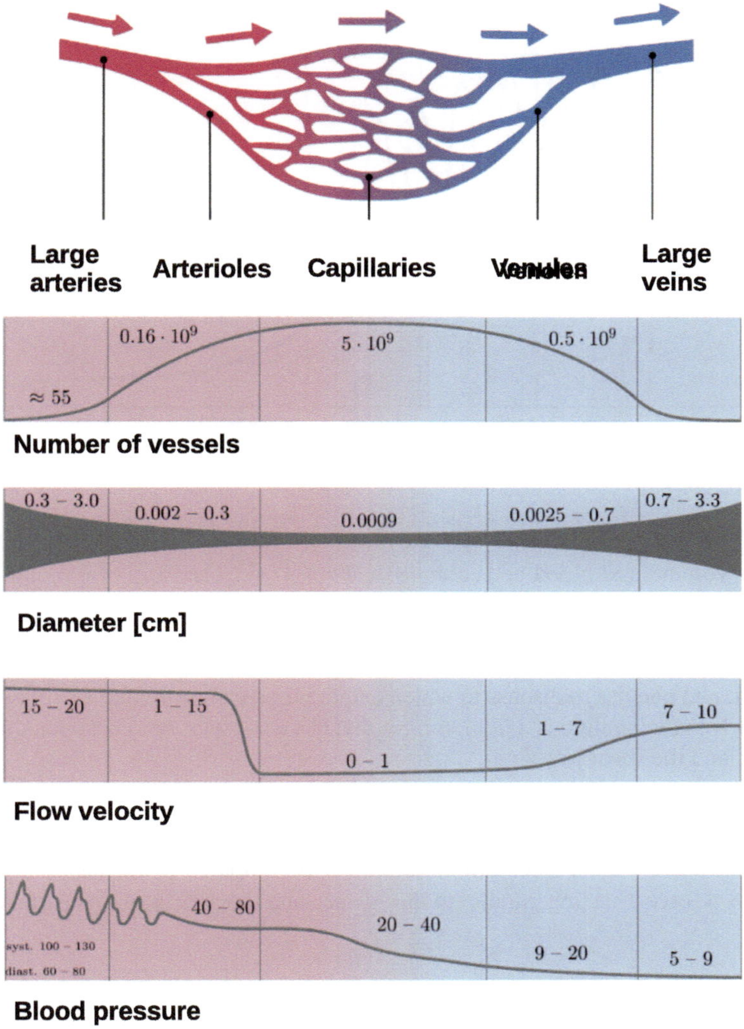

Fig. 1.10 Main components of the blood circulation and their properties [4]

Red blood cells transport oxygen and carbon dioxide, e.g. by attaching oxygen molecules to the red blood cells by means of the protein hemoglobin. The white blood cells support the immune system, while the platelets enable the clotting of blood. By this inner and outer bleedings are prevented. Blood plasma consists to a large extent (approximately 90 %) of water and the rest is given by proteins and ions. As the blood cells, the proteins in the blood plasma have different tasks like substance transport, support of the immune system and maintenance of the pH-value. Overall, blood plasma comprises 55 % of the total blood volume. Concerning the

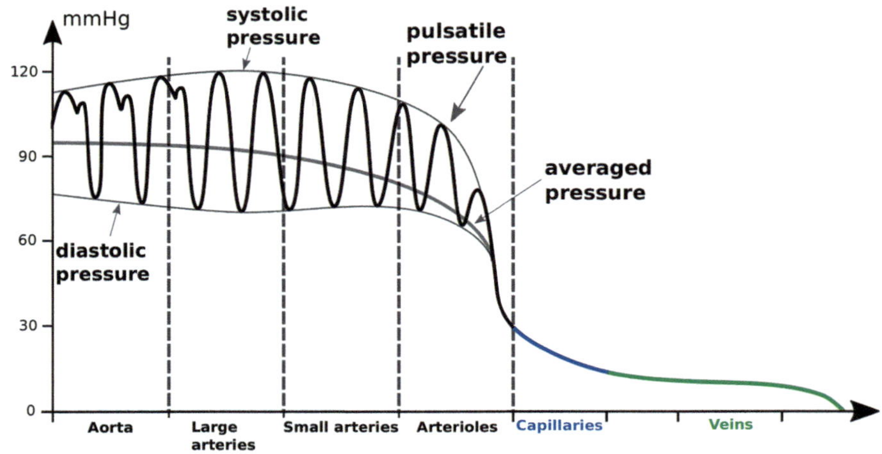

Fig. 1.11 Pressure curves in the different vessel types

cellular volume, red blood cells comprise about 97 %. The ratio of red blood cells and the total blood volume is called the hematocrit value. Thus, at a high hematocrit value the flow behavior is governed by the red blood cells.

Modeling of blood flow, viscous flow behavior of blood has to be taken into account, i.e., one has to discuss to which extent blood can be treated as a Newtonian fluid. A Newtonian fluid is a liquid or gas that has a constant ratio between the shear stress τ and the shear rate $\dot{\gamma}$:

$$\eta = \frac{\tau}{\dot{\gamma}}. \tag{1.3}$$

The shear stress τ on a facet a fluid volume V is computed by means of a quotient given by a force F acting parallel to the surface and the area A of this surface:

$$\tau = \frac{F}{A}. \tag{1.4}$$

The shear stress τ causes a distortion of the fluid volume V and thus generates an angle γ (see Fig. 1.12). To this angle one can assign an angular velocity $\dot{\gamma}$, which is called shear rate. Therefore, the viscosity of a Newtonian fluid is a constant material property. Water is an example of a Newtonian fluid. Since blood plasma consists mainly of water, this component of blood can be considered as a Newtonian fluid. On the other hand, it is observed that viscosity of blood does not increase linearly with increasing hematocrit value. Moreover, temperature, flow conditions and proteins determine the viscoelastic flow behavior. The viscosity of blood is therefore varying in the vascular system, which means that blood exhibits non-Newtonian properties. Thereby, two mechanisms play an important role related to the viscosity of blood:

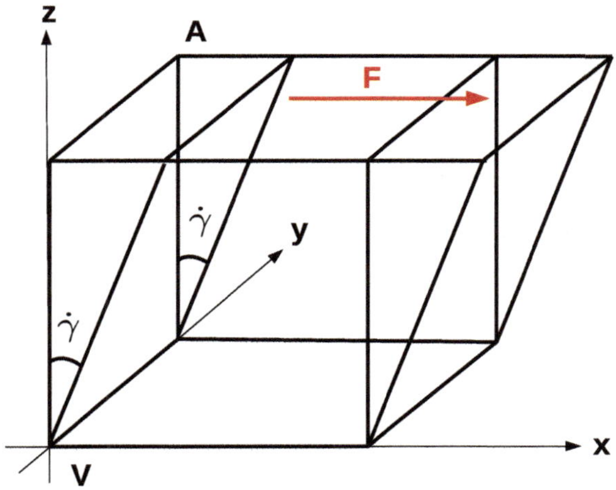

Fig. 1.12 Distortion of the volume element V by the force F

- The first mechanism is the *Fahraeus-Lindqvist effect*. It is characterized by the fact that in smaller vessels (diameter 20–300 μm), the erythrocytes move to the center of the center line of the blood vessel (axial migration), while in the vicinity of the vessel wall a layer of plasma reduces the flow resistances. This is supported by laminar flow in the small vessels. Only in the capillary bed blood viscosity increases again. Since the capillaries have a diameter of about 4–10 μm and the diameter of a red blood cell is about 7 μm, blood cells may have to deform considerably so that they can move through the capillaries. This leads to an increase of the blood viscosity within the capillary bed (see Fig. 1.13).

- The second mechanism is called *reversible aggregation*. If flow is very slow, as it is the case within the smaller vessels, red blood cells may form an aggregation, which can dissolve again. If these aggregations are transported in blood plasma, flow in the smallest vessels can be impeded.

An exact relationship between the relative viscosity η_r and the vessel diameter has been determined in [5] by evaluating different data sets. The relative viscosity η_r results from the quotient of the absolute blood viscosity η and the viscosity of the blood plasma. Depending on the hematocrit value H, the following relationship between the relative viscosity η_r and the scaled vessel diameter $D := D/(1.0\ \mu\text{m})$ is obtained:

$$\eta_r = 1 + (\eta_{0.45} - 1)\,\frac{(1-H)^C - 1}{(1-0.45)^C - 1},$$

$$\eta_{0.45} = 220 \cdot \exp(-1.3D) + 3.2 - 2.44 \cdot \exp\left(-0.06D^{0.645}\right),$$

$$C = (0.8 + \exp(-0.075D)) \cdot \left(-1 + \frac{1}{1 + 10^{-11} \cdot D^{12}}\right) + \frac{1}{1 + 10^{-11} \cdot D^{12}}.$$

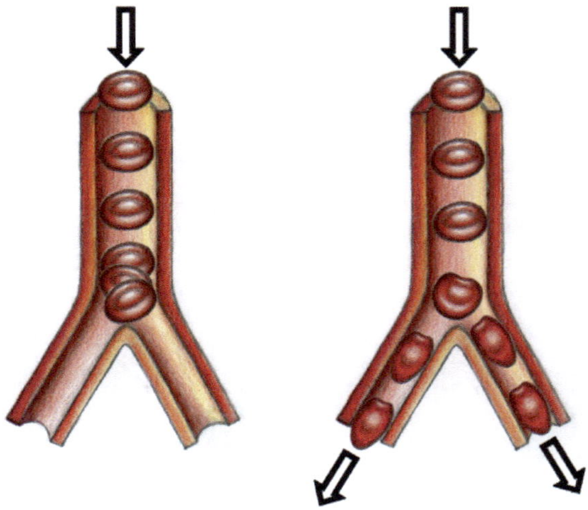

Fig. 1.13 Deformation of red blood cells in small capillaries (right). Blockage of a bifurcation if the red blood cells cannot deform (left)

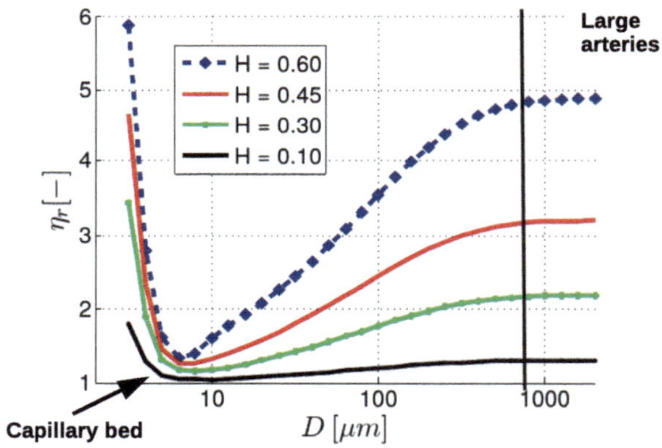

Fig. 1.14 Relative viscosity of blood as a function of the vessel diameter [5]

We observe (see Fig. 1.14) that in the larger vessels η_r is almost constant, i.e. within these vessels the non-Newtonian properties of blood can be neglected [6]. In the other vessel types, however, η_r varies significantly. Due to the Fahraeus-Lindqvist effect, the viscosity decreases until the capillary bed is reached. In the capillary bed η_r is increasing due to red blood cell deformation and reversible aggregation.

1.2 Cardiovascular Diseases

In 2012, a WHO statistical survey showed that 3 out of 10 deaths were caused by cardiovascular diseases. These can be due to a failure of the heart muscle, which in turn can be caused by a bacterial or viral infection. In addition, chronic diseases of the heart valves and vessels are more frequently observed. However, the three most common vascular diseases are arteriosclerosis, aneurysms or genetic malformations (e.g. arterio-venous malformations). Due to their frequent occurrence, these diseases are of particular interest for mathematical modeling [7, 8]. In preparation for the upcoming chapters, this section will provide basic knowledge on this topic.

1.2.1 Arteriosclerosis

Arteriosclerosis results from an accumulation of cholesterol, fatty components, calcium and fibrous elements at or within arterial walls. If such a deposit continues to grow, an affected artery becomes clogged. This can lead to an undersupply of organs. This type of vascular occlusion is also known as *stenosis* (see Fig. 1.15). Symptoms are often not noticeable for decades. Besides slow-growing deposits, rapid occlusions can also occur, primarily due to the formation of thrombi (blood clots) or the detachment of deposits. These objects can block smaller vessels on their way through the vascular tree (embolism) and can cause infarcts. If the coronary vessels are affected, this is referred to as heart attack. A blockage of cerebral vessels (vessels supplying the brain tissue) produces a stroke. In order to better estimate the risk of an embolism, the stability analysis of the deposits is of high interest for mathematical modeling. This is done by computing mechanical quantities such as stresses and displacements at deposits.

Arteriosclerosis usually occurs in larger arteries of the systemic circulation. Within the arteries of the pulmonary circulation and the veins, however, this disease is rarely observed. In this context, one has to note that the deposits are mainly formed at places, where shear stresses between the vessel wall and blood flow exceed a certain threshold. In particular, this is the case at bends of vessels and branches. To determine shear stresses along arterial walls, one requires accurate modeling of the interaction of blood flow with the arterial walls (fluid-structure interaction). This can be used to identify potential locations in a vascular tree where deposits may form.

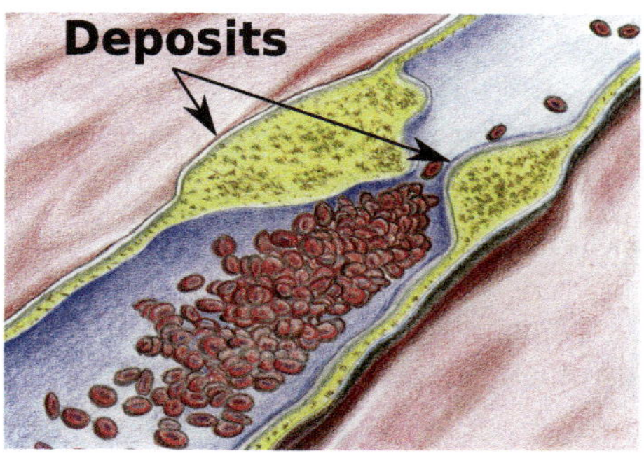

Fig. 1.15 Deposits in a vessel. Due to the narrowing of the blood vessel, the transport of red blood cells is impeded

Furthermore, one can consider the effects of deposits on blood flow and blood pressure. In order to be able to determine whether organs can still be adequately supplied with oxygen and nutrients, it is important to know, how blood pressure and flow rate decrease as the stenosis grows. If the decrease of pressure and flow rate is known, the reduction of oxygen supply can also be estimated.

1.2.2 Aneurysms in Arterial Walls

An aneurysm is a localized arterial dilation caused by a weakening of the arterial wall growing gradually, in some cases on the scale of several years. Regarding their shape (morphology), two types of aneurysms can be distinguished: spindle-shaped and baggy-shaped aneurysms. Spindle-shaped aneurysms are cylindrical dilations caused by a weakening of the entire circumference of the vessel. The main difference related to baggy-shaped aneurysms is that in the case of baggy-shaped aneurysms, the vessel wall is weakened on one side, resulting in the formation of a unilateral bulge. The main reasons of aneurysms are vascular diseases such as arteriosclerosis. Narrowings caused by arteriosclerosis result not only in a decrease of pressure behind the stenosis, but also in an increase of pressure at the inlet of the stenosis. The increase of pressure can cause a weakening of the vessel wall and thus aneurysm formation. However, trauma, infections, inflammations or connective tissue weaknesses can also be responsible for the formation of an aneurysm.

In addition to their shape, aneurysms are also classified according to their location in the arterial vascular tree: Central aneurysms (e.g. in the aorta) and peripheral aneurysms (e.g. in the leg arteries). The greatest danger posed by central aneurysms is their risk of rupture. If aneurysms rupture, e.g. in the brain, the resulting hemorrhage can lead to a stroke or immediate death. For this reason, computer models and mathematical equations are needed to assess the stability of aneurysms. This can be done as in the case of stenoses by computing stresses or displacements. An extremely simple, algebraic law for determining the surface stress σ $[N/m]$ is the Young-Laplace equation:

$$p = \frac{2 \cdot \sigma}{r},$$

where p [Pa] denotes the mean pressure in the aneurysm. Thereby, it is assumed that the aneurysm has the shape of a sphere with radius r [cm]. This equation can be derived as follows:

- Surface of the sphere:

$$A = 4 \cdot \pi r^2.$$

- Volume of a sphere:

$$V = \frac{4}{3} \cdot \pi r^3.$$

- Enlarging or shrinking the radius by dr results in:

$$dA = 8 \cdot \pi r dr, \quad dV = 4 \cdot \pi r^2 dr.$$

- Mechanical work for enlarging or shrinking the surface:

$$dW = \sigma \, dA = 8 \cdot \pi \cdot \sigma \cdot r dr. \tag{1.5}$$

- Mechanical work for enlarging or shrinking the volume:

$$dW = p \, dV = 4 \cdot \pi \cdot p \cdot r^2 dr. \tag{1.6}$$

- Combining (1.5) and (1.6), we obtain:

$$4 \cdot \pi \cdot p \cdot r^2 dr = 8 \cdot \pi \cdot \sigma r \cdot dr \Leftrightarrow p = \frac{2 \cdot \sigma}{r}.$$

More accurate results can be obtained by means of partial differential equations from fluid and structural mechanics (e.g. Navier-Stokes equations coupled with equations governing the deformation of the vessel wall). Peripheral aneurysms endanger the body mainly by thromboembolic vessel occlusions. It can be observed that these aneurysms are partially occluded at the inlet with blood clots, which can detach, e.g., by movements, and cause irreversible vascular occlusions.

1.2.3 Arteriovenous Malformation

Arteriovenous malformation (AVM) is a genetically caused malformation of blood vessels. It is characterized by the fact that small arteries or arterioles are directly connected to the venous vascular system. In this case, the capillary bed is missing at the transition between arteries and veins and blood flows directly into the venous vascular system with high pressure. As a consequence there is a strong expansion of the affected veins, since the veins are have thinner vessel walls compared to the corresponding arteries. Instead of the capillary bed, a tangle of arteries and veins is formed, which are particularly thin-walled and not stable. Partially, even vessel walls are missing inside the vascular tangle. This tangle of vessels is also called *NIDUS*. Due to the features of the veins and the increased blood pressure in a NIDUS, aneurysms are forming and thus internal bleeding can occur if these aneurysms rupture.

This is even more critical, if the NIDUS is located in the brain (cerebral arteriovenous malformation (cAVM)), since hemorraghe in the brain is extremely life-threatening. However, there are further effects that can be threatening for the affected patients. Moreover, it is observed that due to the excessive expansion of veins, oxygen-poor blood is accumulated in the veins and waste in the adjacent tissue

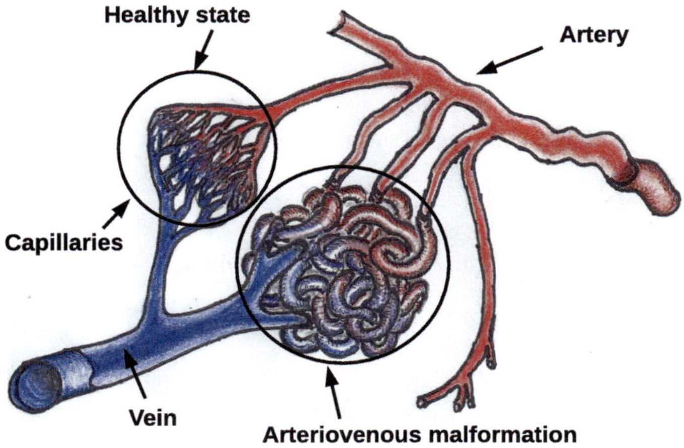

Fig. 1.16 Arteriovenous malformation between veins and arteries

can only be insufficiently removed via the venous vascular system. This is indicated by a dark blue discoloration of the skin. Another consequence of the damaged venous system is a significant increase of the heart volume. Due to the additional volume, the of the heart pumping capacity is increased and blood contained in the veins can be removed in an easier way. This is an immense burden for the heart muscle.

Problems arising in the context of the therapies are manifold. On the one hand, it is very difficult to identify the exact location and shape of the NIDUS. Furthermore, the removal of a NIDUS is dangerous, since the effect of cutting out the NIDUS is not known apriori. Therefore, one tries to remove the NIDUS by irradiation. However, this means that the patient is exposed to high levels of radiation. In addition, accurate measurement of pressures and flow velocities in a NIDUS with an endoscope is often not possible, due to the fact that vessel walls in a NIDUS can be easily damaged during endoscope insertion.

Tackling these difficulties, computer-based methods can help. Improvements in imaging and segmentation techniques may make it easier to determine the location and shape of a NIDUS. A mathematical model for blood flow through arteries, veins, and a NIDUS could reduce the number of invasive measurements.

1.2.4 Therapies for Cardiovascular Diseases

This section contains a brief overview on the most important treatment methods used in the context of cardiovascular disease. In addition to the description of the different treatment methods, it is also discussed, to which extent computer simulations can improve the application of these methods.

- **Insertion of stents:**
  To dilate the stenosis of a vessel, a catheter is inserted through a peripheral vessel and moved towards the stenosis. At the tip of the catheter there is a ballon wrapped with a mesh. When the balloon is inside the stenosis, the balloon is inflated and thus the stenosis is widened. The tubular metal mesh (stent) supports the dilatation after the balloon has been removed and thus keeps the affected vessel open.
  Stents are mostly used to stabilize coronary arteries. However, stents can also be used in other blood vessels, such as the carotid arteries. The main disadvantage associated with the use of stents is the fact that stenoses can quickly form again in the vicinity of stents, as fibers and fatty particles can easily get caught in the metal mesh. Moreover, inflammation occurs in the adjacent tissue if certain materials are used. In context of stent therapies, mathematical modeling focuses on simulating the effects that a stent exerts on blood flow and affected artery wall. The results obtained from these simulations are primarily used in the manufacturing process of stents and thus contribute to the improvement of stent therapy.

- **Insertion of vessel parts:**
  If the stenosis has a very high narrowing degree, it is often no longer possible to insert a stent into the stenosis. Essentially, there are three main options for removing narrow stenoses.
  The first option is to remove the deposit by surgical intervention. However, this requires to cut the affected vessel and to remove the deposit from the vessel. The risk of injuring the vessel wall in the process is very high, as the deposits may have become strongly anchored in the vessel wall.
  Another option for removing a narrow stenosis is to cut out the affected section of the corresponding vessel. The missing piece of the diseased vessel is replaced by a vessel part taken from a dead body and sutured with the healthy ends of the affected vessel. This avoids an injury to the vessel wall, however the risk is that the interfaces of the newly inserted vessel piece and the diseased vessel do not match perfectly.
  The third option is the insertion of bypasses. Vessels are taken from the patient's own body or from dead organisms and are used to bridge the narrowing. Compared to the second therapy option the narrowed part of the vessel is not removed and the substitute vessel is circumventing the stenosis outside of the diseased vessel. If the affected section is too large to be replaced by a biological vessel, it is also possible to use prostheses.
  However, all three options have in common that it is necessary to prevent blood flow through the affected vessel for the duration of the operation. For a physician, it would now be extremely important to know how these interventions affect the cardiovascular system and the supply of organs. Computer simulations could make it easier to assess these risks and facilitate the planning of an operation.

- **Embolization:**
  Embolization is a therapeutic procedure used to occlude vessels. The occlusion of the vessels is usually catheter-guided with microcoils, microparticles or tissue adhesives. Application areas for embolization are, e.g., the occlusion of aneurysms and the treatment of AVMs. In the latter case, one blocks the vessels connected to a NIDUS. To facilitate the planning of such interventions, blockage of specific vessels can be simulated on a computer before surgery and the effects on the cardiovascular system can be studied. This increases the safety of a surgical intervention.

- **Clamping of aneurysms:**
  To prevent baggy-shaped central aneurysms from rupturing, they are clamped at their inlet. This procedure, of course, has a significant effect on the flow behavior in the affected vessel and the adjacent vessels. However, these effects are difficult to estimate and measure. Mathematical models and computer simulations could help at this point to determine these data.

1.2.5 Natural Compensation Mechanisms

A reduced supply due to a clogged artery can be completely or partly prevented by natural compensatory mechanisms. In the following, we describe three different compensatory mechanisms that can be detected in the human body. Modeling and simulation of these mechanisms is of great interest, since in contrast to experiments with living organisms, the compensatory mechanisms can be studied separately and their significance can be assessed in a better way.

- **Arteriogenesis and angiogenesis:**
  The term arteriogenesis refers to the growth of small vessels (shunts) that form connections between larger arteries. These vessels can usually be found between the coronary vessels and the tibial vessels. If one of the larger arteries is completely or partially blocked, e.g., by a stenosis, the missing blood volume can be compensated to a certain extent by these connecting vessels. The connecting or collateral vessels have approximately the same diameter and wall thickness as arterioles. One has to note that growth of the collateral vessels does not primarily refer to the length of these vessels, but to their cross-sectional areas. Experiments with rats show that in a period of about 3 weeks the cross-sectional areas can increase by a factor of 1.5. This significantly reduces the flow resistance of the collateral vessels and makes it possible that a larger volume of blood can bypass the narrowing. The dilatation of the vessels is mainly caused by an increased shear stress on the inner walls of the vessels. As the pressure in the diseased artery decreases, there is a flow through the collateral vessels between the diseased and healthy vessel causing an increased shear stress. The shear stress in turn stimulates certain neurons to transmit signals to the vascular muscles such that they can contract. All in all one can note that these vessels form a natural bypass. Their conductivity can be

further increased by a stimulation with drugs and physical activities.
A disadvantage of these vessels is that the cross-sectional areas decrease towards the center and then increase again towards the other end. In addition, the increase in the cross-sectional areas causes the formation of curves in the vessel. Both features ensure that flow resistance increases. Nevertheless, experiments with rats and simulations show that a sufficient number of collateral vessels can contribute significantly to the reconstruction of blood flow [13, 14].
In contrast to arteriogenesis, angiogenesis is directly caused by a lack of oxygen. During an angiogenesis process, new capillary vessels are generated by sprouting or splitting processes with respect to pre-existing capillaries. This is a complex process in which the endothelial cells and pericytes that are required to stabilize the vessel walls are stimulated by various angiogenic growth factors, such as fibroblast growth factor (FGF) and vascular endothelial growth factor (VEGF). After these cells are stimulated they can enlarge or replicate themselves. If this process is repeated, a denser capillary network results, which ensures that the diffusion distances for oxygen and other molecules are reduced again. As a result, a local lack of oxygen can be compensated again.

- **Vasodilation and vasoconstriction:**
  Vasodilation and vasoconstriction are physiological mechanisms increasing and decreasing the cross-section of vessels. By this the flow resistance in the vessels can be adjusted so that more or less blood volume can be transported through the corresponding vessels. In particular, the arterioles have strong ring muscles (see Fig. 1.10), which can effectively regulate blood pressure and flow velocity. Similar to arteriogenesis, the vascular musculature can be activated by increased shear stress. In most cases, however, these mechanisms are a response to a decrease of oxygen concentrations in tissue due to narrowing of an artery. Other factors are increased carbon dioxide concentrations or nitric oxide concentrations. Due to reduced blood flow, carbon dioxide and nitric oxide molecules cannot be removed by the venous system to a sufficient extent.

- **Circle of Willis:**
  The brain is an organ that requires a huge amount of blood volume. About 15% to 20% of the blood volume pumped from the heart into the systemic circulation is supplying the brain with an adequate amount of oxygen and nutrients. For this reason, it is important to prevent an undersupply of tissue with blood. This is achieved, among other things, by the fact that there are a total of four larger arteries transporting blood into the head. In addition to the left and right carotid arteries, these are two other vessels called the left and right vertebral arteries. The latter vessels are linked and connected to the basilar artery. In contrast, the two carotid arteries are split into an external and internal carotid artery. The internal carotid arteries, together with the basilar artery, are connected to a circular vascular system called *Circle of Willis* (see Fig. 1.17). Six outgoing vessels provide blood for the brain tissue. If one or two of the three suppliers are partially or completely occluded, blood flow can partially be diverted and pumped through the Circle of

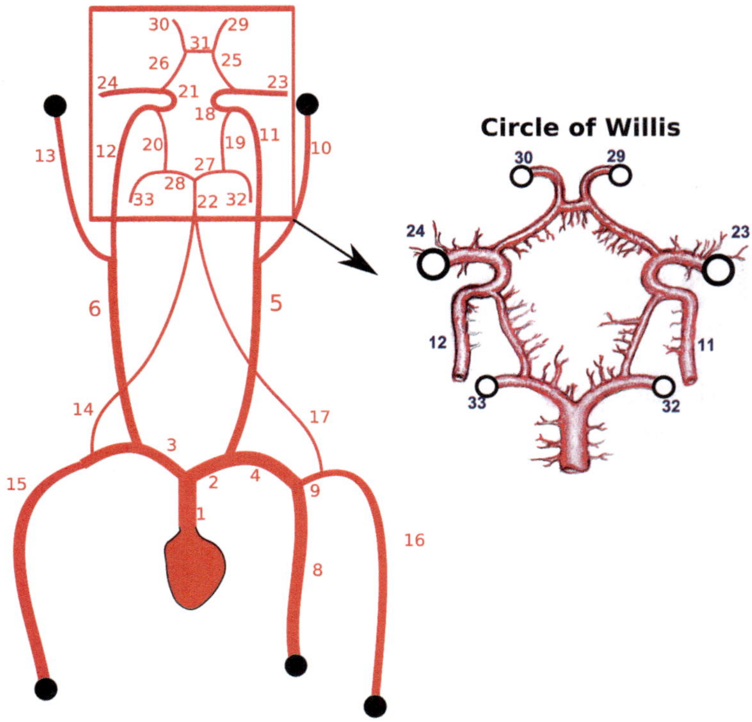

Fig. 1.17 The left side of the figure shows the main arteries branching out of the heart into the arms, head, and lower part of the body. Vessels 5 and 6 denote the left and right carotid arteries, respectively, while the indices 14 and 17 are representing the left and right vertebral arteries, respectively. The suppliers of the CoW consist of the basilar artery (Vessel 22) as well as the left and right internal carotid arteries (Vessels 16 and 17). The CoW has a total of six outlets that supply the brain tissue. The outlets are marked by the label CO (Cerebral Outflow) [11, 12]

Willis. The ring structure of the CoW ensures that blood can also flow to brain regions that should actually be supplied by the diseased vessels. However, the vessels of the CoW have limited conductivity, such that the lack of blood flow can only be partially compensated. In addition, the CoW may be incomplete in more than half of the people (see Fig. 1.19) [11, 12]. This means that in these cases the ring structure of the CoW can not be exploited.

1.3 Further Literature

Many of the information we have provided in this chapter are taken from the first chapter of the book "Cardiovascular Mathematics" by L. Formaggia, A. Quarteroni and A. Veneziani [3]. This book discusses the essential aspects from physiology

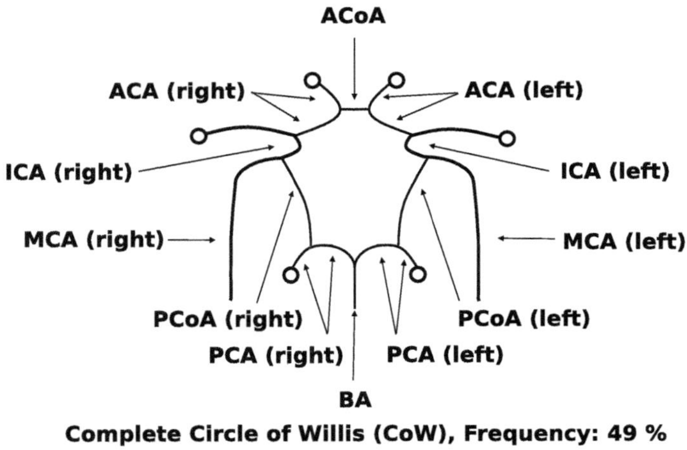

Fig. 1.18 Complete CoW and its components. BA: Basilar artery, PCA: Posterior cerebral artery, MCA: Middle cerebral artery, ACA: Anterior cerebral artery, PCoA: Posterior communicating artery, ACoA: Anterior communicating artery, ICA: Internal carotid artery

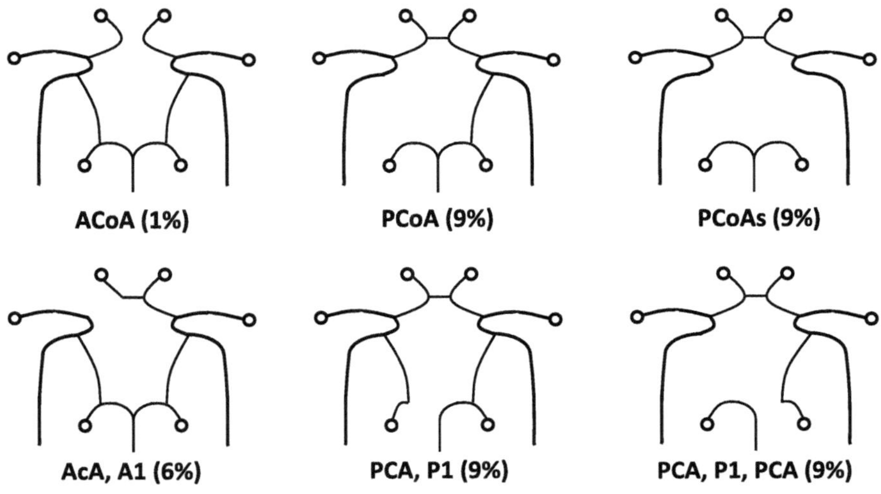

Fig. 1.19 Malformations of the CoW and their frequency. The abbreviations below the different configurations denote the missing vessels and their frequency (see also Fig. 1.18)

that are relevant for mathematical modeling of the cardiovascular system. Further information can be found in the book by M. Thiriet "Biology and Mechanics of Blood Flows" [9]. Apart from that there are many textbooks for medical students such as [10].

References

1. Stergiopulos N, Young D, Rogge T (1992) Computer simulation of arterial flow with applications to arterial and aortic stenoses. J Biomech 25(12):1477–1488. Elsevier
2. Wang JJ, Parker KH (2004) Wave propagation in a model of the arterial circulation. J Biomech 37(4):457–470. Elsevier
3. Formaggia L, Quarteroni A, Veneziani A (eds) (2010) Cardiovascular mathematics: Modeling and simulation of the circulatory system, vol 1. Springer Science & Business Media
4. Galdi G, Rannacher R, Robertson A, Turek S (2008) Hemodynamical flows: Modeling and analysis and simulation, vol 37. Birkhäuser, Basel, Boston, Berlin
5. Pries AR, Neuhaus D, Gaehtgens P (1992) Blood viscosity in tube flow: Dependence on diameter and hematocrit. Am J Physiol-Heart Circulat Physiol 263(6):H1770–H1778
6. Quarteroni A, Tuveri M, Veneziani A (2000) Computational vascular fluid dynamics: Problems, models and methods. Comput Vis Sci 2:163–197
7. Quarteroni A (2006) What mathematics can do for the simulation of blood circulation. MOX Report
8. Willemet M, Alastruey J (2015) Arterial pressure and flow wave analysis using time-domain 1-D hemodynamics. Ann Biomed Eng 43:190–206
9. Thiriet M (2007) Biology and mechanics of blood flows: Part II: Mechanics and medical aspects. Springer Science & Business Media
10. Heiss HW (1983) Human physiology. Robert F (ed) Schmidt and gerhard thews. Springer-Verlag, Berlin, pp 725, 569 figures, ISBN: 3-540-11669-9
11. DeVault K, Gremaud PA, Novak V, Olufsen MS, Vernieres G, Zhao P (2008) Blood flow in the circle of Willis: Modeling and calibration. Multiscale Model Simul 7(2):888–909
12. Alastruey JPKH, Parker KH, Peiró J, Byrd SM, Sherwin SJ (2007) Modelling the circle of Willis to assess the effects of anatomical variations and occlusions on cerebral flows. J Biomech 40(8):1794–1805
13. Scholz D, Ziegelhoeffer T, Helisch A, Wagner S, Friedrich C, Podzuweit T, Schaper W (2002) Contribution of arteriogenesis and angiogenesis to postocclusive hindlimb perfusion in mice. J Molecular Cellular Cardiol 34(7):775–787
14. Deindl E et al (2003) Receptor-independent role of the urokinase-type plasminogen activator during arteriogenesis. FASEB J 17(9):1174–1176

Chapter 2
Modeling Approaches for the Macrocirculation

Abstract As we noted in the previous chapter, the human vascular system consists of a huge number of vessels with different diameters, lengths and wall thicknesses. In both the venous and arterial vessel trees, subnetworks with different length scales can be identified. There are essentially three different types of length scales: Macrocirculation, mesoscale and microcirculation [1, Chap. 1].

Macrocirculation generally contains arterial or venous networks consisting of the largest arteries or veins. The widths and lengths of these networks are on the scale of meters. At the mesoscale one can find individual branches composed of vessels, which are several centimeters long. The vessel diameters are below one centimeter. At the level of the microcirculation, arterioles, venules or capillaries can be found. Small arterioles, venules and capillaries have a diameter in the range of micrometers. Usually, microvascular networks consist of a large number of individual vessels and fill volumes of a few cubic millimetres. A microvascular network in Fig. 2.1 (right) consists of about 20000 segments and fills a box of 1 mm $\times$ 1 mm $\times$ 2 mm.

In the following chapters of this book, we deal mainly with modeling and simulation of blood flow and transport processes on the level of macrocirculation. The main difficulty in modeling blood flow at this scale is that due to the proximity to the heart pulsating and turbulent flow has to be simulated (see Table 1.1). Furthermore, in case of larger arteries, the Windkessel effect (see Fig. 1.9) and in case of veins, the high deformability of the thin vessel walls must be taken into account. Modeling blood flow, the Navier-Stokes equations are usually used, yielding a velocity and a pressure field. In order to be able to represent turbulence or other variables such as the shear stresses on the vessel walls with sufficient accuracy, the vessel geometry may have to be meshed with a very fine grid [5]. This generates a huge amount of data, especially if large networks have to be meshed. Besides the handling of large amounts of data, one is confronted with a high numerical effort to solve the Navier-Stokes equations (Fig. 2.2). Furthermore, to model the Windkessel effect or the deformation of thin-walled veins, the flow equations must be coupled with mathematical models describing the deformation of the vessel wall [8, Chap. 9] [10, Chap. 4]. For this purpose, it is necessary to formulate physically meaningful boundary conditions at the interface between blood flow and vessel wall. This evokes new issues in terms of

T. Köppl and R. Helmig, *Dimension Reduced Modeling of Blood Flow in Large Arteries*, Mathematical Engineering, https://doi.org/10.1007/978-3-031-33087-2_2

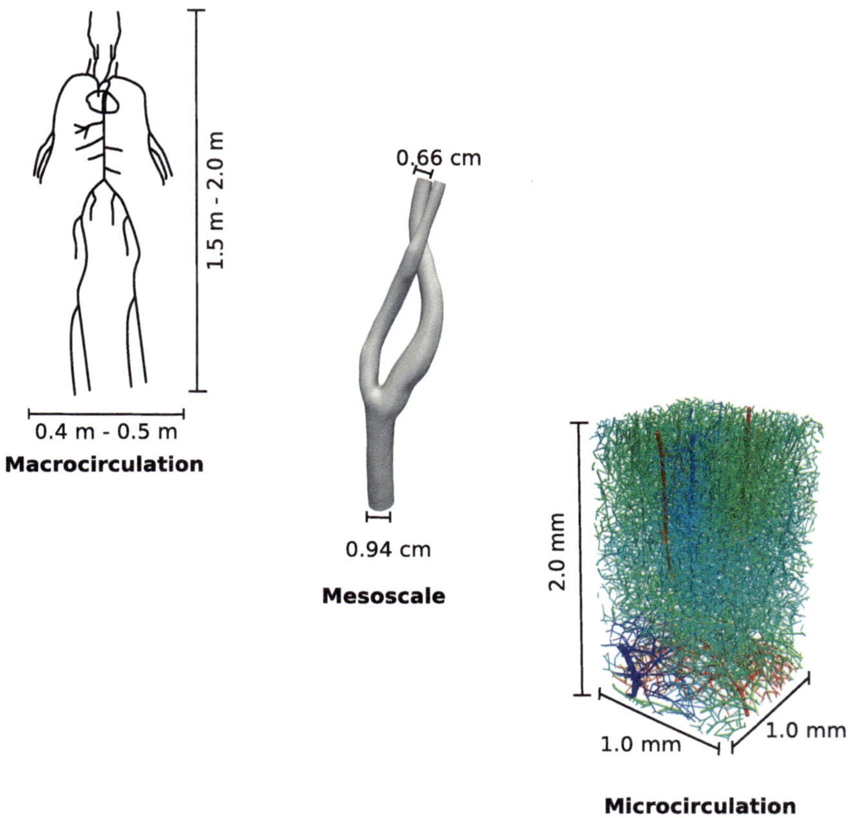

Fig. 2.1 Representation of different vascular systems at the level of the **macrocirculation**, **mesoscale** and **microcirculation**. **Left:** Vascular system consisting of the 55 largest arteries branching out of the heart [6]. Vascular systems of this type connect the heart to the individual organs and therefore they have a height of 1.5–2.0 m and a width of about 0.5 m. **Middle:** Bifurcation in the right carotid artery. The main vessel has an average diameter of 0.94 cm, while the branches have an average diameter of 0.94 cm [6]. **Right:** Capillary network with penetrating arterioles (red) and outgoing venules (blue) in a tissue block of dimension: 1 mm × 1 mm × 2 mm [7] (With permission from Journal of Cerebral Blood Flow & Metabolism, SAGE Journals)

mathematical modeling, since the boundary conditions must be formulated in such a way that the coupled model for the fluid-structure interaction (FSI) is well-posed, i.e., the resulting model equations should have an unique solution depending continuously on the input parameters of the model [5]. Furthermore, robust and efficient methods must be developed for the numerical treatment of these models [9]. In order to reduce the high computational effort, which arises e.g. in the context of optimization problems [11, 12], one considers *multi-dimension* models or multiscale models (*multiscale models*). In contrast to *one-dimension* models, this modeling approach combines models of different spatial dimensions [2][10, Chap. 11]. A *one-dimension*

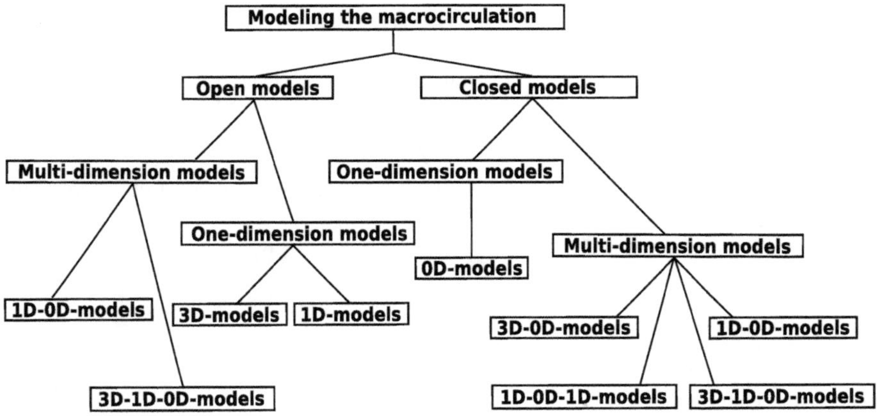

Fig. 2.2 Macrocirculation models can be considered as either **open** or **closed** models. Modeling the macrocirculation, a distinction is made between **one-dimension** models and **multi-dimension** models

model has the feature, that e.g. only 3D or 1D PDE systems are used to simulate blood flow. Quite often, however, it is not even necessary to resolve an entire network with expensive 3D FSI models. It can be sufficient to simulate only a small part of a network by means of 3D FSI models [3, 4, 13–17].

Simulating blood flow through the bifurcation between the external and internal carotid arteries (see Fig. 2.3) in more detail, one represents only the bifurcation i.e. a small part of the arterial network by a 3D geometry, while the remaining part of the network is modeled by means of simplified 1D models [18, 19]. Thereby, the 3D FSI models are coupled with 1D models at the inlets and outlets of the 3D geometry, to provide realistic boundary conditions for the 3D FSI model. At the 1D outlets, the Windkessel effect of the smaller arteries, which are not included in the network, must be taken into account to simulate realistic blood pressures and velocities. In addition, the influence of the arterioles must also be included into the model, since the arterioles are responsible for decreasing the blood pressure before blood enters the capillary bed (see Fig. 1.10). Modeling these mechanisms, a 1D PDE that is assigned to a vessel with an outlet is coupled with ordinary differential equations (ODEs), i.e., with 0D models [20–22]. Overall, a 3D-1D-0D coupled model is obtained.

To derive the 0D models, one uses analogies between hydraulic flows in networks and electrical circuits. The resistance of the vascular network between the arterioles and the outlet of the 1D artery can be described by an ohmic resistance R_1. Similarly, the flow resistance of the arterioles can be considered as a further ohmic resistance R_2. The ability of vessels to store blood volume can be compared to the capacitance C of a capacitor (see Fig. 2.3). From Kirchhoff's and Ohm's laws, as well as the differential equations governing the charging and discharging processes of capacitors, a system of algebraic equations and ODEs (i.e., a 0D model) can be derived in order to account for the Windkessel effect of the omitted vessels. One also refers to this R_1CR_2-model

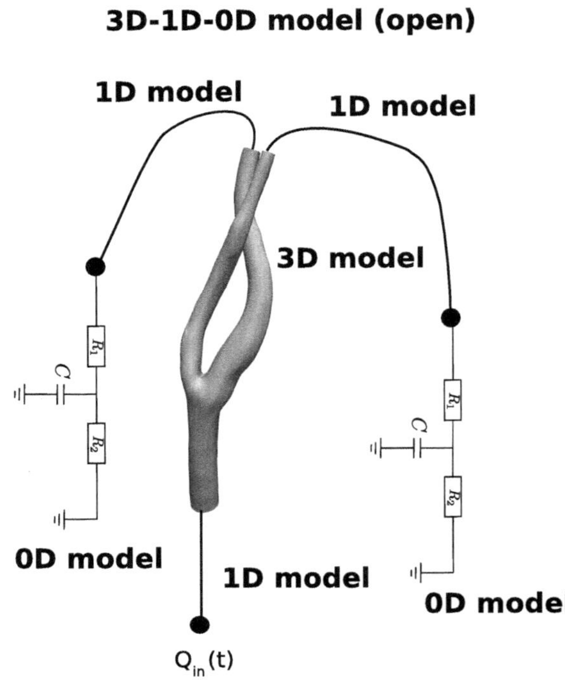

Fig. 2.3 Example for an **open** 3D-1D-0D model. The 3D vessel section, which can be seen in the figure, represents the branching of the right carotid artery into the internal and external carotid arteries. The arteries adjacent to this branch are modeled by 1D models. At the inlet of the 3D geometry, a flow rate Q_{in} is prescribed, whereas at the outflow edges, two 0D models are coupled with the corresponding 1D models to account for the Windkessel effect of the omitted vessels. All in all, this modeling approach results in an **open** loop model, since the Windkessel models are not connected to the inlet of the bifurcation

as 3-element Windkessel model. Further details on this modeling approach can be found in Chap. 3, Sect. 3.6.2. Regarding 3D-1D-0D models, a further distinction is made between *open* and *closed* models (see Figs. 2.3 and 2.4). In case of the closed models, the remaining part of the cardiovascular system is subdivided into further components or compartments. A possible subdivision into different compartments could include two compartments for the systemic circulation, as well as the superior and inferior vena cava. In addition to that, the heart can be divided into its four cavities and the pulmonary circulation is represented by further compartments for the different vessel types such as arteries, veins and capillaries. To each compartment an average pressure is assigned and for each connection between the compartments an averaged flow rate is computed.

In order to determine these variables, a 0D model can be designed for each chamber having a similar structure as the 3-element Windkessel model [23]. Coupling the different compartments with each other and the 1D models, a closed 3D-1D-0D model for the whole circulation results. Thereby, realistic boundary conditions are provided for the higher dimensional parts of this model without the need to describe in a detailed way the capillary networks and other parts of the vascular tree. Overall, this results in a model that exhibits sufficient accuracy in the vessels that are not included in the 3D model. At the same time, the complexity is kept as low as possible, since the omitted parts of the cardiovascular system are modeled in a simplified way.

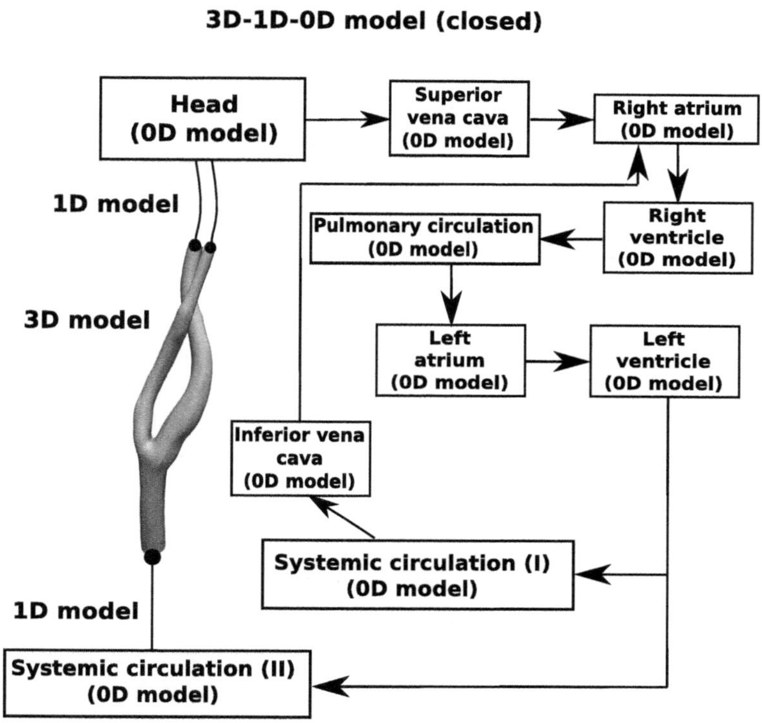

Fig. 2.4 Outline of a **closed** 3D-1D-0D model. The 1D models adjacent to the 3D bifurcation are coupled with 0D models for the rest of the cardiovascular system, in contrast to the open 3D-1D-0D model in Fig. 2.3. Overall, this modeling approach results in a **closed** blood circulation

If one is only interested in modeling bloodflow within networks of larger arteries and not in the remainder of the cardiovascular system, open models are used. As in the case of the closed model, the outlets are coupled with Windkessel models (0D-models). However, contrary to the closed models, the Windkessel models and not coupled with the inlets of the macrocirculation. At the inlets analytical profiles or measurements of the pressure or flow rate are used to compute suitable boundary conditions. Another possibility is to use simplified models for the left ventricle, if the inlet of the aorta is the only inflow boundary. In Fig. 2.6 the open ends of the $R_1 C R_2$-Windkessel models are indicated by a grounding.

For application areas in which 3D quantities such as turbulence or wall shear stresses do not have to be calculated, it is sufficient to consider 1D-0D coupled models, i.e. all the vessels in the network are considered as 1D objects. In terms of this model class, one can again identify open and closed models (see Figs. 2.5 and 2.6). As in case of 3D-1D-0D models, the part of the cardiovascular system that is not resolved by 1D models is described by a combination of algebraic and ordinary differential equations. A closed model that can be assigned to the class of 1D-0D

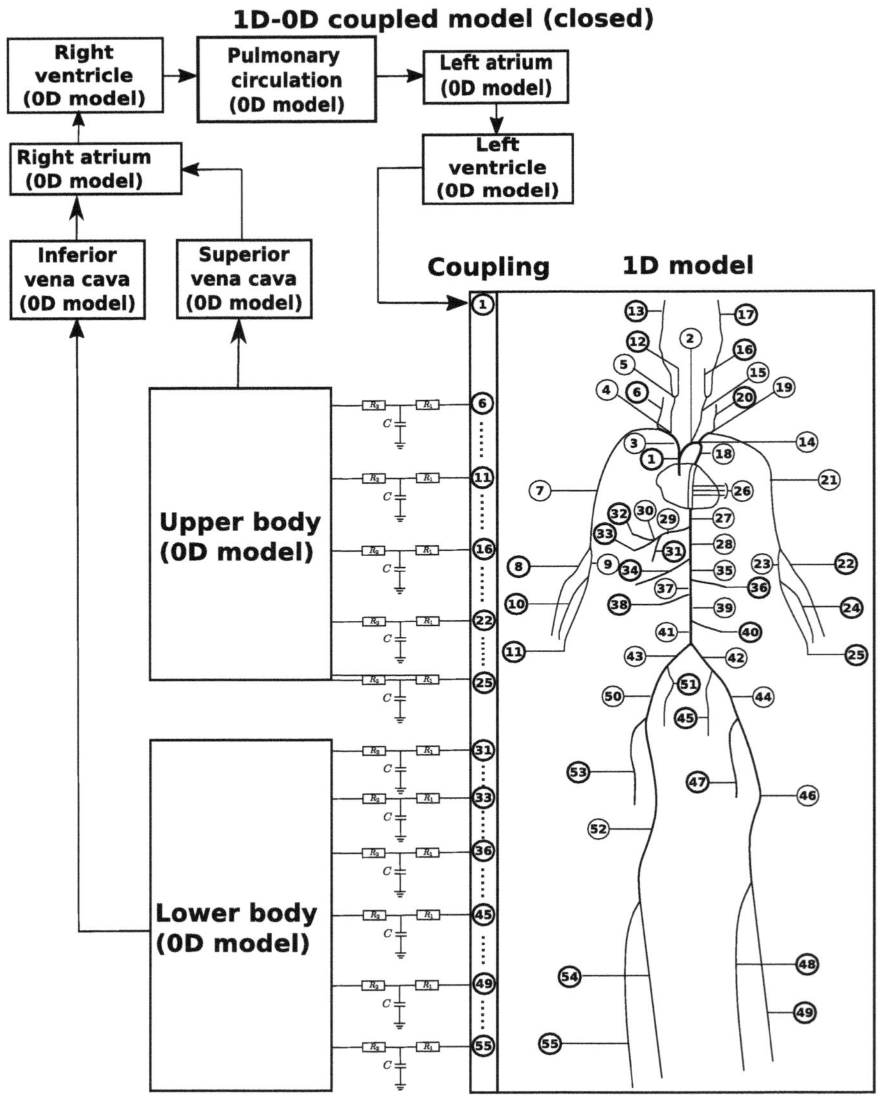

Fig. 2.5 Outline of a **closed** 1D-0D model. The main arteries of the systemic circulation are described by 1D models, while the rest of the cardiovascular system is accounted for by 0D models. Including the Windkessel effect, the outlets of the thick-rimmed 1D arteries are coupled with three-element R_1CR_2-models. The 0D models for the omitted part of the cardiovascular system are coupled with the inlet of the aorta (Vessel 1). All in all, this modeling approach results in a **closed** model

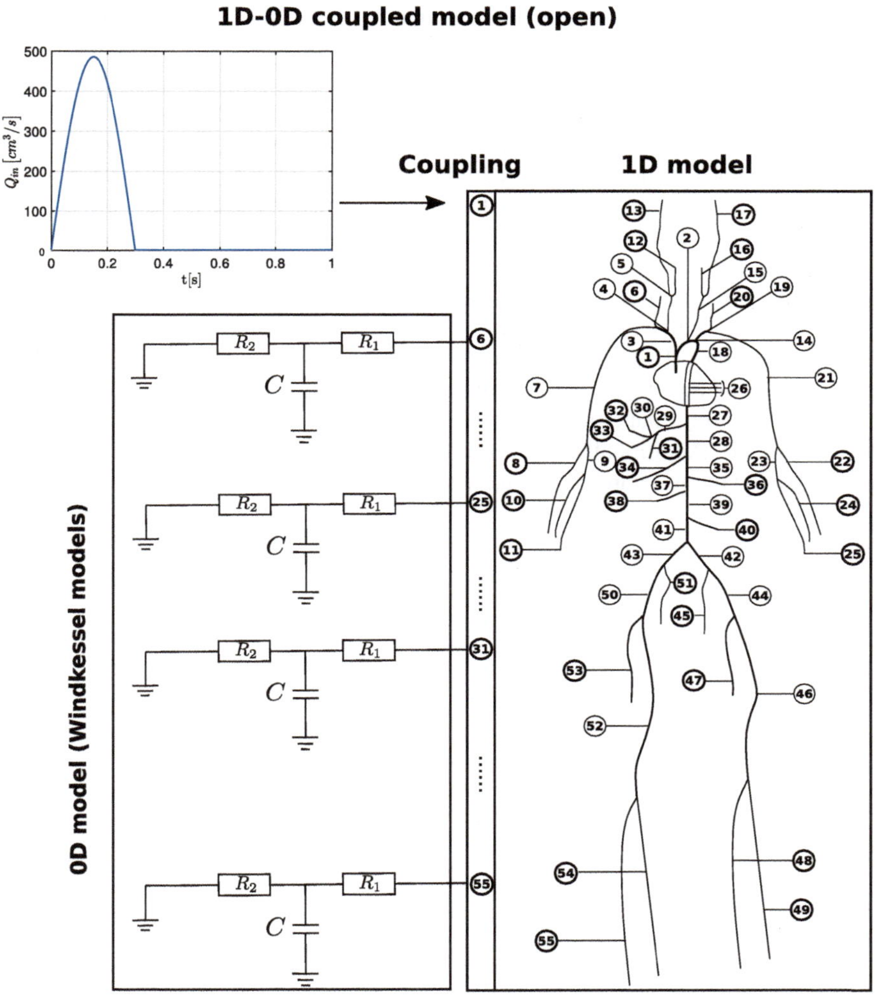

Fig. 2.6 Outline of an **open** 1D-0D model. The main arteries are modeled by means of 1D models, while the influence of the vessels beyond the 1D outlets is included by 0D Windkessel models. Both model types are coupled at the outlets of the thick-rimmed 1D arteries. At the inlet of the aorta (Vessel 1) a flow rate Q_{in} $\left[\mathrm{cm}^3/\mathrm{s}\right]$ is prescribed. Contrary to a **closed** 1D-0D model the R_1CR_2-Windkessel models for the outlets are not connected to the inlet of the aorta. Thus this modeling approach yields an **open** model

coupled models is a 1D-0D-1D coupled model [24]. This type of model is mainly used when a venous and an arterial network are to be simulated at the macrocirculation level [25]. The networks on the macrocirculation level, are represented by 1D models as in the previous modeling approaches. Both 1D networks have corresponding inlets and outlets, which are coupled with each other by 0D models. The intermediate 0D models account for the arterioles, capillary networks, and organs that are located between the respective inlets and outlets of the venous and arterial networks.

Another important modeling class is given by closed 0D models, which belongs to the group of one-dimension models [26, 27]. These models consist of algebraic and ordinary differential equations and allow a fast and inexpensive simulation of a closed cardiovascular system, since one can use optimized solvers for differential-algebraic systems. However, this model provides only mean pressures and flows. Other modeling approaches for the macrocirculation are listed in Fig. 2.2.

In the following chapters of this book, we will mainly focus on modeling of open 1D-0D models. Additionally, we will discuss application areas from physiology in which 1D-0D models can provide realistic and meaningful data.

References

1. Ambrosi D, Quarteroni A, Rozza G (eds) (2012) Modeling of physiological flows, vol 5. Springer Science & Business Media
2. Blanco P, Feijóo R et al (2010) A 3D-1D-0D Computational model for the entire cardiovascular system. In: Dvorking E, Goldschmit M, Storti M (eds) Computational Mechanics, vol 29, pp 5887–5911
3. Blanco P, Feijóo R (2013) A dimensionally-heterogeneous closed-loop model for the cardiovascular system and its applications. Med Eng Phys 35(5):652–667. Elsevier
4. Blanco P, Feijóo R, Urquiza S (2007) A unified variational approach for coupling 3D–1D models and its blood flow applications. Comput Methods Appl Mech Eng 196(41–44):4391–4410. Elsevier
5. Deuflhard P, Weiser M (2012) Adaptive numerical solution of PDEs. de Gruyter, Adaptive Numerical Solution of PDEs
6. Stergiopulos N, Young D, Rogge T (1992) Computer simulation of arterial flow with applications to arterial and aortic stenoses. J Biomech 25(12):1477–1488. Elsevier
7. Reichold J, Stampanoni M, Keller A, Buck A, Jenny P, Weber B (2009) Vascular graph model to simulate the cerebral blood flow in realistic vascular networks. J Cereb Blood Flow Metab 29(8):1429–1443. SAGE Publications Sage, London, England
8. Formaggia L, Quarteroni A, Veneziani A (eds) (2010) Cardiovascular mathematics: modeling and simulation of the circulatory system, vol 1. Springer Science & Business Media
9. Wick T (2011) Adaptive finite element simulation of fluid-structure interaction with application to heart-valve dynamics
10. Quarteroni A, Manzoni A, Vergara C (2017) The cardiovascular system: mathematical modelling, numerical algorithms and clinical applications. Acta Numerica 26:365–590. Cambridge University Press
11. D'Apice C, D'Arienzo M, Kogut P, Manzo R (2018) On boundary optimal control problem for an arterial system: Existence of feasible solutions. J Evol Eqn 1–42. Springer
12. Marsden A (2014) Optimization in cardiovascular modeling. Ann Rev Fluid Mech 46:519–546. Annual Reviews

13. Lantz J, Renner J, Karlsson M (2011) Wall shear stress in a subject specific human aorta–influence of fluid-structure interaction. Int J Appl Mech 3(04):759–778. World Scientific
14. Malossi A, Blanco P, Crosetto P, Deparis S, Quarteroni A (2013) Implicit coupling of one-dimensional and three-dimensional blood flow models with compliant vessels. Multiscale Model Simul 11(2):474–506. SIAM
15. Marsden A, Esmaily-Moghadam M (2015) Multiscale modeling of cardiovascular flows for clinical decision support. Appl Mech Rev 67(3):030804. American Society of Mechanical Engineers
16. Passerini T, De Luca M, Formaggia L, Quarteroni A, Veneziani A (2009) A 3D/1D geometrical multiscale model of cerebral vasculature. J Eng Math 64(4):319. Springer
17. Urquiza S, Blanco P, Vénere M, Feijóo R (2006) Multidimensional modelling for the carotid artery blood flow. Comput Methods Appl Mech Eng 195(33–36):4002–4017. Elsevier
18. Čanić S, Kim E (2003) Mathematical analysis of the quasilinear effects in a hyperbolic model blood flow through compliant axi-symmetric vessels. Math Methods Appl Sci 26(14):1161–1186. Wiley Online Library
19. Payne S (2004) Analysis of the effects of gravity and wall thickness in a model of blood flow through axisymmetric vessels. Med Biol Eng Comput 42(6):799–806. Springer
20. Alastruey J, Parker K, Peiró J, Sherwin S (2008) Lumped parameter outflow models for 1-D blood flow simulations: Effect on pulse waves and parameter estimation. Commun Comput Phys 4(2):317–336
21. Canuto D, Chong K, Bowles C, Dutson E, Eldredge J, Benharash P (2018) A regulated multiscale closed-loop cardiovascular model, with applications to hemorrhage and hypertension. Int J Numer Methods Biomed Eng e2975. Wiley Online Library
22. Shi Y, Lawford P, Hose R (2011) Review of zero-D and 1-D models of blood flow in the cardiovascular system. Biomed Eng 10(1):33. BioMed Central
23. Liang F, Oshima M, Huang H, Liu H, Takagi S (2015) Numerical study of cerebroarterial hemodynamic changes following carotid artery operation: A comparison between multiscale modeling and stand-alone three-dimensional modeling. J Biomech Eng 137(10):101011. American Society of Mechanical Engineers
24. Mynard J, Smolich J (2015) One-dimensional haemodynamic modeling and wave dynamics in the entire adult circulation. Ann Biomed Eng 43(6):1443–146. Springer
25. Acosta S, Puelz C, Rivière B, Penny D, Brady K, Rusin C (2017) Cardiovascular mechanics in the early stages of pulmonary hypertension: A computational study. Biomech Model Mechanobiol 16(6):2093–2112. Springer
26. Liang F, Liu H (2005) A closed-loop lumped parameter computational model for human cardiovascular system. JSME Int J Ser C Mech Syst Mach Elem Manufac 48(4):484–493. The Japan Society of Mechanical Engineers
27. Liang F, Liu H (2006) Simulation of hemodynamic responses to the Valsalva maneuver: An integrative computational model of the cardiovascular system and the autonomic nervous system. J Physiol Sci 56(1):45–65. Physiological Society of Japan

Chapter 3
Dimension Reduced Models for the Macrocirculation

Abstract In this chapter, we design a mathematical model that can be used to simulate flow and transport processes in a vascular system at the macro level (see Chap. 2). The impact of the heart and the omitted vessels is accounted for by suitable boundary conditions and simplified models. Despite of the restriction to the macrocirculation, modeling of flow and transport processes still remains very complex. This is due to the fact that in vessels of the macrocirculation there is a convection-dominated flow which interacts with the vessel wall. The basis of the models discussed in this chapter is a domain decomposition approach (see Fig. 3.1). In a first step of this approach, inlets and outlets of a given network are defined. Next, one decomposes the network into its individual vessels. Flow and transport processes in the individual vessels are governed by dimension-reduced (1D or 0D) models, which are assigned to the individual vessels. This makes it possible to simulate larger networks with less computational effort. As a result, physicians can plan surgeries before the actual medical intervention, since the respective results are available after a short simulation time compared to more complex 3D models. Moreover, simulations could reduce the risks in terms of medical interventions without danger for the patient. The models of the different vessels are linked at the bifurcations of the networks by suitable coupling equations. At the inlets and outlets appropriate boundary conditions for the heart and omitted vessels are prescribed. This approach is reflected in the structure of this chapter: In the first sections, we discuss some technqiues for the derivation of dimension-reduced models. The remaining sections describe the coupling of these models at a bifurcation or junction. Furthermore the incorporation of boundary conditions is discussed.

3.1 Blood Flow Within a Single Vessel

To describe flows in a time interval $[0, t_{end}]$, we use the incompressible Navier-Stokes equations in the following form [1]:

T. Köppl and R. Helmig, *Dimension Reduced Modeling of Blood Flow in Large Arteries*, Mathematical Engineering, https://doi.org/10.1007/978-3-031-33087-2_3

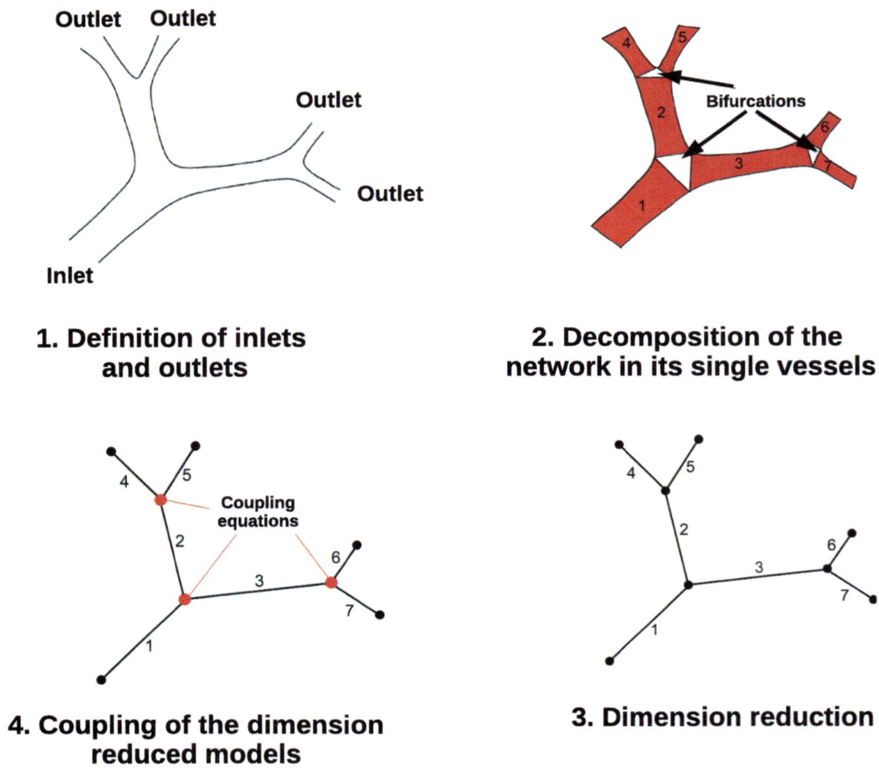

Fig. 3.1 Domain decomposition approach for a given blood vessel network

$$\frac{\partial \mathbf{u}}{\partial t} + (\mathbf{u} \cdot \nabla)\,\mathbf{u} + \frac{\nabla P}{\rho} - \mathbf{div}\left(\frac{2}{\rho}\mu\left(|\mathbf{D}|^2\right)\mathbf{D}\right) = \mathbf{f}, \text{ in } \Omega,\ t > 0, \tag{3.1a}$$

$$\operatorname{div}(\mathbf{u}) = 0, \text{ in } \Omega,\ t > 0. \tag{3.1b}$$

t_{end} is the end time of our simulation period and $\Omega(t) \subset \mathbb{R}^3$ denotes the lumen of a blood vessel. For simplicity, we assume that the main axis of the vessel coincides with the z-axis of the Cartesian coordinate system and that the cross-sectional areas are circular (see Fig. 3.2). If the vessel is curved, the corresponding domain is transformed such that the vessel is straight and its main axis coincides with the z-axis. Using cylindrical coordinates, a vessel of length l can be described as follows:

$$\Omega(t) = \left\{(r, \theta, z) \in \mathbb{R}^3 \mid 0 \leq z \leq l,\ 0 \leq r \leq R(z,t),\ 0 \leq \theta < 2\pi \right\},$$

where $R(z, t)$ is the radius of the vessel at the position z and at time t. The relationship between the cartesian and cylindrical coordinates is given by:

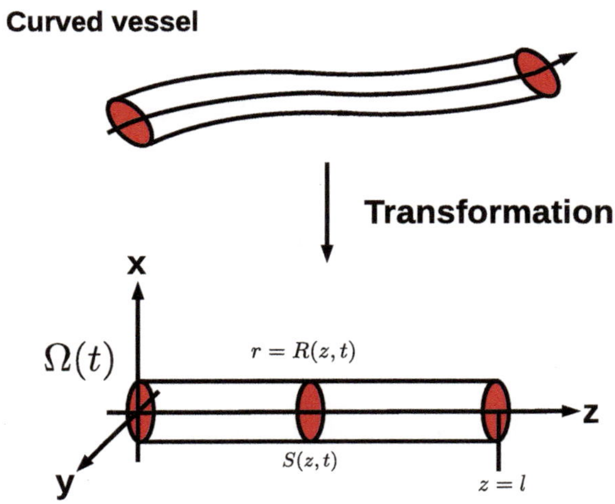

Fig. 3.2 Transformation of a vessel to a simplified geometry

$$\begin{pmatrix} x \\ y \\ z \end{pmatrix} = \begin{pmatrix} r\sin(\theta) \\ r\cos(\theta) \\ z \end{pmatrix}, \quad r = \sqrt{x^2 + y^2}.$$

Equation (3.1a) can be derived from the conservation of momentum wrt. a control volume and consists of three sub-equations, where each sub-equation is associated with a spatial coordinate. The second Eq. (3.1b) is called continuity equation and can be determined by considering mass balancing wrt. a control volume. Here, it is assumed that blood is incompressible, i.e., that its density ρ is constant [2]. The velocity field $\mathbf{u}$ with the components u_1, u_2 and u_3, as well as the pressure P are the solution variables of the Navier-Stokes equations:

$$\mathbf{u} : \Omega(t) \to \mathbb{R}^3, \quad \begin{pmatrix} x \\ y \\ z \end{pmatrix} \mapsto \begin{pmatrix} u_1(x,y,z) \\ u_2(x,y,z) \\ u_3(x,y,z) \end{pmatrix}, \quad P : \Omega(t) \to \mathbb{R}, \quad \begin{pmatrix} x \\ y \\ z \end{pmatrix} \mapsto P(x,y,z).$$

$\mathbf{f}$ can be used to include external forces such as the gravitational force. In our model, however, we set $\mathbf{f}$ to zero: $\mathbf{f} \equiv 0$. The convective component of the flow is described by the term $(\mathbf{u} \cdot \nabla)\,\mathbf{u}$. It is defined as follows:

$$(\mathbf{u} \cdot \nabla)\,\mathbf{u} = \begin{pmatrix} \langle \mathbf{u}, \nabla u_1 \rangle \\ \langle \mathbf{u}, \nabla u_2 \rangle \\ \langle \mathbf{u}, \nabla u_3 \rangle \end{pmatrix},$$

i.e. one determines the gradient of each component u_i and computes the scalar product between the vector $\mathbf{u}$ and ∇u_i. Viscoelastic and diffusive features of the flow are accounted for by the term

$$\mathbf{div}\left(\mu\left(|\mathbf{D}|^2\right)\mathbf{D}\right).$$

D denotes the deformation tensor and is given by:

$$\mathbf{D}(\mathbf{u}) = \frac{1}{2}\left(\nabla\mathbf{u} + \nabla\mathbf{u}^T\right), \ \nabla\mathbf{u} = \begin{pmatrix} \nabla u_1^T \\ \nabla u_2^T \\ \nabla u_3^T \end{pmatrix}.$$

The squared absolute value $|\mathbf{D}|^2$ of the two-stage tensor **D** is given by the sum over the squared entries $\mathbf{D}_{ij}$ of **D**:

$$|\mathbf{D}|^2 = \sum_{i=1}^{3}\sum_{j=1}^{3} \mathbf{D}_{ij}^2.$$

By a suitable choice of the function $\mu(\cdot)$ viscoelastic effects can be integrated into the model. In order to represent non-Newtonian effects, the function $\mu(\cdot)$ can be chosen according to the Carreau model [3]:

$$\mu\left(|\mathbf{D}|^2\right) = \eta_\infty + (\eta_0 - \eta_\infty) \cdot \left(1 + \kappa\,|\mathbf{D}|^2\right)^n.$$

$\kappa > 0$ and $n \in (-0.5, 0)$ are model parameters. η_0 and η_∞ represent the viscosity for very small and very large shear rates, respectively. In case of blood, it is well accepted to choose the parameters as follows [3]:

$$\eta_0 = 65.7 \cdot 10^{-3}\ \text{Pa} \cdot \text{s}, \ \eta_\infty = 4.45 \cdot 10^{-3}\ \text{Pa} \cdot \text{s}, \ \kappa = 212.2\ s^2 \text{ und } n = -0.325.$$

If one is interested in modeling a Newtonian fluid, one simply sets:

$$\mu\left(|\mathbf{D}|^2\right) = \eta_\infty.$$

The divergence operator in the continuity equation is defined as follows:

$$\operatorname{div}(\mathbf{u}) = \frac{\partial u_1}{\partial x} + \frac{\partial u_2}{\partial y} + \frac{\partial u_3}{\partial z}.$$

If the pressure P and the velocity **u** are given, the Cauchy stress tensor can be determined as follows:

$$\mathbf{T} = -p\mathbf{I} + \mathbf{S}, \ \mathbf{S} = 2\mu\left(|\mathbf{D}|^2\right)\mathbf{D}.$$

and

$$\text{WSS} = \langle \mathbf{t}_{blut}, \mathbf{T}n \rangle$$

are the shear stresses at the vessel wall. *WSS* abbreviates the term *wall shear stress*. The vector $\mathbf{t}_{blood}$ denotes the tangential vector to a flow line through a cross-sectional area, while n is the outer normal of this cross-sectional area.

In order to describe the interaction of blood flow and vessel wall, models from structural mechanics are required to compute the deformation of the vessel wall. The link between blood flow and the deformation of the vessel wall is accomplished by fluid-structure interaction (FSI) models. In the following parts of this book, we will restrict ourselves to simple algebraic models (see Sect. 3.2 or (3.18)). Much more complex models and coupling algorithms can be found, e.g., in the following references: [4–8, Chap. 9]. The FSI models provide suitable boundary conditions for the Navier-Stokes equations along the vessel wall. If rigid walls are considered, *no-slip* boundary conditions on the vessel wall can be imposed:

$$\mathbf{u} = 0.$$

At the inlets, one can prescribe a time-dependent velocity or pressure profile, represented by a mathematical function (such as a parabola) or by measured data. Velocity or pressure profiles as well as stress values in normal direction (e.g. $\mathbf{T} \cdot n = 0$) can be considered at the outlets. In addition to the boundary conditions, an initial condition $\mathbf{u_0}$ for $t = 0$ is required:

$$\mathbf{u_0} : \Omega(0) \to \mathbb{R}, \quad \mathbf{x} \mapsto \mathbf{u_0}(\mathbf{x}), \quad \mathbf{u}(\mathbf{x}, 0) = \mathbf{u_0}(\mathbf{x}).$$

The initial condition $\mathbf{u_0}$ could be chosen in an arbitrary way. However, this might cause problems in terms of numerical solution schemes. If $\mathbf{u_0}$ is not chosen appropriately, iterative solvers for the discrete system may not converge. An appropriate initial condition for the velocity can be obtained by solving a stationary Stokes problem:

$$\begin{aligned} -\mathbf{div}\left(\mu \mathbf{D}\left(\mathbf{u_0}\right)\right) + \nabla \mathbf{P} &= \mathbf{f}, \text{ in } \Omega, \\ \operatorname{div}\left(\mathbf{u_0}\right) &= 0, \text{ in } \Omega. \end{aligned}$$

The transport of a substance, such as oxygen or a drug, is determined by computing the corresponding concentration in blood. Usually it is denoted by $c\ \left[mmol/cm^3\right]$ and governed by an advection-diffusion equation:

$$\frac{\partial c}{\partial t} + \operatorname{div}\left(c \cdot \mathbf{u} - D\nabla c\right) = 0. \tag{3.2}$$

In the remainder of this book we denote a PDE such as (3.2) as transport equation. A transport equation can be derived by formulating a balance equation for the number of particles $N\ [mmol]$ (see e.g. [2]). $\mathbf{u}$ results from the momentum and continuity equation. Thus there is a one-sided coupling of (3.2) and the Navier-Stokes equations. Other types of couplings could be introduced, if the diffusivity $D\ \left[\mathrm{cm^2/s}\right]$ is depending on stress [9]. A transport process can be characterized by the *Peclét-number*:

$$Pe = \frac{UL}{2D}.$$

Thererby, U is an averaged velocity and L is a characteristic length like the average length of a vessel or an averaged diameter of a vessel. Thus, high Peclét numbers represent a convection dominated flow, whereas small Peclét numbers indicate a diffusion dominated flow. Typically convection-dominated flow can be observed in larger and medium-sized arteries, while in the capillary bed and tissue diffusion-dominated flow [6, Chap. 2] is present. In order to obtain a well-posed problem the time-dependent problem (3.2) has to be equipped with an initial condition:

$$c\left(\mathbf{x},t\right)=c_0\left(\mathbf{x}\right)\ \text{ for } t=0 \text{ and } \mathbf{x}\in\Omega(0)$$

and suitable boundary conditions. If $\Omega(t)$ represents only a small part of a vessel network with artificial inlets and outlets, one imposes a constant concentration Γ_{in} at the inlet. At the outlet, one can prescribe, e.g., a homogeneous Neumann boundary condition: $\frac{\partial c}{\partial n}=0$ to model a diffusion-free flow across the respective outlet. At the interior surface of the vessel wall, one can impose a Robin boundary condition that prevents both convective and diffusive transport across the vessel wall:

$$c\,\mathbf{u}\cdot n-D\frac{\partial c}{\partial n}=0.$$

However, if one wants to describe the transport of a substance into the vessel wall, a modified Robin boundary condition on the corresponding part of the boundary can be imposed. By this the supply of muscle tissue in a vessel wall with nutrients and oxygen could be modeled:

$$c\,\mathbf{u}\cdot n-D\frac{\partial c}{\partial n}=\mathcal{P}\left(c-c_i\right). \tag{3.3}$$

$\mathcal{P}=\mathcal{P}\left(\left|\langle\mathbf{t}_{blood},\mathbf{T}n\rangle\right|\right)$, denotes the permeability of the endothelial cell layer. Endothelial cells are cells lining the interior of a blood vessel, i.e., the lumen of all healthy vessels are equipped with a layer of endothelial cells (see Sect. 1.1.2). By increasing wall shear stress along this cell layer, the endothelial cells can be stimulated, which causes a change of the permeability $\mathcal{P}$. The constant c_i $\left[\text{mmol/cm}^3\right]$ in (3.3) represents the concentration of a substance (e.g. oxygen) inside the vessel wall (e.g. in the vessel muscles).

3.2 Derivation of the 1D Equations

Usually, the 3D models (Navier-Stokes equations and advection-diffusion equation) are used to simulate blood flow and transport in a small part of the vessel system. In order to simulate the corresponding processes within a larger vessel network by means of a standard computer, it is necessary to decrease the complexity of the given problem. This can be achieved, e.g., by means of dimension reduced models.

However, using dimension reduced models, one has to note that vortices or a spatial distribution of wall shear stresses can not be represented. On the other hand a dimension reduced model enables us to compute averaged pressures, flow rates or concentrations of substances after a relatively short simulation time.

There are different approaches to derive a 1D model for an incompressible fluid in a compliant vessel. One approach is based on the assumption that the radius R of the vessel is small in comparison to the length L of the vessel i.e. $R/L \ll 1$. An asymptotic analysis and neglecting higher order terms depending on R/L that occur in the resulting expansions, yields a suitable 1D model [10]. Another way to derive a 1D blood flow and transport model is based on the assumption that the blood vessels can be represented by a cylindrical geometry as the one that can be seen in Fig. 3.3. Integrating the Navier-Stokes equations and the advection diffusion Eq. (3.2) wrt. the cross sections, one obtains a simplified model for flow and transport [11–13]. In this section, we discuss a third way for the derivation of a dimenison reduced model, which is based on simplifying the Reynolds equations [5, Chap. 10][14, 15]. For this purpose, we consider a small section V_t of a vessel, whose vessel wall can be deformed (see Fig. 3.3).

For simplicity it is assumed that the main axis of the vessel is aligned with the z-axis of the Cartesian coordinate system (see Fig. 3.2). The boundary of V_t is composed of the vessel wall $\partial V_{t,w}$ and the section areas S_1 and S_2 at the inlet and outlet of the considered vessel part. According to Reynold's transport theorem, it holds for a sufficiently smooth function $f = f(\mathbf{x}, t)$:

$$\frac{d}{dt}\int_{V_t} f \, dV = \int_{V_t} \frac{\partial f}{\partial t} \, dV + \int_{\partial V_t} f \, \mathbf{u}_b \cdot n \, d\sigma, \tag{3.4}$$

where $\mathbf{u}_b$ is the velocity field deforming the boundary of V_t. Thereby, we assume that the normal component of $\mathbf{u}_b$ is equal to zero on the cross sections S_1 and S_2. On

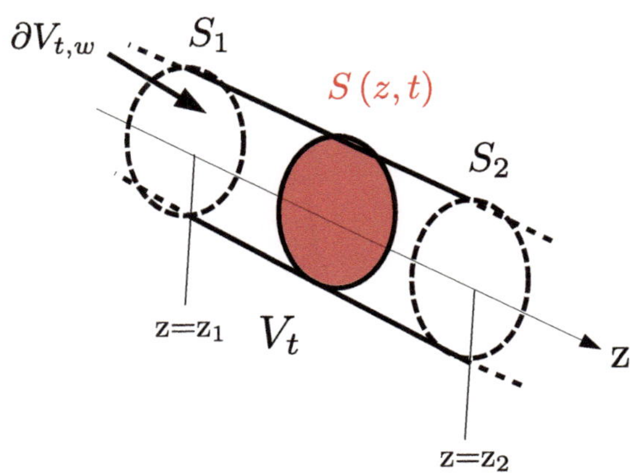

Fig. 3.3 Notation describing the different parts of the vessel portion

the vessel wall $\partial V_{t,w}$, $\mathbf{u}_b$ is denoted by $\mathbf{u}_w$. The relative velocity $\mathbf{w}$ between $\mathbf{u}_w$ and the velocity flield $\mathbf{u} = (u_1, u_2, u_3)^T$ is defined as follows:

$$\mathbf{w} = \mathbf{u}_w - \mathbf{u}.$$

Based on these considerations, one obtains for the boundary integral:

$$\int_{\partial V_t} f\ \mathbf{u}_b \cdot n\ d\sigma = \int_{\partial V_{t,w}} f\ \mathbf{u}_w \cdot n\ d\sigma = \int_{\partial V_{t,w}} f\ \mathbf{w} \cdot n\ d\sigma + \int_{\partial V_{t,w}} f\ \mathbf{u} \cdot n\ d\sigma. \tag{3.5}$$

In a next step, we introduce averaged quantities wrt. S. We denote the average value of f wrt. S by $\overline{f}$. It is given by:

$$\overline{f} = \frac{1}{A} \int_S f\ d\sigma.$$

A is the area of the cross section S:

$$A = A(z,t) = \int_{S(z,t)} d\sigma.$$

By this, the volume integral wrt. V_t can be represented as follows:

$$\int_{V_t} f\ dV = \int_{z_1}^{z_2} \left[\int_S f\ d\sigma\right] dz = \int_{z_1}^{z_2} A \cdot \overline{f}\ dz, \tag{3.6}$$

where z_1 and z_2 $(z_1 < z_2)$ are the z-coordinates corresponding to S_1 und S_2. Assuming that z_1 and z_2 are fixed, the first summand of (3.4) can be reformulated as follows:

$$\int_{V_t} \frac{\partial f}{\partial t}\ dV = \int_{z_1}^{z_2} \frac{\partial}{\partial t} \left(A \cdot \overline{f}\right)\ dz.$$

For the surface integral in the second summand, we obtain:

$$\begin{aligned}\int_{\partial V_{t,w}} f\ \mathbf{u} \cdot n\ d\sigma &= \int_{\partial V_t} f\ \mathbf{u} \cdot n\ d\sigma - \int_{S_1} f\ \mathbf{u} \cdot n\ d\sigma - \int_{S_2} f\ \mathbf{u} \cdot n\ d\sigma \\ &= \int_{\partial V_t} f\ \mathbf{u} \cdot n\ d\sigma + \int_{S_1} f \cdot u_3\ d\sigma - \int_{S_2} f \cdot u_3\ d\sigma.\end{aligned}$$

u_3 is the z-component of the velocity field. Using the Gaussian Theorem (see Sect. 7.9) it results:

$$\int_{\partial V_{t,w}} f\ \mathbf{u} \cdot n\ d\sigma = \int_{V_t} \nabla \cdot (f\mathbf{u})\ dV + \int_{S_1} f \cdot u_3\ d\sigma - \int_{S_2} f \cdot u_3\ d\sigma.$$

By means of (3.5) and the averaging operator, we obtain:

$$\int_{\partial V_{t,w}} f\ \mathbf{u}_w \cdot n\ d\sigma = \int_{\partial V_{t,w}} f\ \mathbf{w} \cdot n\ d\sigma - \int_{z_1}^{z_2} \frac{\partial}{\partial z}\left[A\overline{(f \cdot u_3)}\right] dz$$
$$+ \int_{V_t} \nabla \cdot (f\mathbf{u})\ dV. \tag{3.7}$$

Combining (3.4) with (3.6) and (3.7) yields:

$$\begin{aligned}\int_{z_1}^{z_2} \frac{\partial}{\partial t}\left(A\overline{f}\right) dz &= \int_{z_1}^{z_2} \left(\int_S \frac{\partial f}{\partial t} d\sigma\right) dz + \int_{z_1}^{z_2} \left(\int_{\partial S} f\mathbf{w} \cdot n\ d\gamma\right) dz \\ &- \int_{z_1}^{z_2} \frac{\partial}{\partial z}\left[A\overline{(f \cdot u_3)}\right] dz + \int_{z_1}^{z_2} \left(\int_S \nabla \cdot (f\mathbf{u})\, d\sigma\right) dz \\ &= \int_{z_1}^{z_2} \left(\int_S \frac{\partial f}{\partial t} d\sigma + \int_{\partial S} f\mathbf{w} \cdot n\ d\gamma \right.\\ &\left. - \frac{\partial}{\partial z}\left[A\overline{(f \cdot u_3)}\right] + \int_S \nabla \cdot (f\mathbf{u})\, d\sigma\right) dz. \end{aligned} \tag{3.8}$$

Since z_1 and z_2 are chosen in an arbitrary way, the Fundamental Lemma of Variational Valculus can be applied (see Sects. 7.4 and 7.5). This means that the integrands of the integrals on the left and the right hand side of the equation are equal. By this, we obtain a 1D Version of the Reynolds transport theorem:

$$\frac{\partial}{\partial t}\left(A\overline{f}\right) + \frac{\partial}{\partial z}\left[A\overline{(f \cdot u_3)}\right] = \int_S \left[\frac{\partial f}{\partial t} + \nabla \cdot (f\mathbf{u})\right] d\sigma + \int_{\partial S} f\mathbf{w} \cdot n\ d\gamma. \tag{3.9}$$

Please note that this formula holds both for compressible and incompressible fluids. Using (3.9) a mass and momentum conservation equation wrt. V_t can be derived:

Mass conservation

A mass conservation equation follows directly from (3.9), by choosing $f = \rho$. If we further assume that the fluid is incompressible i.e.

$$\nabla \cdot \mathbf{u} = 0 \text{ and } \rho \equiv \text{ const.}$$

we obtain:

$$\frac{\partial A}{\partial t} + \frac{\partial}{\partial z}\left(A\overline{u_3}\right) = \int_{\partial S} \mathbf{w} \cdot n\ d\gamma. \tag{3.10}$$

Thereby, the term on the right hand side of the equation can be considered as a term accounting for the exhange processes across the vessel wall. It has the unit $cm^3/(s \cdot cm)$.

Momentum conservation

We choose $f = u_3$ in (3.9). Assuming that the fluid is incompressible, it follows:

$$\frac{\partial}{\partial t}\left(A\overline{u_3}\right) + \frac{\partial}{\partial z}\left[A\overline{u_3^2}\right] = \int_S \left[\frac{\partial u_3}{\partial t} + \mathbf{u}\cdot\nabla u_3\right] d\sigma + \int_{\partial S} u_3 \mathbf{w}\cdot n \, d\gamma \tag{3.11}$$

Using the notation $\frac{D}{Dt} = \frac{\partial}{\partial t} + \mathbf{u}\cdot\nabla$ it holds:

$$\frac{\partial}{\partial t}\left(A\overline{u_3}\right) + \frac{\partial}{\partial z}\left[A\overline{u_3^2}\right] = \int_S \frac{Du_3}{Dt} d\sigma + \int_{\partial S} u_3 \mathbf{w}\cdot n \, d\gamma.$$

In general, for a velocity field $\mathbf{u}$ and a volume element V_t, the momentum equation reads as follows:

$$\int_{V_t} \frac{D}{Dt}(\rho\mathbf{u}) \, dV = \int_{V_t} \rho \mathbf{f}^b \, dV + \int_{\partial V_t} \mathbf{T}\cdot n \, d\sigma.$$

The two summands on the right hand side of the previous equation represent the volume forces $\mathbf{f}^b$ acting on V_t (e.g., gravity force) and the frictional forces acting on the boundary of V_t. The frictional forces are described by the Cauchy stress tensor $\mathbf{T}$. Using $\rho \equiv const$ and the Gaussian theorem we obtain:

$$\int_{V_t} \frac{D\mathbf{u}}{Dt} \, dV = \int_{V_t} \mathbf{f}^b \, dV + \frac{1}{\rho}\int_{V_t} \nabla\cdot\mathbf{T} \, dV.$$

A possible choice for the stress tensor is given by:

$$\mathbf{T} = -p\mathbf{I} + \mathbf{D}.$$

By p we denote the pressure, $\mathbf{I}$ is the unit tensor and $\mathbf{D}$ takes into account the viscous properties of the fluid. Setting $\nabla\cdot\mathbf{D} = \mathbf{d}$, we obtain:

$$\nabla\cdot\mathbf{T} = -\nabla p + \mathbf{d}.$$

Thus it holds:

$$\int_{z_1}^{z_2} \left(\int_S \frac{D\mathbf{u}}{Dt} d\sigma\right) dz = \int_{z_1}^{z_2} \left(\int_S \left[\mathbf{f}^b + \frac{1}{\rho}(-\nabla p + \mathbf{d})\right] d\sigma\right) dz.$$

A comparison of the two integrands yields for the z- and third components, respectively:

$$\int_S \frac{Du_3}{Dt} d\sigma = \int_S \left[\mathbf{f}_3^b + \frac{1}{\rho}\left(-\frac{\partial p}{\partial z} + \mathbf{d}_3\right)\right] d\sigma.$$

Together with (3.11) it follows:

$$\frac{\partial}{\partial t}\left(A\overline{u_3}\right)+\frac{\partial}{\partial z}\left[A\overline{u_3^2}\right]=\int_S\left[\mathbf{f}_3^b+\frac{1}{\rho}\left(-\frac{\partial p}{\partial z}+\mathbf{d}_3\right)\right]d\sigma+\int_{\partial S}u_3\mathbf{w}\cdot n\,d\gamma$$

Introducing the notation for averages within the first summand, we obtain:

$$\frac{\partial}{\partial t}\left(A\overline{u_3}\right)+\frac{\partial}{\partial z}\left[A\overline{u_3^2}\right]=\frac{A}{\rho}\left(\rho\overline{\mathbf{f}_3^b}-\frac{\partial\overline{p}}{\partial z}+\overline{\mathbf{d}_3}\right)+\int_{\partial S}u_3\mathbf{w}\cdot n\,d\gamma.$$

To simplify the integral wrt. the pressure derivative, one requires the parametrization of S using polar coordinates and the Leibniz rule (see Sect. 7.1):

$$\begin{aligned}
\int_{S(z,t)}\frac{\partial p}{\partial z}\,d\sigma &= \int_0^{2\pi}\int_0^{R(z)}\frac{\partial p}{\partial z}r\,drd\theta\\
&= \int_0^{2\pi}\frac{\partial}{\partial z}\left(\int_0^{R(z)}pr\,dr\right)-p\left(z,R\left(z\right),\theta,t\right)R\left(z\right)\frac{\partial R}{\partial z}\left(z\right)d\theta\\
&= \frac{\partial}{\partial z}\left(\int_0^{2\pi}\int_0^{R(z)}pr\,drd\theta\right)-\int_0^{2\pi}p\left(z,R\left(z\right),\theta,t\right)R\left(z\right)\frac{\partial R}{\partial z}\left(z\right)d\theta\\
&= \frac{\partial}{\partial z}\left(A\overline{p}\right)-\int_0^{2\pi}p\left(z,R\left(z\right),\theta,t\right)R\left(z\right)\frac{\partial R}{\partial z}\left(z\right)d\theta\\
&= A\frac{\partial\overline{p}}{\partial z}+\frac{\partial A}{\partial z}\overline{p}-\int_0^{2\pi}p\left(z,R\left(z\right),\theta,t\right)R\left(z\right)\frac{\partial R}{\partial z}\left(z\right)d\theta\\
&= A\frac{\partial\overline{p}}{\partial z}+2\pi R\frac{\partial R}{\partial z}\overline{p}-\frac{1}{2\pi}\int_0^{2\pi}p\left(z,R\left(z\right),\theta,t\right)\,d\theta\cdot 2\pi R\frac{\partial R}{\partial z}.
\end{aligned}$$

Since we have assumed vessels, with $R/L=\varepsilon\ll 1$, we can make the following assumption (see Sect. 7.2):

$$\frac{\partial p}{\partial r}=0$$

i.e. p is constant with respect to the radial component. From this it follows:

$$\overline{p}\left(z,t\right)=\frac{1}{R^2\left(z,t\right)\pi}\int_0^{2\pi}\int_0^{R(z)}p\left(r,z,\theta,t\right)r\,drd\theta=\frac{1}{2\pi}\int_0^{2\pi}p\left(R\left(z\right),z,\theta,t\right)\,d\theta.$$

Thus, the following holds:

$$\int_{S(z,t)}\frac{\partial p}{\partial z}\,d\sigma=A\frac{\partial\overline{p}}{\partial z}.$$

By introducing a factor α (Coriolis coefficient), a relation between $\overline{u_3^2}$ and $\overline{u_3}$ can be determined for circular cross-sectional areas:

$$\overline{u_3^2}=\frac{1}{A}\int_S u_3^2\,d\sigma=\alpha\cdot\overline{u_3}^2. \tag{3.12}$$

A flat velocity profile for u_3 results in $\alpha = 1$. For a parabolic velocity profile, we obtain $\alpha = 4/3$. The term modeling the viscous forces in the Newtonian case can be described by a function f_r [11, 14, 16]:

$$\frac{A}{\rho}\overline{\mathbf{d}_3} = f_r(z,t) = 2\mu\pi R(z,t)\left[\frac{\partial u_3}{\partial r}\right]_{r=R},$$

where we denote by $\mu\left[{}^{\mathrm{cm}^2}/_{\mathrm{s}}\right]$ the viscosity of the fluid and $R\,[cm]$ the mean radius of the surface S. u_3 is given by the following profile:

$$u_3(z,r,\theta,t) = \overline{u_3}(z,t)\cdot\frac{\gamma+2}{\gamma}\left[1-\left(\frac{r}{R(z,t)}\right)^{\gamma}\right]. \tag{3.13}$$

The parameter γ governs the shape of the velocity field. For $\gamma = 2$ we obtain a parabolic profile and thus a Poiseuille-like flow. The parameter $\gamma = 9$ produces a profile that decreases fast towards the boundary and is flat in the center of the vessel (see Fig. 3.4). Moreover it holds:

$$\frac{A}{\rho}\overline{\mathbf{d}_3} = -2\mu\pi\cdot(\gamma+2)\cdot\overline{u_3}$$

and finally we obtain:

$$\frac{\partial}{\partial t}(A\overline{u_3}) + \frac{\partial}{\partial z}\left(\alpha A\overline{u_3}^2\right) = A\overline{\mathbf{f}_3^b} - \frac{A}{\rho}\frac{\partial\overline{p}}{\partial z} - K_R\overline{u_3} + \int_{\partial S} u_3\mathbf{w}\cdot n\,d\gamma, \tag{3.14}$$

where

$$K_R = 2\mu\pi\cdot(\gamma+2). \tag{3.15}$$

Note that the viscosity μ in K_R is assumed to be constant, i.e., we assume that blood can be considered a Newtonian fluid. At the macrocirculation level, this can

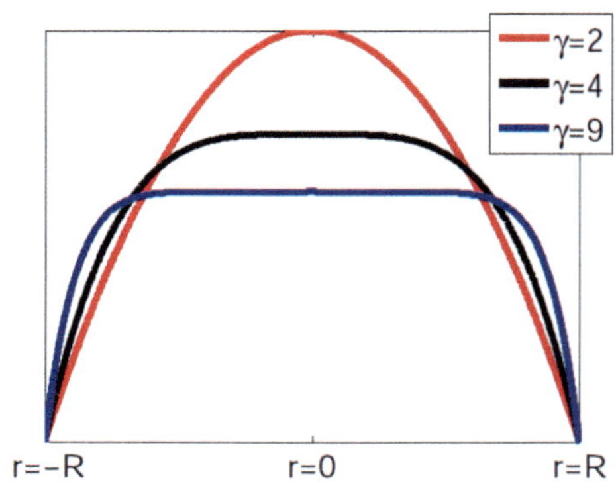

Fig. 3.4 Radial velocity profiles as a function of γ

be justified since measurements reveal that the viscosity is almost constant in vessels of the macrocirculation (see Fig. 1.14). A dimensional reduced blood flow model that also accounts for non-Newtonian effects can be found, e.g., in [17].

According to [18, Sect. 2], the following relationship exists between the parameters α and γ:

$$\gamma = \frac{2-\alpha}{\alpha-1} \quad \text{for } \alpha > 1. \tag{3.16}$$

Counting the unknowns in (3.10) and (3.14), we find that the number of unknowns (p, A and u_3) exceeds the number of equations. For this reason, one needs a further equation to close the system of equations. Up to now, only the conservation of mass and momentum have been taken into account, whereas the interaction of flow with the vessel wall has not yet been integrated into the model. Since the interaction of blood flow with the vessel wall can not be neglected at the level of the macrocirculation, the missing equation should result from an FSI model.

In the following, we present appropriate FSI models from literature (see e.g. [63, Sect. 2.1.3]). Before we investigate these models, some notation has to be introduced: The initial pressure $p(z,0)$ is denoted by $p_0(z)$ and $A_0(z)$ is the corresponding section area. The symbol p_{ext} represents an external pressure at the outer wall of the blood vessel.

Simplified FSI-models

(FSI 1)

The simplest FSI model, is based on the assumption that the cross-sectional area A depends linearly on the internal pressure p and that there are only small deformations:

$$A(z,t) = A_0(z) + C(z,t)\left(p(z,t) - p_0(z)\right),$$

where

$$\left|C(z,t)\left(p(z,t) - p_0(z)\right)\right| \ll A_0(z).$$

The factor

$$C(z,t) = \left.\frac{\partial A}{\partial p}\right|_{A=A_0}$$

can be considered as a compliance parameter of the vessel wall. These equations together with experimentally determined values for the parameter C can be found, e.g., in [19].

(FSI 2)

The FSI model (FSI 1) can be regarded as a Taylor expansion of A, which is truncated after the linear term. A more complex FSI model can be derived by extending the Taylor expansion. In [20] the following equation can be found:

$$A(z,t) = A_0(z)\left[1 + C_0(p(z,t) - p_0(z)) + C_1(p(z,t) - p_0(z))^2\right],$$

where the parameters C_0 and C_1 have to be determined by experiments.

(FSI 3)

In [6, 21, 22] and [23, Abschnitt 4.2.2] a FSI model has been derived, which is based on the assumption that the vessel walls are thin compared to the cross-sectional areas of the blood vessels and that the deformation of the vessel wall is in accordance with a linear-elastic law.

In addition, only deformations in the radial direction are considered. Deformations in the longitudinal direction are are neglected. This means that longitudinal and radial anisotropies cannot be represented by the elasticity module. Neglecting the longitudinal deformations yields an independent ring model [24, Chap. 11], which can be described by means of ODEs or algebraic equations. If the longitudinal deformations are neglected one would obtain a further partial differential equation of second order in space and time [25]. Upon closer inspection, one would realize that this PDE has the shape of a wave equation. In order to avoid the incorporation of another PDE in addition to the PDEs for mass and momentum conservation, longitudinal deformations are negelected in the remainder of this book. In addition to that, we consider the vessel wall as a homogeneous medium (see Fig. 3.5), i.e., the individual vessel layers such as the intermediate membranes, muscle tissue, and the endothelial layer are not considered separately but are homogenized such that they represent a uniform vessel wall.

Using Hooke's law [26], the following relationship between the stress σ [Pa] and the strain ϵ [-] is obtained:

$$\sigma = E\epsilon. \tag{3.17}$$

The constant E $[Pa]$ is known as elasticity module. For simplicity, we assume a constant elasticity module for each individual vessel. To determine σ and ϵ, a cross section with radius R and wall thickness h_0 is taken into account. Making use of the previous assumptions, the strain in (3.17) can be computed as follows:

$$\epsilon = \frac{1}{1-\nu^2}\frac{R-R_0}{R_0},$$

where ν $[-]$ is the Poisson ratio. By means of the Young-Laplace equation for cylindrical objects and assuming there is no external pressure, we obtain for a cylinder with wall thickness h_0:

$$p = \frac{h_0\sigma}{\pi R}.$$

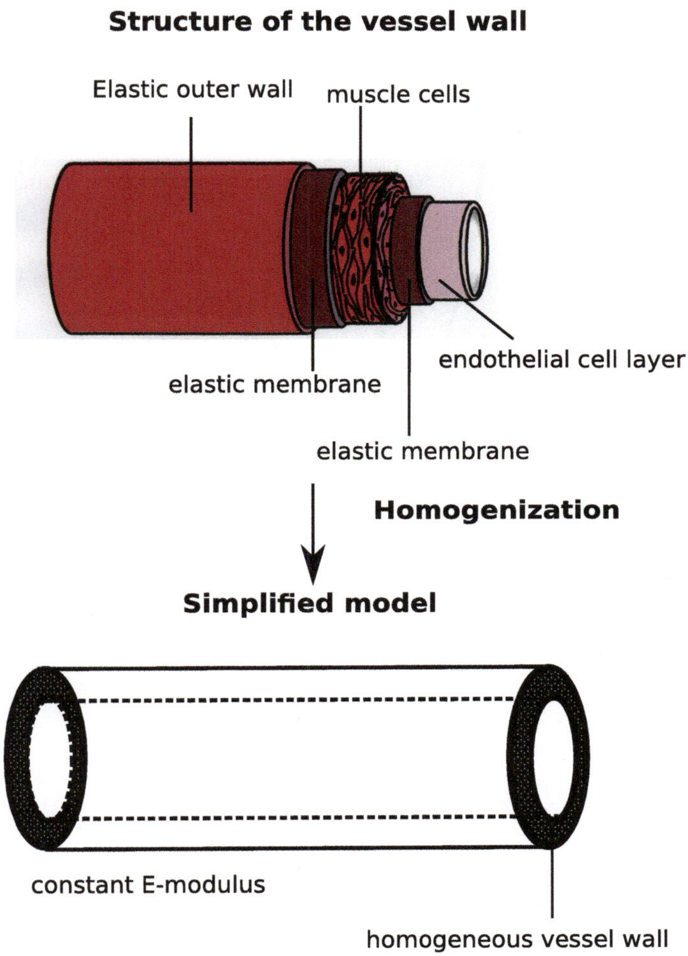

Fig. 3.5 This figure presents the structure of an arterial vessel wall. For the derivation of an FSI model, the vessel wall is homogenized, i.e. instead of the multilayered vessel wall, we only consider in the following homogeneous vessel walls with a constant E-modulus

This equation can be derived similarly to the Young-Laplace equation for spheres (see Sect. 1.2.2) . Combining Hooke's law and the Young-Laplace equation for cylindrical objects, we finally obtain the FSI-model from [11, Sect. 2.5]:

$$p\,(A) = G_0 \left(\sqrt{\frac{A}{A_0}} - 1 \right), \tag{3.18}$$

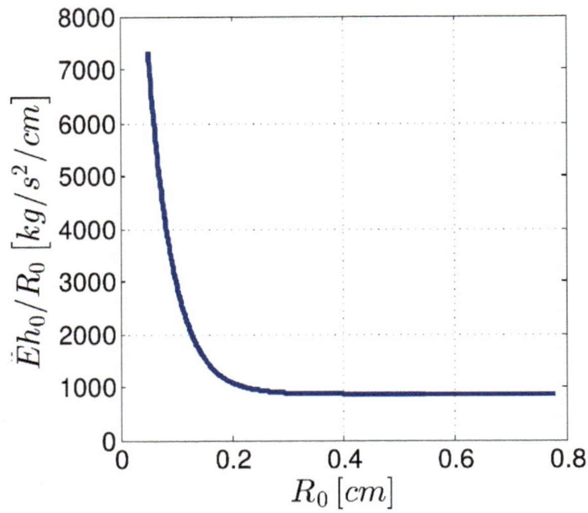

Fig. 3.6 Material parameter $\frac{Eh_0}{R_0}$ as a function of radius R_0

where the elasticity parameter is given by:

$$G_0 = \frac{\sqrt{\pi} E h_0}{\left(1 - \nu^2\right) \sqrt{A_0}}$$

Setting $\nu = 1/2$ it results:

$$G_0 = \frac{4\sqrt{\pi} E h_0}{3\sqrt{A_0}}$$

To be able to determine the material parameter $\frac{Eh_0}{R_0}$ in a more convenient way, Westerhof and his co-authors [27] determined the following relationship from measured data (see Fig. 3.6):

$$\frac{Eh_0}{R_0} = k_1 \exp\left(k_2 R_0\right) + k_3,$$

where the model parameters are given by:

$$k_1 = 2.0 \cdot 10^4\ kg \cdot s^{-2} \cdot cm^{-1},\ k_2 = -22.53\ cm^{-1} \text{ and } k_3 = 8.65 \cdot 10^2\ kg \cdot s^{-2} \cdot cm^{-1}.$$

We observe that the expression $E \cdot h_0/R_0$ increases rapidly for small radii while for vessels with a large radius it is almost constant. Thus, smaller arteries, unlike large arteries, have a larger elasticity parameter G_0. Therefore, the radial extention of the smaller vessels is much smaller than the radial extention of the larger arteries.

(FSI 4)

A generalization of (FSI 3) can be found in [28]. In this publication, the relationship between the area of the cross section and the pressure reads as follows:

$$p(z,t) = \Psi(A, K, A_0),$$

where

$$\Psi(A, K, A_0) = K(z)\,\phi(A, A_0)$$

and

$$\phi(A, A_0) = \left[\left(\frac{A(z,t)}{A_0(z)}\right)^m - \left(\frac{A(z,t)}{A_0(z)}\right)^n\right].$$

m and n are two model parameters. $K(z)$ denotes a elasticity function. Setting $m = \frac{1}{2}$ and $n = 0$, as well as

$$K(z) = \sqrt{\frac{\pi}{A_0(z)}}\left(\frac{h_0 E}{1-\nu^2}\right), \tag{3.19}$$

we obtain (FSI 3). For more information on generalized FSI models, one can consider: [29, 30].

(FSI 5)

The two previous FSI models are based on the assumption that the deformation behavior of the vessel walls exhibits purely elastic features. However, a closer look reveals that the deformation of blood vessels is subject to viscoelastic effects. If some stress is imposed on a vessel wall one can observe that it continues to deform after pressure has been imposed. This means that besides a pure elastic deformation, which ends after a short time period, there is also a retarded deformation caused by viscoelasticity. A well-known viscoelastic FSI-model is the Kelvin-Voigt model [31, 32]. It represents the stress σ by the following relationship [33]:

$$\sigma(t) = E\epsilon(t) + \phi\frac{\partial \epsilon(t)}{\partial t}, \tag{3.20}$$

where ϕ is a coefficient quantifying the viscoelasticity of the vessel wall. Note that for $\phi = 0$ the linear elasticity Eq. (3.17) results. If a constant stress σ_0 is imposed on the viscoelastic material, the deformations approach the deformation σ_0/E of a pure elastic material:

$$\epsilon(t) = \frac{\sigma_0}{E}\left(1 - e^{-\lambda t}\right),$$

where the parameter λ can be considered as a relaxation rate:

$$\lambda = \frac{E}{\phi}.$$

This means that λ incorporates the retarded deformation of the vessel wall into the model. Using (3.18) the following FSI-models are obtained:

$$p = G_0 \left(\sqrt{\frac{A}{A_0}} - 1 \right) + \nu_s \frac{\partial A}{\partial t}.$$

ν_s is the viscosity parameter of the vessel wall. It is determined as follows [34]:

$$\nu_s = \frac{\sqrt{\pi} \phi h_0}{2 \left(1 - \nu^2\right) \sqrt{A_0 A}}.$$

Other FSI models can be found, e.g., in [35]. In the remainder of this book, we use the FSI model (FSI 4) given by Eq. (3.18).This model has successfully been applied in a series of publications [11, 12, 22, 29, 36, 37].

***AQ*-system**

To further simplify the notation, we introduce the variable Q $\left[{}^{\mathrm{cm}^3}/_{\mathrm{s}}\right]$ for the volumetric flux through the cross section S:

$$Q = A \cdot \overline{u_3}.$$

In addition, it is assumed that there is no blood volume exchange neither with tissue of the vessel wall nor through the vessel wall, i.e., it holds that: $\mathbf{w} \cdot n = 0$. The external forces are neglected. Thus we set: $\overline{\mathbf{f}_3^b} = 0$. Combining (3.10) and (3.14), it follows:

$$\begin{aligned}
&\frac{\partial A}{\partial t} + \frac{\partial Q}{\partial z} = 0, \\
&\frac{\partial Q}{\partial t} + \frac{\partial}{\partial z} \left(\alpha \frac{Q^2}{A} \right) + \frac{A}{\rho} \frac{\partial p}{\partial z} = -K_R \frac{Q}{A} = -2\pi \mu \frac{\alpha}{\alpha - 1} \frac{Q}{A}.
\end{aligned}$$

Using (3.18) for the pressure, one obtains a 1D system for the fluid variables (A, Q):

$$\frac{A}{\rho} \frac{\partial p}{\partial z} = \frac{A}{\rho} \frac{G_0}{2\sqrt{A \cdot A_0}} \frac{\partial A}{\partial z} = \frac{\partial}{\partial z} \left(\frac{G_0 A^{3/2}}{3\rho \sqrt{A_0}} \right)$$

$$\begin{aligned}
&\frac{\partial A}{\partial t} + \frac{\partial Q}{\partial z} = 0, \\
&\frac{\partial Q}{\partial t} + \frac{\partial}{\partial z} \left(\alpha \frac{Q^2}{A} + \frac{G_0 A^{3/2}}{3\rho \sqrt{A_0}} \right) = -K_R \frac{Q}{A} = -2\pi \mu \frac{\alpha}{\alpha - 1} \frac{Q}{A}.
\end{aligned} \tag{3.21}$$

Thereby, $G_0 \equiv \text{const}$ and $A_0 \equiv \text{const}$ holds.

Representation of the 1D fluid equations for flat velocity profiles ($\alpha = 1$)

In literature [11, 18] different versions of the 1D fluid Eq. (3.21) with flat velocity profile, i.e., $\alpha = 1$, are considered. It can be seen at a first glance that for $\alpha = 1$ the viscosity parameter K_R is not defined. In order to circumvent this issue, one neglects the viscosity parameter: $K_R = 0$. This is acceptable, since for a flat velocity profile there is no boundary layer with friction on the vessel walls. From this, one obtains an inviscid version of the fluid equations:

$$\begin{aligned} \frac{\partial A}{\partial t} + \frac{\partial Q}{\partial z} &= 0, \\ \frac{\partial Q}{\partial t} + \frac{\partial}{\partial z}\left(\frac{Q^2}{A} + \frac{G_0 A^{3/2}}{3\rho\sqrt{A_0}}\right) &= 0. \end{aligned} \tag{3.22}$$

Besides the frictionless formulation of the 1D fluid equations one can often find in literature [38] a mixed formulation. In this context, the parameter α is chosen to be one, while in the viscous parameter K_R, $\alpha > 1$ is used:

$$\begin{aligned} \frac{\partial A}{\partial t} + \frac{\partial Q}{\partial z} &= 0, \\ \frac{\partial Q}{\partial t} + \frac{\partial}{\partial z}\left(\frac{Q^2}{A} + \frac{G_0 A^{3/2}}{3\rho\sqrt{A_0}}\right) &= -K_R\frac{Q}{A} = -2\pi\mu\frac{\alpha}{\alpha - 1}\frac{Q}{A}. \end{aligned} \tag{3.23}$$

Despite of the fact that (3.23) differs from (3.21) and is not in accordance with the derivation of the 1D model, it is frequently used for bloodflow simulations. However, it turns out that even a wrong weighting of the convective term Q^2/A yields reasonable pressure and velocity curves [22, 39, 40]. A further attractive feature of (3.23) is that the mathematical analysis of this system is more convenient [11, 25]. In Sects. 3.6.3 and 4.4 this issue is investigated more detailed.

Transport processes

System (3.21) governs the one-dimensional velocity field in a vessel and thus it can be regarded as the counterpart to the three-dimensional model Eqs. (3.1a) and (3.1b). In order to model the transport of a substance in 1D, we require as in case of the fluid equations a 1D version of the corresponding 3D Eq. (3.2). Deriving this equation, we choose in (3.9): $f = c$, where c $\left[\frac{mmol}{cm^3}\right]$ denotes the volumetric concentration in (3.2). The averaged concentration variable is defined as follows:

$$\overline{c} = \frac{1}{A}\int_S c \, d\sigma =: \frac{\Gamma}{A}.$$

Next we assume that the boundary integral on the right hand side of Eq. (3.9), which determines the loss of transported mass across the vessel wall, is set to zero. This is motivated by the fact that the vessel walls of large arteries are impermeable for blood flow. In addition, the amount of a certain substance like oxygen that is transported from the lumen of a vessel into the vessel walls is considered to be negligible. The second term on the right hand side of Eq. (3.9) can be set to zero for the derivation of a dimension reduced transport equation. According to Table 1.1 there are higher Reynolds numbers in the larger vessels. Therefore, it is obvious that a convection-dominated flow can be detected in this type of vessels. This means that in the 3D Eq. (3.2) the diffusive terms can be neglected. Combined with the modeling assumption that blood is an incompressible fluid, we obtain:

$$\frac{\partial c}{\partial t} + \nabla \cdot (\mathbf{u}\, c) = 0 \quad \text{i.e.} \quad \int_S \left[\frac{\partial c}{\partial t} + \nabla \cdot (\mathbf{u}\, c) \right] d\sigma = 0.$$

Simplifying the second term at the left hand side, we assume as in [12, Equation (2.14)] that the profile of the concentration on a cross section $S(z,t)$ can be described by a radial symmetric function with two shape parameters κ und δ:

$$c(z, r, \theta, t) = \overline{c}(z,t) \cdot \frac{\kappa + 2}{\kappa + 2(1-\delta)} \left(1 - \delta \left(\frac{r}{R}\right)^{\kappa}\right).$$

By this one obtains:

$$\begin{aligned}
A\, \overline{c \cdot u_3}(z,t) &= A \cdot \frac{1}{A} \int_{S(z,t)} c \cdot u_3 \, d\sigma = \int_{S(z,t)} c \cdot u_3 \, d\sigma \\
&= \overline{u_3} \cdot \frac{\gamma + 2}{\gamma} \int_{S(z,t)} c \cdot \left(1 - \left(\frac{r}{R}\right)^{\gamma}\right) d\sigma \\
&= \frac{Q}{A} \cdot \overline{c} \cdot \frac{\gamma+2}{\gamma} \frac{\kappa+2}{\kappa+2(1-\delta)} \int_{S(z,t)} \left(1 - \delta\left(\frac{r}{R}\right)^{\kappa}\right) \cdot \left(1 - \left(\frac{r}{R}\right)^{\gamma}\right) d\sigma \\
&= \frac{Q}{A} \cdot \Gamma \cdot \frac{\gamma+2}{\gamma} \frac{\kappa+2}{\kappa+2(1-\delta)} \frac{1}{A} \int_{S(z,t)} \left(1 - \delta\left(\frac{r}{R}\right)^{\kappa}\right) \cdot \left(1 - \left(\frac{r}{R}\right)^{\gamma}\right) d\sigma \\
&= \frac{Q}{A} \cdot \Gamma \cdot \underbrace{\frac{\gamma+2}{\gamma} \frac{\kappa+2}{\kappa+2(1-\delta)} 2 \left[1 - \frac{\delta}{\kappa+2} - \frac{1}{\gamma+2} + \frac{\delta}{\kappa+\gamma+1}\right]}_{=:\omega} = \omega \frac{Q}{A} \cdot \Gamma.
\end{aligned}$$

Summarizing the previous considerations, the following 1D equation for the averaged concentration variable Γ results:

$$\frac{\partial \Gamma}{\partial t} + \frac{\partial}{\partial z}\left(\omega \frac{Q\Gamma}{A}\right) = 0. \tag{3.24}$$

Introduction of flux functions

Summarizing the fluid Eq. (3.21) and the transport Eq. (3.24), one obtains the following first order system:

$$\frac{\partial A}{\partial t} + \frac{\partial Q}{\partial z} = 0, \ z \in (0, L), \ t > 0,$$
$$\frac{\partial Q}{\partial t} + \frac{\partial}{\partial z}\left(\alpha \frac{Q^2}{A} + \frac{G_0 A^{3/2}}{3\rho\sqrt{A_0}}\right) = -K_R \frac{Q}{A}, \ z \in (0, L), \ t > 0,$$
$$\frac{\partial \Gamma}{\partial t} + \frac{\partial}{\partial z}\left(\omega \frac{\Gamma \cdot Q}{A}\right) = 0, \ z \in (0, L), \ t > 0. \tag{3.25}$$

The three unknowns A, Q und Γ form a vector $\mathbf{U} = (A, Q, \Gamma)^T$ and the expressions associated with the spatial derivative define a flux function $\mathbf{F}$:

$$\mathbf{F}(\mathbf{U}) = \begin{pmatrix} Q \\ \alpha \frac{Q^2}{A} + \frac{G_0 A^{3/2}}{3\rho\sqrt{A_0}} \\ \omega \frac{\Gamma \cdot Q}{A} \end{pmatrix}.$$

The right hand side of the above system is described by a source term $\mathbf{S}$:

$$\mathbf{S}(\mathbf{U}) = \begin{pmatrix} 0 \\ -K_R \frac{Q}{A} \\ 0 \end{pmatrix}.$$

Overall a compact form of the model is obtained:

$$\frac{\partial \mathbf{U}}{\partial t} + \frac{\partial \mathbf{F}}{\partial z}(\mathbf{U}) = \mathbf{S}(\mathbf{U}), \ z \in (0, L), \ t > 0, \tag{3.26}$$
$$\mathbf{U}(0) = \mathbf{U_0}, \ z \in (0, L), \ t = 0.$$

The vector $\mathbf{U_0} = (A_0, 0, \Gamma_0)^T$ governs the initial condition of the problem. On closer examination one can observe that the first two components do not depend on the concentration Γ. As a consequence a subsystem for the fluid variables A and Q consisting of the flux function

$$\mathbf{F_f}(\mathbf{U_f}) = \begin{pmatrix} Q \\ \alpha \frac{Q^2}{A} + \frac{G_0 A^{3/2}}{3\rho\sqrt{A_0}} \end{pmatrix}, \tag{3.27}$$

the source term

$$\mathbf{S_f}(\mathbf{U_f}) = \begin{pmatrix} 0 \\ -K_R \frac{Q}{A} \end{pmatrix} \tag{3.28}$$

and the vector $\mathbf{U_f} = (A, Q)^T$ can be defined. Combined with a scalar equation for the concentration the following system results:

$$\frac{\partial \mathbf{U_f}}{\partial t} + \frac{\partial \mathbf{F_f}}{\partial z} (\mathbf{U_f}) = \mathbf{S_f} (\mathbf{U_f}), \ z \in (0, L), \ t > 0, \tag{3.29a}$$

$$\frac{\partial \Gamma}{\partial t} + \frac{\partial}{\partial z} \left(\omega \frac{Q}{A} \Gamma \right) = 0, \ z \in (0, L), \ t > 0. \tag{3.29b}$$

The initial conditions are given by:

$$\mathbf{U_f}(0) = (A_0, 0)^T, \ \Gamma(0) = \Gamma_0, \ z \in (0, L), \ t = 0.$$

3.3 Classification of the Model Equations

In the upcoming sections, we want to investigate the regularity of solutions resulting from the blood flow model and the transport equation. For this purpose, it is required to classify the model equations in a mathematical way. Considering the system of Eq. (3.25), one can recognize that it is a coupled and non-linear system of partial differential equations (PDEs). Since only partial derivatives of first order occur, we have a first order system.

Among PDEs of first order, different classes can be identified. One of these classes is denoted as hyperbolic. Besides elliptic and parabolic PDEs, hyperbolic PDEs belong to the three most important classes of PDEs [41, Sect. 1.2]. Hyperbolic PDEs of first order with a single spatial variable can be represented in the following form:

$$\frac{\partial \mathbf{U}}{\partial t} + \frac{\partial \mathbf{F}}{\partial z} (\mathbf{U}) = \mathbf{S} (\mathbf{U}). \tag{3.30}$$

$\mathbf{U} \in \mathbb{R}^n$ is the vector of unknowns, where n is the dimenion of the system:

$$\mathbf{U} : \mathbb{R} \times [0, \infty) \to \mathbb{R}^n, \ (z, t) \mapsto \begin{pmatrix} U_1(z, t) \\ \vdots \\ U_n(z, t) \end{pmatrix}, \ n \geq 1.$$

The functions

$$\mathbf{F} : \mathbb{R}^n \to \mathbb{R}^n, \ \mathbf{U} \mapsto \mathbf{F}(\mathbf{U}) \text{ and } \mathbf{S} : \mathbb{R}^n \to \mathbb{R}^n, \ \mathbf{U} \mapsto \mathbf{S}(\mathbf{U})$$

are called flux function and source term. Defining (3.30) on $\mathbb{R}$ and adding an initial condition:

$$\mathbf{U}(z, 0) = \mathbf{U_0}(z), \ z \in \mathbb{R}, \tag{3.31}$$

we obtain a *Cauchy problem*. If the components of $\mathbf{F}$ are partially differentiable, (3.30) can be provided in a quasi-linear form:

$$\frac{\partial \mathbf{U}}{\partial t} + \mathbf{H}(\mathbf{U})\frac{\partial \mathbf{U}}{\partial z} = \mathbf{S}(\mathbf{U}), \tag{3.32}$$

the matrix $\mathbf{H}(\mathbf{U})$ is the Jacobian of $\mathbf{F}$: $\mathbf{H}(\mathbf{U}) = \frac{\partial \mathbf{F}}{\partial \mathbf{U}}$. By means of the Jacobian, we can define the term *hyperbolic PDE* [41, Sect. 1.3].

Definition 3.1 The partial differential Eq. (3.32) is called hyperbolic for a vector $\mathbf{U}$, if the matrix $\mathbf{H}(\mathbf{U})$ has only real eigenvalues and if $\mathbf{H}(\mathbf{U})$ is diagonalizable, i.e., there is a basis of eigenvectors.

The name *hyperbolic* is motivated by the fact that by means of a Jacobian $\mathbf{H}$ derived from a hyperbolic PDE, *hyperbolas* can be constructed [42]. At the end of this section, we provide some examples of first order hyperbolic PDEs:

(i) *Linear transport equation* with a constant velocity $a \in \mathbb{R}$:

$$\frac{\partial u}{\partial t} + a\frac{\partial u}{\partial z} = 0. \tag{3.33}$$

In this case it holds: $\mathbf{U} = u,\ \mathbf{F}(\mathbf{U}) = a \cdot u$ and $\mathbf{S}(\mathbf{U}) = 0$.

(ii) *Burgers equation:*

$$\frac{\partial u}{\partial t} + \frac{\partial}{\partial z}\left(\frac{u^2}{2}\right) = 0 \tag{3.34}$$

Here we have: $\mathbf{U} = u,\ \mathbf{F}(\mathbf{U}) = \frac{u^2}{2}$ and $\mathbf{S}(\mathbf{U}) = 0$.

(iii) *Wave equation* with the speed $c \in \mathbb{R}$ and source term f:

$$\frac{\partial^2 u}{\partial t^2} - c^2\frac{\partial^2 u}{\partial z^2} = f. \tag{3.35}$$

Defining $u_1 = \frac{\partial u}{\partial z}$ and $u_2 = \frac{\partial u}{\partial t}$, one obtains for the vector

$$\mathbf{U} = \begin{pmatrix} u_1 \\ u_2 \end{pmatrix}$$

and the source term

$$\mathbf{S}(\mathbf{U}) = \begin{pmatrix} 0 \\ f \end{pmatrix}$$

the following equation:

$$\frac{\partial \mathbf{U}}{\partial t} + A\frac{\partial \mathbf{U}}{\partial z} = \mathbf{S}(\mathbf{U}). \tag{3.36}$$

The matrix A is given by

$$A = \begin{pmatrix} 0 & -1 \\ -c^2 & 0 \end{pmatrix}$$

and the flux function reads as follows:

$$\mathbf{F}(\mathbf{U}) = A \cdot \mathbf{U}.$$

The eigenvalues λ_1 and λ_2 are given by: $\lambda_1 = c$ and $\lambda_2 = -c$. Further considerations show that the matrix A is diagonalizable. Thus system (3.36) is hyperbolic.

3.3.1 The Blood Flow Model is a System of Hyperbolic PDEs

In a first step, we transform the PDE system (3.29a) into its quasi-linear form:

$$\frac{\partial \mathbf{U}_\mathbf{f}}{\partial t} + H(\mathbf{U}_\mathbf{f}) \frac{\partial \mathbf{U}_\mathbf{f}}{\partial z}(\mathbf{U}_\mathbf{f}) = \mathbf{S}_\mathbf{f}(\mathbf{U}_\mathbf{f}), \ z \in (0, L), \ t > 0. \tag{3.37}$$

For the Jacobian one obtains:

$$H(\mathbf{U}_\mathbf{f}) = \begin{pmatrix} 0 & 1 \\ -\alpha \frac{Q^2}{A^2} + \frac{G_0 A^{1/2}}{2\rho\sqrt{A_0}} & 2\alpha \frac{Q}{A} \end{pmatrix}.$$

Next we decompose the matrix $H(\mathbf{U}_\mathbf{f})$ into a product that consists of a diagonal matrix containing the eigenvalues, the matrix of eigenvectors and its inverse. These matrices are used to transform the PDE system locally into a system of characteristic variables. Computing the eigenvalues λ_1 and λ_2 of the Jacobian $H(\mathbf{U}_\mathbf{f})$, we obtain:

$$\lambda_1(\mathbf{U}_\mathbf{f}) = \alpha \cdot v_{AQ} - c_\alpha \text{ und } \lambda_2(\mathbf{U}_\mathbf{f}) = \alpha \cdot v_{AQ} + c_\alpha.$$

The variables v_{AQ} and c_α denote the velocity of blood flow and the characteristic velocity of the blood vessel:

$$v_{AQ} = \frac{Q}{A} \text{ und } c_\alpha = \sqrt{c_{AQ}^2 + \alpha(\alpha - 1) v_{AQ}^2}, \ c_{AQ} = \sqrt{\frac{G_0}{2\rho}\left(\frac{A}{A_0}\right)^{\frac{1}{4}}}. \tag{3.38}$$

According to [39], it holds for realistic blood flows: $v_{AQ} \ll c_\alpha$. This implies: $\lambda_1 < 0$ and $\lambda_2 > 0$ i.e., the eigenvalues of the Jacobian have a different sign. Computing the left eigenvectors l_1 and l_2 we have to consider the following equation:

$$\left[H^T(\mathbf{U_f}) - \lambda_1 \mathbf{I}\right] \cdot l_1 = 0$$

$$\left[\begin{pmatrix} 0 & -\alpha\frac{Q^2}{A^2} + \frac{G_0\sqrt{A}}{2\rho\sqrt{A_0}} \\ 1 & 2\alpha\frac{Q}{A} \end{pmatrix} - \begin{pmatrix} \alpha v_{AQ} - c_\alpha & 0 \\ 0 & \alpha v_{AQ} - c_\alpha \end{pmatrix}\right] \cdot l_1 = 0$$

$$\begin{pmatrix} -\alpha v_{AQ} + c_\alpha & -\alpha v_{AQ}^2 + c_{AQ}^2 \\ 1 & \alpha v_{AQ} + c_\alpha \end{pmatrix} \cdot l_1 = 0.$$

Solving this system of equations and repeating the same procedure for l_2, one obtains:

$$l_1 = \xi_1 \begin{pmatrix} -c_\alpha - \alpha v_{AQ} \\ 1 \end{pmatrix} \text{ and } l_2 = \xi_2 \begin{pmatrix} c_\alpha - \alpha v_{AQ} \\ 1 \end{pmatrix},$$

where the parameters $\xi_1 \in \mathbb{R}$ and $\xi_2 \in \mathbb{R}$ can be chosen in an arbitrary way. Here, we set $\xi_1 = -1/A$ and $\xi_2 = 1/A$. By means of these matrices the Jacobian $H(\mathbf{U_f})$ can be decomposed as follows:

$$H(\mathbf{U_f}) = L^{-1}(\mathbf{U_f})\,\Lambda(\mathbf{U_f})\,L(\mathbf{U_f}).$$

The matrices L^{-1}, L and Λ are given by

$$L^{-1}(\mathbf{U_f}) = \frac{A}{2c_\alpha}\begin{pmatrix} 1 & 1 \\ -c_\alpha + \alpha v_{AQ} & c_\alpha + \alpha v_{AQ} \end{pmatrix}, \quad L(\mathbf{U_f}) = \frac{1}{A}\begin{pmatrix} c_\alpha + \alpha v_{AQ} & -1 \\ c_\alpha - \alpha v_{AQ} & 1 \end{pmatrix}$$

and

$$\Lambda(\mathbf{U_f}) = \begin{pmatrix} \alpha v_{AQ} - c_\alpha & 0 \\ 0 & \alpha v_{AQ} + c_\alpha \end{pmatrix} = \begin{pmatrix} \lambda_1(\mathbf{U_f}) & 0 \\ 0 & \lambda_2(\mathbf{U_f}) \end{pmatrix}.$$

Thus, the Jacobian matrix can be diagonalized i.e. the blood flow model can be described by a hyperbolic PDE system.

A crucial step required in terms of the analytical and numerical treatment is the transformation of the blood flow model into its characteristic form. For this purpose, a matrix-valued differential equation is established for the characteristic variables:

$$\frac{\partial \mathbf{W_f}}{\partial \mathbf{U_f}} = L(\mathbf{U_f}), \quad \mathbf{W_f}(A_0, 0) = \begin{pmatrix} \sqrt{\frac{G_0}{2\rho}} \\ 0 \end{pmatrix}. \tag{3.39}$$

Using the chain rule, we obtain the characteristic form of the blood flow model:

$$\frac{\partial \mathbf{W_f}}{\partial t} + \Lambda(\mathbf{U_f})\frac{\partial \mathbf{W_f}}{\partial z} = \mathbf{S_{W_f}}, \quad z \in (0, L), \quad t > 0,$$

where

$$\mathbf{W_f} = \begin{pmatrix} W_1 \\ W_2 \end{pmatrix}, \quad \mathbf{S}_{\mathbf{W_f}} = L\left(\mathbf{U_f}\right)\mathbf{S_f}\left(\mathbf{U_f}\right).$$

Recalling the signs of the eigenvalues, it can be concluded by means of the characteristic system that the solution $\mathbf{U_f}$ can be locally composed by a wave W_1 propagating in the negative z direction and a wave W_2 propagating in the positive z direction. This is due to the assumption that the eigenvalue $\lambda_1\left(\mathbf{U_f}\right)$ is negative and the eigenvalue $\lambda_2\left(\mathbf{U_f}\right)$ is positive. Therefore at each of the two boundaries $z = 0$ or $z = L$ a characteristic variable leaves the simulation domain $(0, L)$. To determine the entering characteristic variable, a boundary condition must be imposed in each case, i.e. we need exactly one boundary condition for the well-posedness of the blood flow model. The influence of flow inside the vessel is determined by the outgoing characteristic variables.

3.3.2 *The Transport Model is a Hyperbolic PDE*

Obviously, the quasi-linear form of the PDE (3.29b) reads as follows:

$$\frac{\partial \Gamma}{\partial t} + \left(\omega \frac{Q}{A}\right)\frac{\partial \Gamma}{\partial z} = 0, \ z \in (0, L), \ t > 0. \tag{3.40}$$

i.e. the Jacobian $H(\Gamma)$ is just a number and independent of the solution variable Γ:

$$H(\Gamma) = \omega \frac{Q}{A}.$$

Thus $H(\Gamma)$ is diagonalizable, if $\omega \cdot v_{AQ} = \omega \cdot Q/A \neq 0$ and the eigenvalue is also equal to $\omega \cdot v_{AQ}$. From this it follows that the incorporation of boundary conditions is depending on the sign of the velocity field.

3.4 Existence of Classical Solutions for Flat Velocity Profiles

A solution triple (A, Q, Γ) of the system of Eq. (3.25) is denoted as classical solution, if

$$(A, Q, \Gamma) \in C^1\left((0, L) \times \mathbb{R}^+\right)^3$$

holds. Since the equation for Γ depends on A and Q and due to the fact that A and Q do not depend on Γ, the blood flow model (3.29a) can be analyzed separately from the transport model (3.24).

We begin by examining the blood flow model. For simplicity a "flat" velocity profile is considered. This leads us to the parameter choice $\alpha = 1$ and a homogeneous system of equations. From a physical point of view, a model for frictionless flow is resulting. Using the dimensionless system of equations, it can be shown that the viscous source term is not significant for long arteries with larger cross-sections. Thus, the theoretical results related to the homogeneous system are also representative for this type of artery. This can also be observed in numerical experiments (see [11, Sect. 5.2]). The inhomogeneous blood flow model is given by:

$$\begin{aligned}
&\frac{\partial A}{\partial t} + \frac{\partial Q}{\partial z} = 0, \\
&\frac{\partial Q}{\partial t} + \frac{\partial}{\partial z}\left(\alpha \frac{Q^2}{A}\right) + \frac{A}{\rho}\frac{\partial p}{\partial z} = -2\pi\mu \frac{\alpha}{\alpha - 1}\frac{Q}{A}.
\end{aligned} \tag{3.41}$$

Deriving the dimensionless form, we introduce some reference variables and dimensionless parameters as well as solution variables. To the dimensionless variables a tilde is attached, while the reference variables have the index ref:

$$\begin{aligned}
&A = \tilde{A} \cdot R_{ref}^2, \;\; p = \tilde{p} \cdot \rho \cdot v_{ref}^2, \;\; z = \tilde{z} \cdot L_{ref}, \\
&v = \tilde{v} \cdot v_{ref}, \;\; t = \tilde{t} \cdot \frac{L_{ref}}{v_{ref}}, \;\; Q = \tilde{Q} \cdot v_{ref} \cdot R_{ref}^2, \;\; \mu = \tilde{\mu} \cdot \mu_{ref}.
\end{aligned}$$

Substituting these quantities into the above equation, we obtain:

$$\begin{aligned}
&\frac{\partial \tilde{A}}{\partial \tilde{t}} + \frac{\partial \tilde{Q}}{\partial \tilde{z}} = 0, \\
&\frac{\partial \tilde{Q}}{\partial \tilde{t}} + \frac{\partial}{\partial \tilde{z}}\left(\alpha \frac{\tilde{Q}^2}{\tilde{A}}\right) + \tilde{A}\frac{\partial \tilde{p}}{\partial \tilde{z}} = -2\pi \cdot \frac{\mu_{ref} L_{ref}}{v_{ref} R_{ref}^2} \cdot \tilde{\mu}\frac{\alpha}{\alpha - 1}\frac{\tilde{Q}}{\tilde{A}}.
\end{aligned}$$

Using the reference variables [11]:

$$\mu_{ref} = 3.2 \cdot 10^{-2} \mathrm{cm}^2/\mathrm{s}, \;\; R_{ref} = 0.82 \text{ cm}, \;\; L_{ref} = 10.0 \text{ cm}, \;\; \text{und } v_{ref} = 10.0 \text{ cm/s},$$

results in the following value for the prefactor in the second component of the source term:

$$\frac{\mu_{ref} L_{ref}}{v_{ref} R_{ref}^2} \approx 0.048.$$

However, this means that the viscous term is at least an order of magnitude smaller compared to the other terms and can be neglected for the upcoming analysis.

Existence of Global Classical Solutions

In order to prove the existence of classical solutions for hyperbolic differential equations, we consider the characteristic form of (3.41). Neglecting the source term in (3.41) and choosing $\alpha = 1$, we obtain:

$$\begin{aligned} \frac{\partial W_1}{\partial t} + \lambda_1 (W_1, W_2) \frac{\partial W_1}{\partial z} &= 0, \\ \frac{\partial W_2}{\partial t} + \lambda_2 (W_1, W_2) \frac{\partial W_2}{\partial z} &= 0, \end{aligned} \tag{3.42}$$

defined in the space-time domain:

$$D = \mathbb{R} \times (0, \infty) = \{(z, t) |\ t > 0,\ 0 < z < \infty\}.$$

The characteristic variables in (3.42) can be explicitly determined by a componentwise integration of the differential Eq. (3.39):

$$W_1 = -v_{AQ} + 4c_{AQ} \text{ und } W_2 = v_{AQ} + 4c_{AQ}. \tag{3.43}$$

Thus we obtain for the eigenvalues $\lambda_1 (W_1, W_2)$ and $\lambda_2 (W_1, W_2)$:

$$\begin{aligned} \lambda_1 (W_1, W_2) &= v_{AQ} - c_1 = -\frac{5}{8} W_1 + \frac{3}{8} W_2, \\ \lambda_2 (W_1, W_2) &= v_{AQ} + c_1 = -\frac{3}{8} W_1 + \frac{5}{8} W_2. \end{aligned}$$

For a further analysis of the blood flow model with a “flat” velocity profile, we reduce the system of partial differential Eq. (3.42) to a system of ordinary differential equations. Therefore, we define two curves z_1 and z_2 in the space-time domain D. These curves are defined as follows:

$$\begin{aligned} z_1' (t) &= \lambda_1 (W_1 (z_1 (t), t), W_2 (z_1 (t), t)), \\ z_2' (t) &= \lambda_2 (W_1 (z_2 (t), t), W_2 (z_2 (t), t)). \end{aligned}$$

In literature the curves z_1 and z_2 are referred to in literature as *characteristic curves* of the equation system (3.42). Recalling the assumption that the eigenvalues have the following signs: $\lambda_1 < 0$ and $\lambda_2 > 0$, we can conclude that the curve z_1 is moving into negative direction in time, while the curve z_2 is heading into positive direction. Restricting W_1 and W_2 to z_1 and z_2 and computing the total derivative wrt. t, we obtain combined with (3.42):

$$\begin{aligned} \frac{dW_1}{dt} (z_1 (t), t) &= \frac{\partial W_1}{\partial t} (z_1 (t), t) + z_1' (t) \frac{\partial W_1}{\partial z} (z_1 (t), t) = 0, \\ \frac{dW_2}{dt} (z_2 (t), t) &= \frac{\partial W_2}{\partial t} (z_2 (t), t) + z_2' (t) \frac{\partial W_2}{\partial z} (z_2 (t), t) = 0. \end{aligned}$$

This means that the characteristic variables are constant on the respective curves (wrt. to the time variable).

Moreover, it can be seen that W_1 moves into the negative spatial direction due to the negative sign of z_1' and W_2 moves into the positive space direction due to the positive sign of z_2'. Therefore, on the boundary $z = 0$, we require a boundary condition for the entering characteristic variable W_2 that we denote by g. On the boundary $t = 0$ two functions $W_{1,0}$ and $W_{2,0}$ have to be prescribed for W_1 and W_2. They can be considered as initial conditions for the time dependent problem (3.42). All in all, it holds:

$$W_1(z,t) = W_{1,0}(z), \; W_2(z,t) = W_{2,0}(z), \text{ for } t = 0, \; 0 \le z < \infty,$$
$$W_2(0,t) = g(t, W_1), \text{ for } 0 \le t < \infty. \tag{3.44}$$

To prove the existence of a global classical solution, we formulate the following assumptions regarding the given parameters and variables:

(A1)
Smoothness of the data: $W_{1,0}, \; W_{2,0}$ and g are C^1-functions.

(A2)
Boundary conditions for $t = 0$: The norms $\|W_{1,0}\|_{C(\mathbb{R}^+)}$ and $\|W_{2,0}\|_{C(\mathbb{R}^+)}$ are bounded i.e. there is a constant $C > 0$ such that: $\|W_{1,0}\|_{C(\mathbb{R}^+)} < C$ and $\|W_{2,0}\|_{C(\mathbb{R}^+)} < C$. Moreover it may hold that $W_{1,0}'(z) \le 0$ and $W_{2,0}'(z) \ge 0$.

(A3)
Boundary conditions for $z = 0$: The boundary condition g and the variable W_1 fulfill the inequalities: $\frac{\partial g}{\partial W_1} \ge 0$ and $\frac{\partial g}{\partial t} \le 0$.

(A4)
Compatibility of the boundary conditions:

$$W_{2,0}(0) = g\left(0, W_{1,0}(0)\right),$$
$$-\lambda_2\left(W_{1,0}(0), W_{2,0}(0)\right) W_{2,0}'(0) = \frac{\partial g}{\partial t}\left(0, W_{1,0}\right)$$
$$-\frac{\partial g}{\partial W_1}\left(0, W_{1,0}\right) \lambda_1\left(W_{1,0}(0), W_{2,0}(0)\right) W_{1,0}'(0).$$

(A5)
Eigenvalues: $\lambda_1(z,t) < 0$ and $\lambda_2(z,t) > 0$ for $(z,t) \in D$. Moreover it may hold:

$$\frac{\partial \lambda_1}{\partial W_1} < 0 \text{ and } \frac{\partial \lambda_2}{\partial W_2} > 0.$$

There is a constant $M > 0$, such that $\lambda_2(z,t) - \lambda_1(z,t) > M$.

Using the definitions just introduced, a statement on the existence of classical solutions can be formulated.

Theorem 3.1 *Let the assumptions (A1)–(A5) be fulfilled. Then problem* (3.42) *combined with* (3.44) *has a global classical solution* (W_1, W_2) *in* D.

Proof To prove the statement, we follow the proof in [11, Theorem 3.1]. Let $Z_2(t)$ be forward moving characteristic curve, starting at $z = 0$ and $t = 0$. The curve $Z_2(t)$ divides the space-time domain in two parts D_1 and D_2:

$$D_1 = \{(z,t) |\, z \geq Z_2(t),\ t \geq 0\},\ D_2 = D \setminus D_1.$$

In D_1 the existence of a classical solution has already been shown in [43]. Furthermore in [43] the following inequality has been derived:

$$\frac{dW_1}{dt}(Z_2(t), t) \leq 0. \tag{3.45}$$

Thus, we concentrate in the following on D_2. We show that for a fixed $T_0 > 0$ and $0 < T \leq T_0$ the C^1-norm of the solution is bounded in the domain

$$D^T = \{(z,t) |\, z \leq Z_2(t),\ 0 \leq t \leq T\}$$

independent of T. This means that there is a constant $C(T_0) > 0$, such that the C^1-norms $\|W_1\|_{C^1(D^T)}$ and $\|W_2\|_{C^1(D^T)}$ are bounded by $C(T_0)$. For this purpose, we consider an arbitrary point $(z,t) \in D^T$. In order to analyze the behavior of the solution in the point (z,t), we trace back the solutions and the corresponding derivates along the characteristic curves through this point. Tracing back the forward moving curve $z_2(t)$, one notes that it hits the boundary $z = 0$ in the point $(0, \delta(z,t))$. On the other hand the backward moving curve $z_1(t)$ hits the curve $Z_2(t)$ in the point $(\eta(z,t), \beta(z,t))$ (see Fig. 3.7). Denoting the solution of W_1 on $Z_2(t)$ with $W_{1,Z_2}(t)$, we obtain:

$$W_1(z,t) = W_{1,Z_2}(\beta(z,t)) \text{ and } W_2(z,t) = g(\delta(z,t), W_1(0, \delta(z,t))).$$

Due to $\beta(z,t) \leq t$ it holds:

$$\|W_1\|_{C(D^T)} \leq C(T_0) \text{ with } 0 < T < T_0,$$

since in [43] it has been shown that $W_{1,0}(t)$ is bounded on the curve $Z_2(t)$. From this, it follows together with (A1) and the inequality $\delta(z,t) \leq t$:

$$\|W_2\|_{C(D^T)} \leq C(T_0) \text{ with } 0 < T < T_0.$$

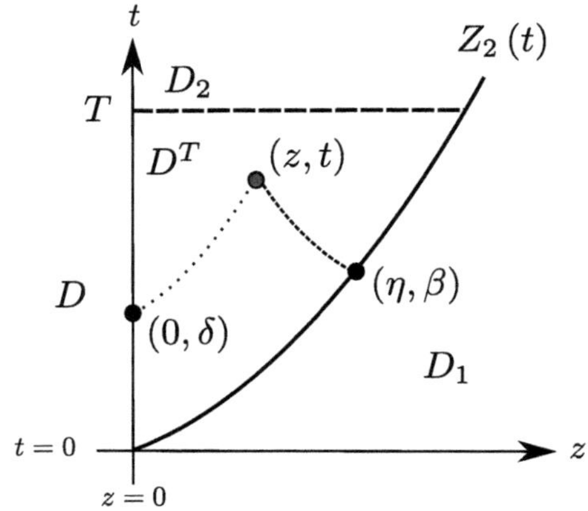

Fig. 3.7 Decomposition of D into D_1 and D_2

Thus the norms $\|W_1\|_{C(D^T)}$ and $\|W_1\|_{C(D^T)}$ are bounded by a constant independent of T. In a next step, we derive upper bounds for the derivatives of the solutions and start with the derivative $\partial W_1/\partial z$. Therefore, we consider the function:

$$v = e^{k(W_1,W_2)} \frac{\partial W_1}{\partial z},$$

where k is defined as follows:

$$\frac{\partial k}{\partial W_2} = -\frac{1}{\lambda_2 - \lambda_1} \frac{\partial \lambda_1}{\partial W_2}.$$

The function v solves on the backward moving curve $z_2(t)$ the following ordinary differential equation:

$$\frac{dv}{dt} = \frac{\partial v}{\partial t} + \lambda_1(W_1, W_2) \frac{\partial v}{\partial z} = -e^{-k(W_1,W_2)} \frac{\partial \lambda_1}{\partial W_1} v^2. \tag{3.46}$$

On the curve $Z_2(t)$ we enforce the following initial condition:

$$v|_{Z_2} = e^{k(W_1,W_2)} \frac{\partial W_1}{\partial z}. \tag{3.47}$$

The initial condition (3.47) can further be simplified. Therefore, we consider the derivative of W_{1,Z_2} on $Z_2(t)$:

$$W'_{1,Z_2}(t) = \frac{\partial W_1}{\partial t} + \lambda_2(W_1, W_2) \frac{\partial W_1}{\partial z}.$$

Together with

$$\frac{\partial W_1}{\partial t} = -\lambda_1 (W_1, W_2) \frac{\partial W_1}{\partial z} \text{ on } Z_2 (t)$$

it results

$$\frac{\partial W_1}{\partial z} = \frac{W'_{1,Z_2}(t)}{\lambda_2 - \lambda_1} \text{ on } Z_2 (t) \text{ and thus } v|_{Z_2} = \frac{e^{k\left(W_{1,0}(t), W_{2,0}\right)}}{(\lambda_2 - \lambda_1)\left(W_{1,0}(t), W_{2,0}\right)} W'_{1,Z_2}(t).$$

By separating the variables and integrating both sides, we obtain the following solution of the initial value problem (3.46) and (3.47):

$$v(z,t) = \frac{e^{k\left(W_{1,0}(\beta), W_{2,0}\right)} W'_{1,Z_2}(t)}{B(t, \beta(z,t))}, \tag{3.48}$$

where

$$\begin{aligned} B(t, \beta(z,t)) &= (\lambda_2 - \lambda_1)\left(W_{1,Z_2}(\beta), W_{2,0}\right) \\ &\quad + W'_{1,Z_2}(\beta) \int_\beta^t e^{-k(W_1(z_1(\tau,\beta),\tau), W_2)} \frac{\partial \lambda_1}{\partial W_1} (W_1(z_1(\tau,\beta),\tau))\, d\tau. \end{aligned}$$

$z_1(\tau, \beta)$ is the backward moving characteristic through the point (η, β).

Next we show that B is bounded from below by a positive constant. According to (A5) the first summand $\lambda_2 - \lambda_1$ is bounded from below by a positive constant. The second summand is positive, since $\partial\lambda_1/\partial W_1 < 0$ holds thanks to (A5) and since $W'_{1,Z_2}(\beta) < 0$ is true due to (A2). From the fact that B is bounded from below by a positive constant, we can conclude that $1/B$ is bounded from above. The term

$$e^{-k(W_1(z_1(\tau,\beta),\tau), W_2)}$$

is bounded, due to the boundedness of W_1 and W_2. From (3.46), it follows that $dv/dt \geq 0$, which means that v is bounded from below by the initial condition. All in all, we obtain that v is bounded from above and below. Thus the spatial derivative of W_1 can be bounded:

$$\left|\frac{\partial W_1}{\partial z}(z,t)\right| \leq C(T_0), \ (z,t) \in D^T, \ 0 < T \leq T_0.$$

Furthermore it holds:

$$\text{sign}\left(\frac{\partial W_1}{\partial z}\right) = \text{sign}\left(W'_{1,Z_2}(t)\right) = -1. \tag{3.49}$$

In order to accomplish the proof, we require an estimate for the derivative $\frac{\partial W_2}{\partial z}$. For this purpose, we determine the sign of $\frac{\partial W_2}{\partial z}$ on the boundary $z = 0$. On this boundary the variable W_2 is governed by the boundary condition g. Thus it holds:

$$\frac{\partial W_2}{\partial z} = -\frac{1}{\lambda_2}\left[\frac{\partial g}{\partial t} - \lambda_1 \frac{\partial g}{\partial W_1}\frac{\partial W_1}{\partial z}\right]. \tag{3.50}$$

Due to (A3), sign $\left(\frac{\partial W_1}{\partial z}\right) = $ sign $\left(W'_{1,z_2}\right) = -1$, $\lambda_1 < 0$ and $\lambda_2 > 0$ we have $\frac{\partial W_2}{\partial z} \geq 0$ on $z = 0$. In a next step we consider the sign of $\frac{\partial W_2}{\partial z}$ in D^T: Since W_2 is constant on the forward moving characteristics z_2, it holds: $W_2(z, t) = W_2(0, \delta(z, t))$. From this follows:

$$\frac{\partial W_2}{\partial z}(z, t) = \frac{\partial W_2}{\partial t}(0, \delta)\frac{\partial \delta}{\partial z}(z, t) = -\lambda_2 \frac{\partial W_2}{\partial z}(0, \delta)\frac{\partial \delta}{\partial z}(z, t).$$

In order to determine the sign of $\frac{\partial W_2}{\partial z}(z, t)$, we recall that $-\lambda_2 < 0$ and $\frac{\partial W_2}{\partial z}(0, \delta) \geq 0$. Using the definition of $\delta(z, t)$ it becomes obvious that $\delta(z, t)$ is decreasing, if z is increasing. This yields $\frac{\partial \delta}{\partial z} \leq 0$ and all in all we have $\frac{\partial W_2}{\partial z}(z, t) \geq 0$.

Now we can prove that the derivative $\frac{\partial W_2}{\partial z}$ is bounded from above. Similar to the estimate of $\frac{\partial W_1}{\partial z}$ we define an auxiliary function that is denoted by u:

$$u = e^{h(W_1, W_2)}\frac{\partial W_2}{\partial z} \text{ with } \frac{\partial h}{\partial W_1} = \frac{1}{\lambda_2 - \lambda_1}\frac{\partial \lambda_2}{\partial W_1}.$$

u solves the following differential equation

$$\frac{du}{dt} = -e^{-h(W_1, W_2)}\frac{\partial \lambda_2}{\partial W_2}u^2$$

combined with the initial condition

$$u(0, \delta) = e^{h(W_1(0,\delta), W_2(0,\delta))}\frac{\partial W_2}{\partial z}(0, \delta) \text{ on the boundary } z = 0.$$

Solving this ODE using the separation of variables, we obtain the following expression for u:

$$u(z, t) = \frac{e^{h(W_1(0,\delta), W_2(0,\delta))}\frac{\partial W_2}{\partial z}(0, \delta)}{E(t, \delta)}, \tag{3.51}$$

where E is defined by:

$$E(t, \delta) = 1 + \frac{\partial W_2}{\partial z}e^{h(W_1, W_2)}\int_\delta^t \frac{\partial \lambda_2}{\partial W_2}\left(W_1\left(z_1(\tau, \delta), \tau\right), W_2(0, \delta)\right)e^{-h(W_1, W_2)}\,d\tau. \tag{3.52}$$

$z_1(\tau, \delta)$ represents the forward moving characteristic through the point $(0, \delta(z, t))$. From $\partial \lambda_2 / \partial W_2 \geq 0$ and $\partial W_2 / \partial z \geq 0$ we can conclude that $E(t, \delta)$ is positive and bounded from below. Thus $1/E(t, \delta)$ is bounded from above. By this, the boundedness of u can be derived from the boundedness of W_1, W_2 and $\frac{\partial W_2}{\partial z}(0, \delta)$ in D^T. In order to prove that $\frac{\partial W_2}{\partial z}(0, \delta)$ is bounded from above, one uses the boundedness of $\frac{\partial W_1}{\partial z}$ and the initial condition (3.44). The existence of a constant $C(T_0) > 0$ such that

$$\left| \frac{\partial W_2}{\partial z}(z, t) \right| \leq C(T_0), \quad (z, t) \in D^T, \ 0 < T \leq T_0,$$

follows from the boundedness of u.

Formation of Shock Waves

As we have shown in Theorem 3.1, Assumptions (A1)–(A5) guarantee the existence a global classical solution. However, it is revealed that simulating application relevant issues, some of these assumptions are too restrictive. In particular (A3) and $\partial g / \partial t \leq 0$ have to be considered. If a velocity profile v_{in} is to be described at the inlet of a vessel, we obtain by means of (3.43) the following expression for g:

$$W_2(0, t) = g(t, W_1) = W_1(0, t) + 2 v_{in}(t).$$

If v_{in} is a non-monotonic decreasing function, then Assumption (A3) may be violated and thus the existence of a global classical solution can no longer be guaranteed. This is e.g. the case, if the pulsation of the heartbeat at the inlet of the aorta has to be simulated. Thus, if there is an alternating sign of the derivative v_{in}', the classical solution may exist only up to a time t_s. Due to $\partial g(t, W_1) / \partial t \leq 0$ (see Assumption (A3)) a global existence of solutions can no longer be guaranteed.

A rough estimate of this time point is of high importance for the choice of a suitable numerical method, since the convergence and stability of a numerical method is strongly depending on the regularity of the analytical solution. In order to derive a formula for the computation of t_s, we consider the expression (3.52) and the solution formula (3.51), which governs the derivative $\partial W_2 / \partial z$.

If $E(t, \delta)$ in (3.52) becomes zero for a time point t_s, the derivative $\partial W_2 / \partial z$ can become infinitely large and a shock in the solution results. Assuming that $W_{1,0}$ and $W_{2,0}$ are given by:

$$W_{1,0}(z) = W_{2,0}(z) = 4 \sqrt{\frac{G_0}{2\rho}},$$

one can show that $W_1(z, t) = W_{1,0}$ is a constant [11, Sect. 4.3]. This means that

$$\begin{aligned} e^{h(W_1(0,\delta), W_2(0,\delta)) - h(W_1(\tau, z_2(\tau,\delta)), W_2(0,\delta))} &= e^{h(W_1(0,\delta), W_2(0,\delta)) - h(W_1(0,\delta), W_2(0,\delta))} \\ &= e^{h\left(W_{1,0}, W_2(0,\delta)\right) - h\left(W_{1,0}, W_2(0,\delta)\right)} = 1. \end{aligned}$$

Combined with $\partial \lambda_2 / \partial W_2 = 5/8$ one obtains:

$$E(t,\delta) = 1 + \frac{5}{8}\frac{\partial W_2}{\partial z}(0,\delta)(t-\delta).$$

From (3.50) it follows for $\frac{\partial W_2}{\partial z}\Big|_{z=0}$:

$$\frac{\partial W_2}{\partial z}\Big|_{z=0} = -2\frac{v'_{in}(t)}{\lambda_2}\Big|_{z=0} + \frac{\lambda_1}{\lambda_2}\frac{\partial W_1}{\partial z}\Big|_{z=0}.$$

Due to $\frac{\partial W_1}{\partial z}\Big|_{z=0} = 0$ it holds:

$$E(t,\delta) = 1 - \frac{5}{4}\frac{v'_{in}(t)}{\lambda_2}(t-\delta).$$

Setting $E(t,\delta) = 0$, it results for t_s:

$$t_s = \delta + \frac{v_{AQ} + \sqrt{\frac{G_0}{2\rho}}\left(\frac{A}{A_0}\right)^{\frac{1}{4}}}{\frac{5}{4}v'_{in}(t)}. \tag{3.53}$$

An estimate for the place at which the shock is formed is given by:

$$z_s = t_s \cdot \lambda_2\left(W_{1,0}, W_{2,0}\right) \tag{3.54}$$

A numerical validation of these estimates can be found in Sect. 4.4.2. In this case, we will consider and analyze simulation results for both a "flat" ($\alpha = 1$) and another velocity profile for $\alpha \neq 1$.

Existence of Weak Solutions

A possible formation of shock waves in terms of application relevant boundary condition shows that the classical solution concept is not sufficient to formulate a well-posed blood flow model. This is due to the fact that a shock causes infinitely large derivatives and thus the solution is not continuous and differentiable. However the resulting jumps in the solution exhibit a finite height, but no singularity. Thus a weaker solution concept is required, which accepts jumps of finite height within the solution. In order to design this solution concept, we consider the space-time domain:

$$D_1^T = \{(z,t) \mid z \in (0,1),\ t \in (0,T)\}$$

and specify the following initial and boundary conditions:

$$
\begin{aligned}
&Q(0,t) = Q(1,t) = 0,\ t \in (0,T),\\
&A(z,0) = A_0(z) \in L^\infty([0,1]),\ Q(z,0) = Q_0(z) \in L^\infty([0,1]),\\
&0 < A_0(z) \le M,\ |Q_0(z)| \le M A_0(z),\ z \in [0,1],\\
&\int_0^1 A_0(z) = \overline{A} > 0.
\end{aligned}
$$

In a next step, we multiply the equations of the blood flow model (3.23) by test functions Ψ from the following function space $V\left(D_1^T\right)$:

$$
V\left(D_1^T\right) = \left\{ \Psi \in C^\infty\left(D_1^T\right) \middle|\ \Psi|_{t=T} = \Psi|_{z=0} = \Psi|_{z=1} = 0 \right\}.
$$

Integrating the two equations in space, one obtains by means of a partial integration:

$$
\begin{aligned}
&\int_0^1 \frac{\partial A}{\partial t} \Psi\, dz + \int_0^1 Q \frac{\partial \Psi}{\partial z}\, dz = 0,\\
&\int_0^1 \frac{\partial Q}{\partial t} \Psi\, dz + \int_0^1 \left(\frac{Q^2}{A} + \frac{G_0 A^{3/2}}{3\rho\sqrt{A_0}} \right) \frac{\partial \Psi}{\partial z}\, dz = -\int_0^1 K_R \frac{Q}{A} \Psi\, dz.
\end{aligned}
$$

A repetition of these steps wrt. the time variable t yields:

$$
\begin{aligned}
&\int_0^T \int_0^1 A \frac{\partial \Psi}{\partial t}\, dz dt + \int_0^T \int_0^1 Q \frac{\partial \Psi}{\partial z}\, dz dt + \int_0^1 A_0(z)\, \Psi|_{z=0}\, dz = 0,\\
&\int_0^T \int_0^1 Q \frac{\partial \Psi}{\partial t}\, dz dt + \int_0^T \int_0^1 \left(\frac{Q^2}{A} + \frac{G_0 A^{3/2}}{3\rho\sqrt{A_0}} \right) \frac{\partial \Psi}{\partial z}\, dz dt + \int_0^1 Q_0(z)\, \Psi|_{z=0}\, dz\\
&\qquad + \int_0^T \int_0^1 K_R \cdot \frac{Q}{A} \cdot \Psi\, dz dt = 0, \qquad (3.55)
\end{aligned}
$$

thereby we have assumed that the integrals with respect to t and z may be interchanged. By means of partial integrations, the regularity requirements of the solution variables A and Q are reduced by deriving the test functions. In [44, Theorem 2.4] it has been shown by means of further assumptions and definitions that (3.55) has a solution and thus a weak solution exists.

Existence of Classical Solutions for the Transport Problem

After studying the solvability of the blood flow model in detail, we consider the transport problem at the end of this section. Since this equation is only linear and not coupled with other equations the existence of classical solutions can be shown in an easier way. Wrt. the fluid variables A, Q and p we reconsider the Assumptions (A1)–(A5). Furthermore we set $\omega = 1$. The characteristic variable of the transport equation

$$
\frac{\partial \Gamma}{\partial t} + \frac{\partial}{\partial z}\left(\frac{Q}{A} \Gamma \right) = 0,\ (z,t) \in D^T,
$$

is given by: $C = \frac{\Gamma}{A}$, due to the fact that this variable is constant along the characteristic curve z_c. For this purpose, let the curve z_c be defined as follows:

$$\begin{aligned} \frac{dz_c}{dt} &= \frac{Q}{A}\left(z_c\left(t\right), t\right) = v_{AQ}\left(z_c\left(t\right), t\right), t \geq 0, \\ z_c\left(0\right) &= \xi \in \mathbb{R}^+. \end{aligned} \tag{3.56}$$

Using (3.56), we compute:

$$\begin{aligned} 0 &= \frac{\partial \Gamma}{\partial t} + \frac{\partial}{\partial z}\left(\frac{Q}{A}\Gamma\right) = \frac{\partial A}{\partial t}C + A\frac{\partial C}{\partial t} + \frac{\partial Q}{\partial z}C + Q\frac{\partial C}{\partial z} \\ &= A\frac{\partial C}{\partial t} + Q\frac{\partial C}{\partial z} = A\left(\frac{\partial C}{\partial t} + v_{AQ}\frac{\partial C}{\partial z}\right) = A\frac{dC}{dt}\left(z_c\left(t\right), t\right). \end{aligned}$$

Due to $A > 0$, we have shown that C is constant along z_c. Combined with initial and boundary conditions

$$\begin{aligned} C\left(z, 0\right) &= C_0\left(z\right), \ z \in \mathbb{R}^+, \ C_0 \in C^1\left(\mathbb{R}^+\right), \\ C\left(z, t\right) &= C_{in}\left(t\right), \ \text{für } z = 0, \ t > 0, \end{aligned} \tag{3.57}$$

and the condition

$$-v_{AQ}\left(W_{1,0}\left(0\right), W_{2,0}\left(0\right)\right) C_0'\left(0\right) = C_{in}'\left(0\right), \tag{3.58}$$

we can define the PDE:

$$\frac{\partial C}{\partial t} + v_{AQ}\frac{\partial C}{\partial z} = 0 \tag{3.59}$$

The velocity field v is depending on W_1 and W_2:

$$v_{AQ} = -\frac{1}{2}W_1 + \frac{1}{2}W_2.$$

Based on some previous publications like [12], we can show the existence of classical solutions for the transport problem.

Theorem 3.2 *Let us consider the problem given by* (3.57)–(3.59). *Moreover let* $\varepsilon > 0$ *be a parameter such that*

$$v_{AQ}\left(W_1\left(z, 0\right), W_2\left(z, 0\right)\right) > \varepsilon, \ z \in \mathbb{R}^+ \ \text{and} \ v_{AQ}\left(W_1\left(0, t\right), W_2\left(0, t\right)\right) > \varepsilon, \ t \in \left(0, T\right).$$

Then there is a global classical solution $C \in C^1\left(D^T\right)$ *for* (3.57)–(3.59).

Proof First we assume that W_2 is constant on a forward moving characteristic z_2. Due to (3.49) it holds $\partial W_1/\partial z \leq 0$ and according to (A5), $\lambda_2 - \lambda_1 > 0$ is true. From this it follows:

$$\begin{aligned}\frac{dv_{AQ}}{dt}\left(W_1\left(z_2\left(t\right),t\right),W_2\left(z_2\left(t\right),t\right)\right) &= -\frac{1}{2}\frac{d}{dt}\left(W_1\left(z_2\left(t\right),t\right)-W_2\left(z_2\left(t\right),t\right)\right)\\ &= -\frac{1}{2}\frac{dW_1}{dt}\left(z_2\left(t\right),t\right) = -\frac{1}{2}\left(\lambda_2-\lambda_1\right)|_{(W_1(z_2(t),t),W_2(z_2(t),t))}\frac{\partial W_1}{\partial z}\left(z_2\left(t\right),t\right)\\ &\geq 0.\end{aligned}$$

Thus v_{AQ} increases along the characteristic curve z_2. Together with the assumptions concerning v_{AQ} it follows: $v_{AQ}(z,t) > \varepsilon$ for $(z,t) \in D^T$ i.e. the curve z_c is moving forward in space and does not intersect itself. Thus the existence of a solution can be guaranteed, since the characteristic curves z_c in the space-time domain are not vertical. Moreover it holds for $(z,t) = (z_c(t), t) \in D^T,\ t > 0$:

$$z_c\left(t\right)-z_c\left(t^*\right) = \underbrace{\int_{t^*}^{t} v_{AQ}\left(z_c\left(\tau\right),\tau\right)\,d\tau}_{>0}.$$

This implies:

$$C\left(z,t\right) = \begin{cases} C_0\left(z-\int_{t^*}^{t} v_{AQ}\left(\left(z_c\left(\tau\right),\tau\right)\,d\tau\right), & \text{if } z_c\left(t^*\right) > 0 \text{ and } t^* = 0,\\ C_{in}\left(t^*\right), & \text{if } z_c\left(0\right) = 0 \text{ and } z_c \text{ intersects the time axis at } t^*.\end{cases}$$

Thus it holds: $C \in C^1\left(D^T\right)$.

3.5 Simplification of the Non-Linear 1D Models

As one can see in Fig. 1.11, pulsating blood flow is turned into a uniform flow. For this reason the vessel walls of the smaller vessels are much less deformed compared to the deformation of the vessel walls contained in the macrocirculation. Moreover, Table 1.1 reveals that the Reynolds numbers in the small arteries and arterioles are small compared to the large arteries. These observations motivate a further simplification of the non-linear blood flow model.

Linearized Models

Due to the small Reynolds numbers, we first neglect the convective term

$$\alpha\frac{Q^2}{A}$$

in Eq. (3.21). The deformation properties of small arteries and arterioles can be integrated into the model by linearizing the non-linear terms around

$$(A, Q)^T = (A_0, 0)^T = \mathbf{U_{f,0}}$$

To do this, we introduce two new variables a and q and represent a general state $(A, p, Q)^T$ as follows:

$$(A, p, Q)^T = (A_0 + a, p, q)^T .$$

Using the notation

$$\mathbf{U_l} = \begin{pmatrix} a \\ q \end{pmatrix},$$

the non-linear flux function in (3.27) can be approximated by a truncated Taylor series as follows:

$$\begin{aligned}\mathbf{F_f}\left(\mathbf{U_f}\right) &= \mathbf{F_f}\left(\mathbf{U_{f,0}} + \mathbf{U_l}\right) \approx \mathbf{F_f}\left(\mathbf{U_{f,0}}\right) + \mathbf{H}\left(\mathbf{U_{f,0}}\right)\left(\mathbf{U_f} - \mathbf{U_{f,0}}\right) \\ &= \begin{pmatrix} 0 \\ \frac{2}{3}c_{AQ,0}^2 A_0 \end{pmatrix} + \begin{pmatrix} 0 & 1 \\ c_{AQ,0}^2 & 0 \end{pmatrix}\begin{pmatrix} a \\ q \end{pmatrix}\end{aligned}$$

The source term is approximated by:

$$\mathbf{S_f}\left(\mathbf{U_f}\right) \approx \mathbf{S_f}\left(\mathbf{U_{f,0}}\right) = \begin{pmatrix} 0 \\ -R_{1D} \cdot q \end{pmatrix},$$

where

$$R_{1D} = \frac{K_R}{A_0} = \frac{2\pi \mu \alpha}{(\alpha - 1)\ A_0}.$$

Linearizing the area pressure relationship (3.18) around $A = A_0$ yields:

$$p(A) \approx p\left(A_0\right) + \left.\frac{\partial p}{\partial A}\right|_{A=A_0}\left(A - A_0\right) = \frac{a}{C_{1D}}.$$

Due to $\frac{\partial \mathbf{F_f}}{\partial z}\left(\mathbf{U_{f,0}}\right) = 0$, we can neglect the constant term $\mathbf{F_f}\left(\mathbf{U_{f,0}}\right)$ and the linearized system with the variables p and q is given by:

$$\frac{\partial}{\partial t}\begin{pmatrix} p \\ q \end{pmatrix} + \frac{\partial}{\partial z}\begin{pmatrix} \frac{q}{C_{1D}} \\ \frac{p}{L_{1D}} \end{pmatrix} = \begin{pmatrix} 0 \\ -R_{1D} \cdot q \end{pmatrix}.$$

The parameters C_{1D} and L_{1D} are given by the following equations:

$$C_{1D} = \left(\frac{\partial p}{\partial A}\left(A_0\right)\right)^{-1} = \frac{2 \cdot A_0}{G_0} \text{ and } L_{1D} = \frac{\rho}{A_0}.$$

Setting $c = \frac{\Gamma}{A}$ in (3.29b) the resulting PDE has the following shape:

$$\frac{\partial c}{\partial t} + \frac{\partial}{\partial z}\left(\frac{q \cdot c}{A}\right) = 0.$$

Introducing the matrix

$$H_l = \begin{pmatrix} 0 & \frac{1}{C_{1D}} \\ \frac{1}{L_{1D}} & 0 \end{pmatrix}$$

and summarizing the previous equations, one obtains the following linearized system for blood flow and transport:

$$\frac{\partial}{\partial t}\begin{pmatrix} p \\ q \end{pmatrix} + H_l \frac{\partial}{\partial z}\begin{pmatrix} p \\ q \end{pmatrix} = \begin{pmatrix} 0 \\ -R_{1D}q \end{pmatrix}, \; z \in (0, L), \; t > 0, \tag{3.60a}$$

$$\frac{\partial c}{\partial t} + \frac{\partial}{\partial z}\left(\frac{q}{A}c\right) = 0, \; z \in (0, L), \; t > 0. \tag{3.60b}$$

Considering the above system, one can see that in the linearized system there is a coupling between the fluid Eq. (3.60a) and the concentration Eq. (3.60b) but not vice versa. In this case (3.60a) provides the velocity field $v = q/A$ for the transport Eq. (3.60b).

Lumped Parameter (0D) Models

If a network consists of vessels whose lengths differ significantly, it is useful to describe blood flow and transport through the smaller vessels by models that are further simplified. This is motivated by the observation that pulse waves travel through the small vessels in a very short time.

An example for such a network is depicted in Fig. 1.17. Besides the carotid arteries and aorta there is also a ring shaped vessel system, whose outlets are supplying the brain tissue. This vessel system can compensate the blockage of a carotid artery (Vessel 5 and 6) by means of the healthy carotid artery and the ring shaped vessel system. This vessel system is denoted by Circle of Willis (see Sect. 1.2.5). Some vessels of the Circle of Willis [12] exhibit lengths that are small compared to lengths of other vessels, e.g., Artery 31 (anterior communicating artery, ACoA) linking the two halves of the brain has a length of 0.3 cm. Since the typical speed of pulse waves ranges from 10 $^{\mathrm{cm}}/_{\mathrm{s}}$ to 20 $^{\mathrm{cm}}/_{\mathrm{s}}$, a pulse wave moves very fast through this vessel.

If the spatial dynamics of blood flow can be neglected, (3.60a) and (3.60b) can be integrated wrt. z over an interval $(0, L)$. Assuming that the parameters C_{1D}, L_{1D} and R_{1D} are constant on $(0, L)$, it results:

$$C_{0D}\frac{d\hat{p}}{dt} + (q_{out} - q_{in}) = 0, \; t > 0, \tag{3.61}$$

$$L_{0D}\frac{d\hat{q}}{dt} + (p_{out} - p_{in}) = -R_{0D}\hat{q}, \; t > 0, \tag{3.62}$$

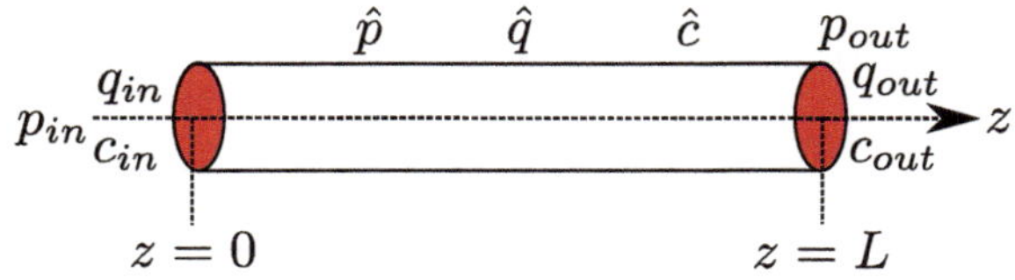

Fig. 3.8 Variables for the lumped parameter model (0D-model)

$$A_{0D} \frac{d\hat{c}}{dt} + (q_{out} \cdot c_{out} - q_{in} \cdot c_{in}) = 0, \ t > 0. \tag{3.63}$$

The averaged variables $\hat{p}$, $\hat{q}$ and $\hat{c}$ are given by the following integral expressions:

$$\hat{p}(t) = \frac{1}{L} \int_0^L p \, dz, \ \hat{q}(t) = \frac{1}{L} \int_0^L q \, dz \ \text{and} \ \hat{c}(t) = \frac{1}{L} \int_0^L c \, dz, \tag{3.64}$$

where the boundary values are defined by (see Fig. 3.8):

$$q_{in}(t) = q(0, t), \ q_{out}(t) = q(L, t), \ p_{in}(t) = p(0, t), \ p_{out}(t) = p(L, t).$$

The concentration variables at the boundaries are given by:

$$c_{in}(t) = c(0, t) \ \text{and} \ c_{out}(t) = c(L, t).$$

Finally, the model parameters read as follows:

$$A_{0D} = A_0 \cdot L, \ C_{0D} = C_{1D} \cdot L, \ L_{0D} = L_{1D} \cdot L \ \text{and} \ R_{0D} = L_{0D} \cdot R_{1D}.$$

Approximating the integrals in (3.61) and (3.62) by a rectangular rule, we have:

$$\hat{p}(t) \approx p_{out}, \ \hat{q}(t) \approx q_{in} \ \text{ or } \ \hat{p}(t) \approx p_{in}, \ \hat{q}(t) \approx q_{out}. \tag{3.65}$$

It can be shown that the discretization of a vessel of length L with N ODE systems (3.61) - (3.63) converges linearly ($\mathcal{O}(L/N)$, $N \to \infty$) to the solutions of (3.60a), (3.60b) [45, 46]. Models such as (3.61)–(3.63) are denoted by 0D-models or lumped parameter models (concentrated on a point), since a vessel with some length is reduced to a system without any spatial dimension. Due to that such model types are denoted as 0D-models. Approximating $\hat{c}$, we have to check the sign of $\hat{q}$:

- **Case 1:** $\hat{q}(t) > 0$. In this case, blood flows into the vessel from the left ($z = 0$) and $c_{in}(t)$ is given by a boundary condition, whereas $\hat{c}$ is approximated by the value c_{out}.
- **Case 2:** $\hat{q}(t) \leq 0$. In this case, the value $c_{out}(t)$ is prescribed as a boundary condition and $\hat{c}$ is approximated by the value c_{in}.

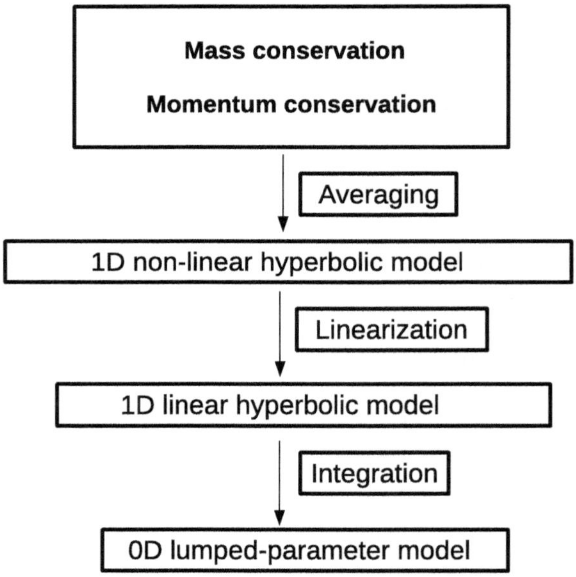

Fig. 3.9 Derivation of dimensional reduced models

The whole model reduction process is summarized in Fig. 3.9. In literature one can find other representations of the fluid Eqs. (3.61) and (3.62). Thereby, one introduces the volume V of a vessel and defines the parameter C_{0D}, as the fraction of the time derivative wrt. V and $\hat{p}$ [47, 48]:

$$C_{0D} = \frac{\frac{dV}{dt}}{\frac{d\hat{p}}{dt}} \Rightarrow \frac{dV}{dt} = C_{0D} \frac{d\hat{p}}{dt}.$$

Integrating the above equation on both sides with respect to time, we obtain:

$$V(t) - V(0) = C_{0D} \left(\hat{p}(t) - \hat{p}(0) \right).$$

The volume for $t = 0$: $V(0) = V_0$ is equivalent to the volume for $\hat{p} = 0$. Setting $\hat{p}(0) = 0$, it results:

$$V(t) - V(0) = C_{0D} \cdot \hat{p}(t). \tag{3.66}$$

Overall, the following $0D$ fluid equations are obtained:

$$\frac{dV}{dt} + q_{out} - q_{in} = 0, \ t > 0, \tag{3.67}$$

$$L_{0D} \frac{dq_{out}}{dt} + p_{out} - p_{in} = -R_{0D} q_{out}, \ t > 0. \tag{3.68}$$

Analogies from Electrical Engineering

ODEs like (3.61) or (3.62) are frequently used to model electrical circuits. Thereby, voltage is compared to pressure, since both voltage and hydraulic pressure are caused by gradients of a potential. Moreover, current can be related to a volumetric flux. By means of current, we determine the amount of electrical charge passing through the cross-section of a wire during a certain time interval. The volumetric flow rate takes into account, how much fluid mass is transported through the cross-section of a vessel. This means that in electrical engineering one is concentrated on the transport of charge, whereas in hydraulic engineering the transport of fluid mass is considered.

As the hydraulic variables, the parameters R_{0D}, C_{0D} and L_{0D} can be compared with parameters from electrical engineering (see also Fig. 3.10 and Table 3.1):

- **Capacity of a capacitor:** The storage capacity of a vessel with a length L, a cross-sectional area A_0 and an elasticity parameter G_0, is described by the parameter

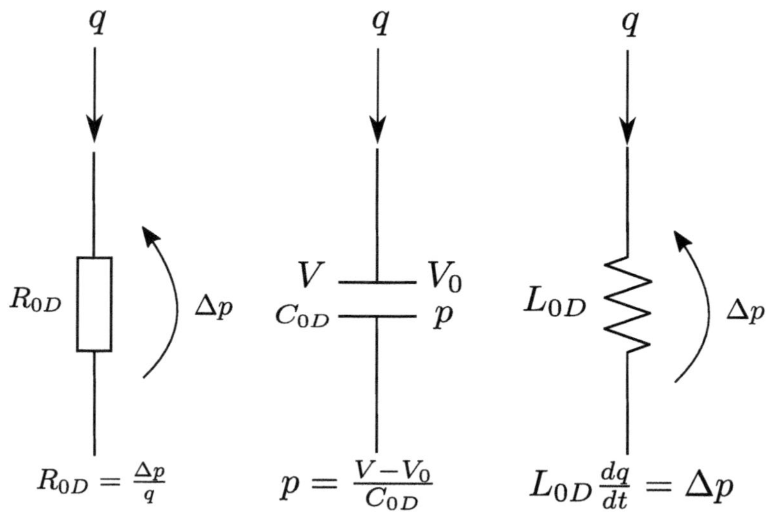

Fig. 3.10 Electrical components and hydraulic variables. To simplify the notation, the hat over the hydraulic variables has been omitted

Table 3.1 Analogies between electrical and hydraulic variables

Hydraulic variables	Electrical variables
Pressure	Voltage
Volumetric flow rate	Current
Vessel parameter R_{0D}	Ohmic resistance R
Vessel parameter L_{0D}	Inductivity L_S
Vessel parameter C_{0D}	Capacity C

$C_{0D} = \frac{2LA_0}{G_0}$. Analyzing this parameter in more detail reveals that vessels with a large length L, a large cross-sectional area A_0 and a small elasticity parameter G_0 exhibit a large capacity C_{0D}. At the same time, it can be seen that vessels with a large C_{0D} can store a larger volume of blood compared to other vessels that do not exhibit these features. A component that plays an important role in electrical engineering is the capacitor. This component consists of two conductors, usually two parallel plates, which are separated by a dielectric material. Depending on the properties of these plates, some amount of electrical charge can be stored on them. For this reason, the capacitance C of a capacitor can be compared to the parameter C_{0D} of a blood vessel. In case of a blood vessel, blood volume is stored, while in case of a capacitor, electrical charges are stored.

- **Inductance:** The parameter $L_{0D} = \frac{\rho L}{A_0}$ is a weighting factor for the inertia term in (3.62). The longer a vessel is and the smaller its cross-sectional area A_0 is and the higher the density ρ of the fluid is, the larger is the inertia parameter of a vessel. If current is passing through a coil contained in an electrical circuit, voltage builds up in the coil, which counteracts the actual current. Obviously it holds that: The longer the wire winding of the coil, the greater the countervoltage. Thus, in an alternating current (AC) circuit, a long wire winding of a coil and a small conductor cross-section yields a higher impedance (AC resistance). Another important influence on the countervoltage is exerted by the material that is surrounded by the coil. Comparing the inductance of a coil with the vessel parameter L_{0D}, we find that the length of the wire winding and the length of the blood vessel as well as the cross-sectional area of the conductor and the blood vessel correspond to each other. In addition, properties of the fluid and the material wrapped by the coil correspond.
- **Ohmic resistance:** The flow resistance of a vessel is described by the parameter $R_{0D} = \frac{\rho K_R L}{A_0^2}$. It is easy to understand that the larger the viscous parameter K_R, the longer the vessel, and the smaller the cross-sectional area, the greater is the flow resistance. In an electrical circuit, this parameter corresponds to the ohmic resistance R, which is given by the ratio of voltage and current. Transferring Ohm's law to fluid mechanics, we find that $R_{0D} = \frac{\Delta p}{q}$, where Δp denotes the pressure drop across the resistor.

The simplified 0D-models and electrical engineering analogies derived in this section will be used in the upcoming sections and chapters to simulate flows and transport processes in small connecting vessels such as Artery 31 in Fig. 1.17. The 0D models are of particular importance for the formulation of boundary conditions at inflow and outflow boundaries. This is due to the fact that 0D models can be used to describe in a simplified way vascular systems or the ventricles of the heart. Based on suitable parameter selection, these models can describe the influence of the heart and the omitted vessels with sufficient accuracy. At the same time, the reduced model complexity keeps the computational effort for the boundary conditions within a reasonable range.

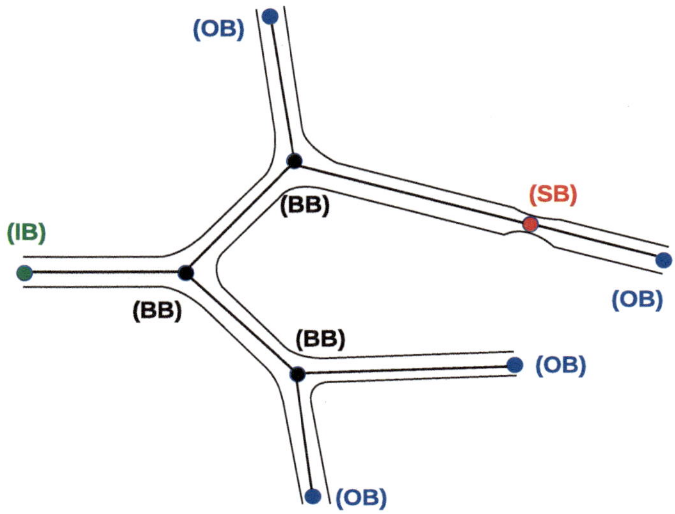

Fig. 3.11 Outline of a network with one inflow boundary (IB), four outflow boundaries (OB), one stenosis (SB), and three bifurcations (BB). The black bold lines indicate the center lines of the network

3.6 Boundary Conditions for a Complete Vascular Network

In the previous section, we studied how boundary conditions can be formulated for the blood flow model and for the transport equation. Thereby, we focused on single vessels represented by an interval $(0, L)$. In a next step, the techniques developed for single vessels will be used to simulate a complete vessel network. Here, we restrict ourselves to the open 1D-0D model discussed in Chap. 2 (see also Fig. 2.6). In order to construct an open 1D-0D model for arterial macroscale networks, it is necessary to derive boundary conditions and coupling constraints that integrate the influence of the heart, unresolved vessels, and branches into the model (see Fig. 3.11). The corresponding boundary and coupling conditions will be discussed in the following subsections.

3.6.1 Modeling of Inflow Boundaries (IB)

In the following chapters, we will perform our simulations based on two arterial networks that have exactly one inflow boundary at the inlet of the aorta. These are, on the one hand, the network consisting of the main arteries (see Fig. 1.7) and, on the other hand, the network containing the Circle of Willis (see Fig. 1.17). In order to perform realistic simulations of blood flow for these two networks, it is necessary to develop a model for flow through the aortic valve. As we noted in Chap. 1, the heart muscle produces pulsatile flow through the aortic valve by means of contraction (systole) and relaxation (diastole) phases. Overall, $5.7 - 6.0$ liters of blood volume per minute are pumped through the systemic circulation. The contraction phase has a duration of approximately $0.15\ s$ and the total heart beat has a duration of approximately $0.85\ s$. At the beginning of the contraction phase, the valves of the left ventricle remain closed, allowing the blood pressure to build up within the ventricle. If the ventricle pressure exceeds the blood pressure in the aorta, the aortic valve can open and blood in the ventricle is pushed into the aorta. Using these information, we propose in this subsection two different models that generate appropriate boundary conditions at the inlet of the aorta.

Explicit Representation of Q_{AV}

The simplest way to model the influence of the heart is to provide a function for the volumetric flow Q. To represent the pulsation of flow, we use a half sine wave to describe Q during the contraction phase. In the remainder of the heart beat, we set Q to zero to simulate the closed aortic valve. This yields the following flow rate Q_{AV} for $t_h \in [0\ \text{s},\ T\ \text{s}]$:

$$Q_{AV}\,(t_h) = \begin{cases} Q_{\max}\ \sin\left(\frac{\pi}{0.30T} t_h\right), & \text{for } 0\ \text{s} \le t_h \le 0.30 \cdot T, \\ 0, & \text{for } 0.30 \cdot T < t_h \le T. \end{cases} \tag{3.69}$$

Setting $Q_{max} = 425.0\ \text{cm}^3/\text{s}$ and $T = 1.0\ \text{s}$ (3.69) one obtains:

$$60 \cdot \int_{0\ \text{s}}^{0.3\ \text{s}} Q_{AV}\,(t)\ dt \approx 5729.58\ \text{cm}^3 \approx 5.73\ \text{l},$$

i.e. using the function in (3.69) in 60 s or one minute about 5.73 l are transported through the aortic valve. This means that the function Q_{AV} is a simple but effective way to emulate the heart beats at the inlet of the aorta. Since Q_{AV} is only defined on $[0, T]$, one has to find a way how it can be computed for an arbitrary time point $t \in [0\ \text{s}, t_{end}]$, where t_{end} is the end of the simulation time period. For this purpose the following mapping of t to $t_h \in [0, T]$ is considered:

$$\tau_r : [0\ \text{s}, t_{end}] \rightarrow [0, T]\,,\ t \mapsto t - T \cdot \text{floor}\left(\frac{t}{T}\right) =: t_h. \tag{3.70}$$

Modeling the Left Ventricle

The model based on Eq. (3.69) has the disadvantage that the aortic valve is assumed to open immediately at the beginning of the heartbeat. However, this does not correspond to the real behavior of the aortic valve. Actually, it takes some time to build up enough pressure such that the aortic valve can be opened. In order to be able to understand this mechanism in a better way, it is required to model the pressure in the left ventricle. In order to compute the pressure P_{LV}, we use the following elasticity model:

$$P_{LV}(t_h) = E_{LV}(t_h)\left(V_{LV}(t_h) - V_{0,LV}\right) + S_{LV}(t_h)\frac{dV_{LV}}{dt_h}, \tag{3.71}$$

$$\frac{dV_{LV}}{dt_h} = -Q_{AV}(t_h). \tag{3.72}$$

V_{LV} $\left[\mathrm{cm}^3\right]$ is the volume of the left ventricle and $V_{0,LV}$ $\left[\mathrm{cm}^3\right]$ corresponds to the volume for $P_{LV} = 0$. $E_{LV}(t_h)$ $[\mathrm{mmHg}/\mathrm{cm}^3]$ denotes the time-dependent elastic modulus of the ventricular wall and Q_{AV} is the volumetric flow through the aortic valve. $S_{LV}(t_h)$ $[\mathrm{mmHg}\cdot\mathrm{s}/\mathrm{cm}^3]$ is a parameter representing the viscoelastic behavior of the ventricular wall. In the following sections and chapters, we assume that this parameter depends linearly on P_{LV}:

$$S_{LV}(t_h) = 5.0 \cdot 10^{-4}\, P_{LV}(t_h)\frac{\mathrm{s}}{\mathrm{cm}^3}.$$

The first part of (3.71) corresponds to the Laplace equation, which is based on the assumption that the internal volumetric forces and the surface forces in the ventricular wall are equal to each other. This is a sufficiently accurate model if the wall of an elastic body is relatively thin and the forces are balance out without a long retardation. Since this is not the case for the ventricle, a further term is added to account for the viscoelastic effects. The differential equation for the change in volume considers only flow of blood through the aortic valve. Flow through the mitral valve, is modeled in a simplified way, since we consider only an open loop model. At the beginning of a heart beat, we set the blood volume V_{LV} to $V_{LV}^{\max}$:

$$V_{LV}(k \cdot T) = V_{LV}^{\max},\ k \in \mathbb{N}.$$

$V_{LV}^{\max}$ is the maximal blood volume in the left ventricle and T [s] denotes the duration of a single heartbeat i.e. we adjust the volume of the ventricle at the start of a new heart cycle. The elasticity function E_{LV} is represented as follows:

$$E_{LV}(t_h) = E_A^{(LV)} e_{LV}(t_h) + E_B^{(LV)}. \tag{3.73}$$

The parameters $E_A^{(LV)}$ $[\mathrm{mmHg}/\mathrm{cm}^3]$ and $E_B^{(LV)}$ $[\mathrm{mmHg}/\mathrm{cm}^3]$, which are independent of t_h, correspond to the peak and base values of the E-module. e_{LV} determines the dynamics of the elasticity parameters E_{LV}:

$$e_{LV}(t_h) = \begin{cases} \frac{1}{2}\left(1 - \cos\left(\frac{\pi t_h}{T_{vc}}\right)\right), & \text{falls } 0 \leq t_h \leq T_{vc}, \\ \frac{1}{2}\left(1 + \cos\left(\frac{\pi (t_h - T_{vc})}{T_{vr}}\right)\right), & \text{falls } T_{vc} < t_h \leq T_{vc} + T_{vr}, \\ 0, & \text{falls } T_{vc} + T_{vr} < t_h \leq T. \end{cases}$$

The first part of the above function describes the contraction phase, while the second and third part represent the relaxation and rest phases. Accordingly, the parameter T_{vc} [s] is the duration of the contraction phase and T_{vr} [s] is the duration of the relaxation phase. As in the case of Q_{AV}, the function is enhanced by a mapping such as (3.70). In order to determine the flow rate Q_{AV} in (3.72), we use a further differential equation:

$$L_{AV} \frac{dQ_{AV}}{dt_h} = O\left(\Delta P_{AV}\right) \Delta P_{AV} - R_{AV} Q_{AV} - B_{AV} Q_{AV} \left|Q_{AV}\right|. \tag{3.74}$$

This equation represents a balance of forces, where the left side corresponds to the inertia term with the inertia parameter L_{AV} $\left[\mathrm{mmHg \cdot s^2/cm^3}\right]$ [1, Chap. 9]. The right hand side contains the forces acting on the aortic valve. One component is the force caused by the pressure difference $\Delta P_{AV} = P_{LV} - P_{Aorta}$ between the ventricular pressure P_{LV} and the aortic pressure P_{Aorta}. To model the opening and closing of the aortic valve, the pressure difference is multiplied by an opening factor $O \in [0, 1]$, where O solves the following ODE:

$$\frac{dO}{dt} = \frac{\tanh\left(C_v \Delta P_{AV}\right) + 1}{2C_d} - \frac{O}{C_d}. \tag{3.75}$$

The parameters C_v and C_d determine the opening speed of the aortic valve. An opening factor $O = 0$ corresponds to a closed heart valve and $O = 1$ corresponds to an open heart valve. Energy losses due to the viscous forces and turbulences are contained in the second and third term of the right hand side in (3.74). The parameter for the viscous forces is given by R_{AV} $[\mathrm{mmHg \cdot s/cm^3}]$, while B_{AV} $\left[\mathrm{mmHg \cdot s^2/cm^6}\right]$ is weighting the turbulence.

The Left Ventricle can be Modeled by an Electrical Schematic

As we have already noted in Chap. 2 and Sect. 3.5, one can find correspondances between electrical and hydraulic engineering. Following this, the 0D model for the left ventricle based on ODEs can be compared to an electrical component. This technique can also be applied to the remaining ventricles and the rest of the cardiovascular system, so that the entire cardiovascular system is represented as an "electrical schematic" [1, Chap. 9] [25–27, 46]. This approach allows for a clear subdivision of the cardiovascular system and due to its compact presentation, a clear overview on the different components of the blood circulation can be provided.

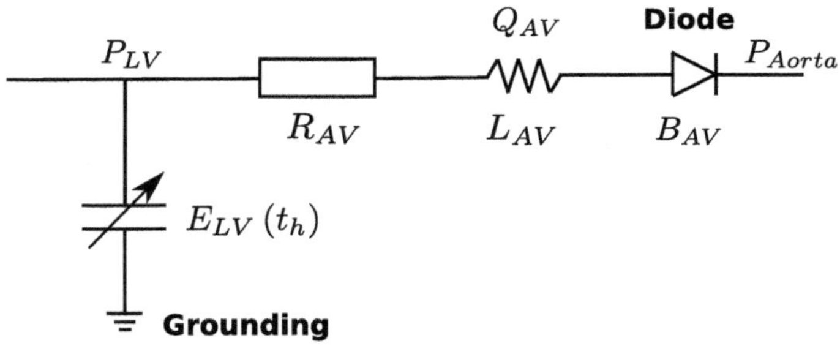

Fig. 3.12 Representation of the left ventricle as an electric schematic

A ventricle can be considered as a capacitor with a variable capacity, where the elasticity E_{LV} is reciprocal to the capacity of a ventricle. The aortic valve is represented by a coil with inductance L_{AV} and an ohmic resistance R_{AV}. A further feature of the aortic valve is that it allows flow out of the left ventricle and prevents backflow from the aorta into the left ventricle. In an electric circuit, a diode is generating a unidirectional current. For this reason we add to the coil and ohmic resistance a diode. All in all, we obtain for the left ventricle an electric schematic that is presented in Fig. 3.12.

Boundary Conditions for the Transport Problem

Simulating e.g. the transport of oxygen, we choose as a boundary condition a fixed concentration value, which corresponds to an averaged volumetric oxygen concentration in arterial blood [12, Chap. 2]:

$$C_{in} = \frac{\Gamma_{in}}{A_{in}} = 8.75\ \mu\text{mol/cm}^3. \tag{3.76}$$

Thereby, Γ_{in} and A_{in} denote the averaged concentration and the section area at the inlet of the aorta.

3.6.2 Modeling of Outflow Boundaries (OB)

Since macroscaled arterial networks (see e.g. Figs. 1.7 and 1.17) are comprising just a small part of the whole cardiovascular system, there is the issue how the influence of the vessels, which are not part of these networks, can be integrated into the model. In Sect. 3.6.1, we modeled the influence of the heart at the inflow boundaries. Besides the inflow boundary connecting a macroscaled arterial network to the heart and the pulmonary circulation, there are further boundaries connecting the network to the remainder of the systemic circulation. These boundaries are denoted as outflow boundaries.

A well-know method for modeling the outflow boundaries is the usage of three-element Windkessel models that are denoted as R_1CR_2-Windkessel models [5, 20, 30, 49]. The concept "Windkessel model" has already been discussed in Sect. 1.1.2. It describes the ability of elastic arteries to store blood volume and to push it towards the organs by means of elastic vessel walls. By this a continuous supply of organs with oxygen rich blood is ensured. The arteries that are not contained in the macroscaled network are usually not considered as arteries with a large elasticity. However, they contain a huge number of middle-sized arteries, which still exhibit some storage capacity. Moreover, some of them exhibit muscles in their vessel walls as the larger arteries. A contraction of these muscles causes some pressure on the blood volume such that it can be transported through the corresponding vessels. Thus in these vessels a Windkessel effect can be recognized.

Modeling the outflow boundaries we use as in case of the heart correspondences from electrical engineering that have already been discussed on Sect. 3.5. Thereby, C represents the capacity of arterial vessels and capillaries attached to the outlet of a 1D vessel (see Fig. 3.13). The resistance R_1 covers the flow resistances of the vessels between the 1D vessel and the capillary bed, while the resistance R_2 summarizes the

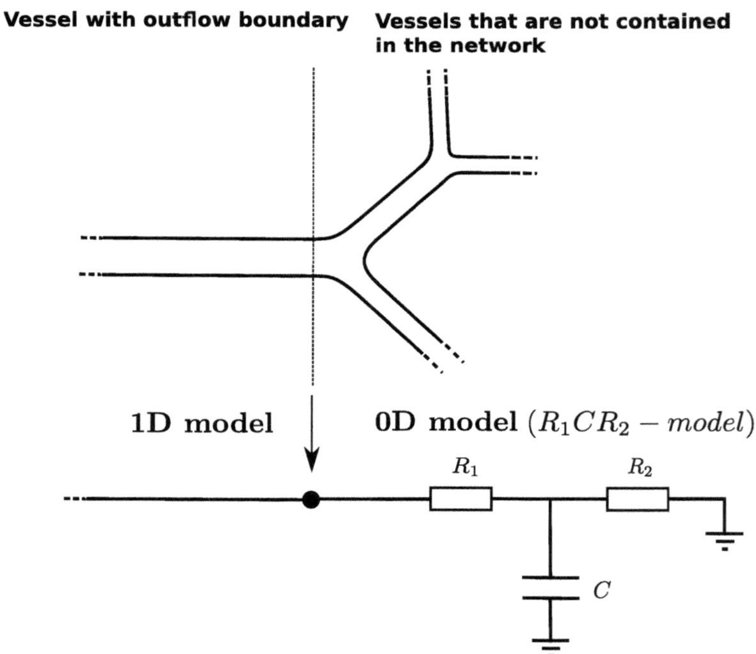

Fig. 3.13 Modeling of an outflow boundary with a three-element outflow Windkessel model. To the terminal vessel of the network we assign a 1D model, while the omitted vessels attached to the outlet of this vessel are represented by an electrical component consisting of two resistances and a capacitor

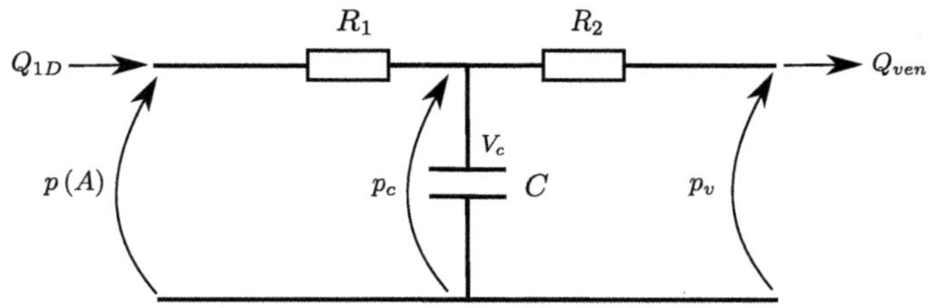

Fig. 3.14 Electric schematic of the R_1CR_2 component modeling the omitted vessels. Besides the different electrical components, the most important parameters and variables that are required for the derivation of the model equations, are listed

flow resistances of the remaining vessels. We assign an averaged pressure p_c and a blood volume V_c. The blood volume V_c is governed by the two flow rates Q_{1D} and Q_{ven}. Thereby Q_{1D} is the flow rate out of this vessel into the omitted arterial vessels and capillaries, while Q_{ven} represents blood flow out of these vessels into the large veins. This yields the following ODE:

$$\frac{dV_c}{dt} = Q_{1D} - Q_{ven}. \tag{3.77}$$

Moreover, it holds analogous to (3.66):

$$V_c - V_{0,c} = C \cdot p_c,$$

where $V_{0,c}$ is equal to V_c for $p_c = 0$. Using Ohm's law with hydraulic variables, the flow rates Q_{1D} and Q_{ven} can be represented as follows (see Fig. 3.14):

$$Q_{1D} = \frac{p\,(A) - p_c}{R_1} \text{ and } Q_{ven} = \frac{p_c - p_v}{R_2}.$$

p_v represents the mean venous pressure and $p\,(A)$ is the pressure at the outlet of the 1D vessel, which is given by (3.18). Combining the last equations, we get:

$$C\frac{dp_c}{dt} = Q_{1D} - \frac{p_c - p_v}{R_2},$$

$$C\frac{d}{dt}\,(p - R_1 Q_{1D}) = Q_{1D} - \frac{p - R_1 Q_{1D} - p_v}{R_2}. \tag{3.78}$$

Finally, we obtain:

$$p - p_v + CR_2\frac{dp}{dt} = (R_1 + R_2)\,Q_{1D} + CR_1R_2\frac{dQ_{1D}}{dt}. \tag{3.79}$$

Solving this differential equation at time points of interest yields one equation for determining the fluid variables. The second equation results from the discretization of the differential Eq. (3.21) (see Chap. 4).

The boundary conditions for transport problem are governed by the sign of the velocity field and an ODE, which yields the mean concentration c in the omitted vessels. If blood is leaving the vessel, the concentration variable Γ_{1D} must be determined by an upwinding procedure as in Sect. 4.2. In the other case, we need another ODE to determine c. For the derivation, we design a balance equation considering the outflow and inflow of particles similar to the case of blood flow:

$$\frac{d}{dt}\left(V_c\left(t\right)c\right) = N_{1D}\left(t\right) - N_{ven}\left(t\right) - r\left(c\left(t\right), t\right).$$

N_{1D} represents the number of particles leaving or entering the 1D vessel in a given time interval. On the other hand, N_{ven} denotes the number of particles moving into the rest of the venous vessels in a given time interval. The term r takes into account the transfer of the substance into the tissue or lymphatic system. V_c is the volume of blood contained in the omitted vessels. As we have already noted, V_c can be computed using (3.77) and N_{1D} is given by a combination of the 1D variables:

$$N_{1D}\left(t\right) = \begin{cases} \frac{\Gamma_{1D}(t) Q_{1D}(t)}{A_{1D}(t)}, & \text{if } Q_{1D}\left(t\right) \geq 0, \\ Q_{1D}\left(t\right) c\left(t\right), & \text{if} Q_{1D}\left(t\right) < 0. \end{cases}$$

N_{ven} results from the following equation:

$$N_{ven}\left(t\right) = Q_{ven}\left(t\right) \cdot c\left(t\right) = \frac{p_c\left(t\right) - p_v}{R_2} \cdot c\left(t\right).$$

All in all one obtains the following ODE:

$$\frac{d}{dt}\left(V_c\left(t\right)c\right) = \begin{cases} \frac{\Gamma_{1D}(t) Q_{1D}(t)}{A_{1D}(t)} - \frac{p_c - p_v}{R_2} \cdot c\left(t\right) - r\left(c\left(t\right), t\right), & \text{if } Q_{1D}\left(t\right) \geq 0, \\ Q_{1D}\left(t\right) c\left(t\right) - \frac{p_c - p_v}{R_2} \cdot c\left(t\right) - r\left(c\left(t\right), t\right), & \text{if } Q_{1D}\left(t\right) < 0. \end{cases} \tag{3.80}$$

Choice of the Parameters R_1, C and R_2

Of high importance is now the issue on how to choose the resistance parameters R_1 and R_2, as well as the capacity parameter C. For the networks in Figs. 1.7 and 1.18, some suitable values for these parameters can be found in [1] and [12], respectively. In these references, the total resistances $R = R_1 + R_2$ and the capacity parameters C are listed for the respective outflow boundaries.

One possible approach to estimate the total resistances R and the parameter C is described in [12]. Thereby, it is assumed that the total resistance of the arterial vasculature is given by $R_T = 1.34 \cdot 10^8$ ${}^{\text{Pa}\cdot\text{s}}/_{\text{m}^3}$ and the total capacity is given by $C_T = 9.45 \cdot 10^{-9}$ ${}^{\text{m}^3}/_{\text{Pa}}$. Next, one tries to roughly estimate how the total blood volume is distributed at the outlets of the network. Accordingly, the total capacity and the total resistance are distributed to the outflow boundaries. Denoting the index set of the vessels with an outflow boundary as I_{out} and the index set of the whole network as I_{in}, the total resistance and total capacity can be determined as follows:

$$\frac{1}{R_T} \approx \sum_{i \in I_{out}} \frac{1}{R_{p,i}} \text{ und } C_T = \sum_{i \in I_{in}} C_{0D,i} + \sum_{i \in I_{out}} C_{p,i}.$$

The resistance $R_{p,i}$ represents the resistances of the peripheral circulation, which is adjacent to the outflow boundary $i \in I_{out}$. Correspondingly, the capacities $C_{p,i}$ summarize the capacities of the omitted vessels adjacent to the outflow boundaries $i \in I_{out}$. However, one has to take into account that the formula for the total resistance R_T does not include the vessels of the 1D network. This is due to the fact that the resistances

$$R_{0D,i} = \frac{\rho K_R L_i}{A_{0,i}^2}, \quad i \in I_{in}, \tag{3.81}$$

of the 1D vessels in a macroscale network are very small compared to the peripheral resistances $R_{p,i}$. Therefore they can be neglected in the computation of R_T. The parameter L_i in (3.81) denotes the length of the vessel and ρ stands for the density of blood. $A_{0,i}$ is the cross-sectional area of the corresponding vessel in the relaxed state and K_R is the viscosity parameter introduced in Eq. (3.15). Since the resistances of the peripheral vessels can be interpreted as resistances connected in parallel, the reciprocal value of the total resistance is determined by the sum of the reciprocal partial resistances. This is due to the fact that the total resistance of a parallel circuit is determined in exactly the same way as in electrical engineering [50, 51].

Computing the total capacity, we have to include the capacities of the network. Since networks on the macroscale contain large deformable arteries having the potential to store a large amount of blood volume the capacity of the network has to be considered. Summing the capacity parameters of the macroscale vessels:

$$C_{0D,i} = \frac{L_i A_{0,i}}{\rho \cdot c_{AQ,i}^2 \left(A_{0,i}\right)}, \quad i \in I_{in},$$

one can estimate the total capacity of the macroscale network by this value.

$$c_{AQ,i}\left(A_{0,i}\right) = \sqrt{\frac{G_0}{2\rho}}$$

is the characteristic wave speed of the i-th 1D vessel (see Eq. (3.38)). The total capacity of the peripheral vessels is determined is the same way as in the case of parallel connected capacitors i.e. we sum up the partial capacities $C_{p,i}$. This is due to the fact that in electrical engineering the total capacity of parallel capacitors is given by the sum of the respective capacities. Together with the total capacitance of the 1D network, we obtain the total capacitance C_T.

It remains to discuss how the partial resistances R_1 and R_2 can be chosen. In [52], the authors propose the splitting $R_1 = 4/5\ R$ and $R_2 = 1/5\ R$. This splitting seems to be quite reasonable, since the resistance R_1 includes the arterioles. The arterioles regulate the flow resistance between the capillary bed and the macrocirculation (see also Chap. 1) so that the flow velocity is sufficiently low. On the other hand, the second resistance R_2 can be considered as the resistance of the veins, which have a much lower flow resistance. Therefore it makes sense to integrate the main part of the total resistance into the resistance R_1. Another well-known strategy to determine the resistances R_1 and R_2 can be found in e.g. [12]. Here, it is assumed that the resistance R_1 is equal to the characteristic impedance of the 1D vessel: $Z_0 = \rho \cdot c_{AQ}(A_0)/A_0$, where A_0 is the cross-sectional area of the 1D vessel. This choice can be found in many references, because it minimizes unphysical reflections at the outflow boundaries.

However, if one has to deal with a network for which the total resistances $R_{p,i}$ and capacitances $C_{p,i}$ are unknown or difficult to estimate, the parameters R_1, C and R_2 can also be determined by a calibration algorithm. Thereby, the parameters are computed by classical optimization methods such that the difference between the simulated systolic and diastolic pressures and the corresponding values from the medical literature are minimized [53]. In addition, the shape of the pressure curves is optimized so that one can observe a strong pressure increase during systole and a slow pressure decrease during diastole (see also Fig. 1.6). Another way to calibrate the three parameters is a data-based approach, which is characterized by the use of neural networks (PINNs: *physics informed neural networks*) and stochastic optimization methods [54].

3.6.3 Modeling of Bifurcations (BB)

Both the arterial and venous vascular systems have a large number of branches to distribute oxygen-rich blood throughout the body and remove oxygen-poor blood. In order to represent flow through a bifurcation in its full complexity, one would have to construct a 3D geometry. However, if the computation of shear stresses on the vessel walls or turbulence is not of interest, it is sufficient to omit the exact modeling of the branching region. In a first step, we simplify the geometric structure in such a way that the branching region is neglected. Thus the effect of the branching angles is neglected for the moment (see Fig. 3.15). In a next step, the bifurcation is decomposed into three 1D vessels, which we denote as the main vessel Ω_m and daughter vessels Ω_{b_1} and Ω_{b_2}. Moreover, we assume that the 1D main axes z_m, z_{b_1}, and z_{b_2} of the three vessels are oriented in such a way that the boundaries $z_m = L_m$, $z_{b_1} = 0$, and $z_{b_2} = 0$ are adjacent to the branching region. L_m is the length of Ω_m. Now there is the issue how to couple the 1D fluid equations at the bifurcation. In order to derive the coupling conditions, we are guided by two basic principles: *conservation of mass* and

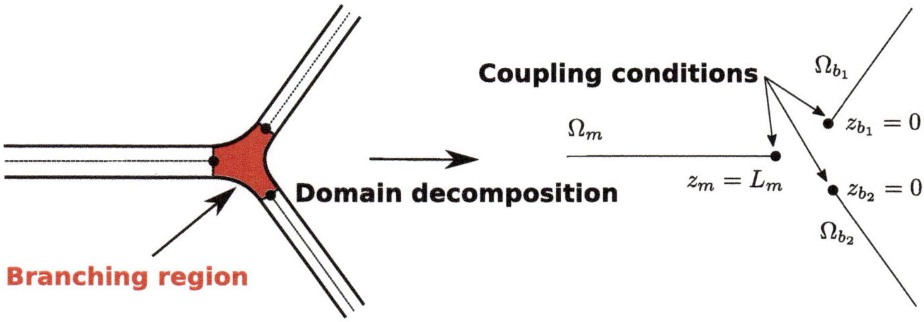

Fig. 3.15 1D domain decomposition approach for a bifurcation

conservation of energy. The conservation of mass at the bifurcation can be enforced by a simple relationship between the corresponding flow rates:

$$Q_m(L_m) = Q_{b_1}(0) + Q_{b_2}(0). \tag{3.82}$$

Integrating this equation on both sides over an arbitrary time period, one obtains that the fluid volume transported into the bifurcation during this time period is also leaving the bifuration. In order to formulate a coupling equation which guarantees that the energy of the system is not accumulated, we derive in a first step an energy balance equation for the blood flow model (3.23). Thereby, we use the same arguments as in [55].

Energy Balance Equation for Blood Flow

Multiplying the second equation in (3.23) by $v = Q/A$ and integrating over the spatial domain $(0, L)$ yields four different terms that are analyzed in the following:

- **Term 1:**

$$\begin{aligned} T_1 &= \int_0^L \frac{\partial Q}{\partial t} v \, dz = \int_0^L \frac{\partial (Av)}{\partial t} v \, dz = \frac{1}{2}\int_0^L A \frac{\partial v^2}{\partial t} dz + \frac{1}{2}\int_0^L \frac{\partial A}{\partial t} v^2 dz \\ &= \frac{1}{2}\frac{d}{dt}\int_0^L A v^2 \, dz + \frac{1}{2}\int_0^L \frac{\partial A}{\partial t} v^2 dz. \end{aligned}$$

- **Term 2:**

$$\begin{aligned} T_2 &= \int_0^L \frac{\partial}{\partial z}\left(Av^2\right) v \, dz = \int_0^L \frac{\partial}{\partial z}(Av)\, v^2 \, dz + \int_0^L A v^2 \frac{\partial v}{\partial z} dz \\ &= \frac{1}{2}\int_0^L \frac{\partial}{\partial z}(Av)\, v^2 \, dz + \frac{1}{2}\int_0^L \frac{\partial A}{\partial z} v^3 \, dz + \frac{3}{2}\int_0^L A v^2 \frac{\partial v}{\partial z} dz \\ &= \frac{1}{2}\int_0^L \frac{\partial Q}{\partial z} v^2 \, dz + \frac{1}{2}\int_0^L \frac{\partial \left(Av^3\right)}{\partial z} dz. \end{aligned}$$

Using the first equation in (3.23), we obtain for T_2:

$$T_2 = \frac{1}{2}\left[-\int_0^L \frac{\partial A}{\partial t} v^2 \, dz + A v^3\big|_0^L\right].$$

- **Term 3:** By means of the non-linear relationship between the pressure p and the cross-sectional area A:

$$p(A) = G_0\left(\sqrt{\frac{A}{A_0}} - 1\right),$$

we obtain:

$$T_3 = \int_0^L \frac{A}{\rho}\frac{\partial p}{\partial z} v \, dz = \frac{1}{\rho}\left[-\int_0^L \frac{\partial Q}{\partial z} p \, dz + Qp\big|_0^L\right].$$

Substituting the first equation from (3.23) into this equation, the following expression can be derived for T_3:

$$T_3 = \frac{1}{\rho}\left[\int_0^L \frac{\partial A}{\partial t} p \, dz + Qp\big|_0^L\right] = \frac{1}{\rho}\left[\frac{d}{dt}\int_0^L \Psi(A) \, dz + Qp\big|_0^L\right],$$

where $\Psi(A)$ is defined as follows:

$$\Psi(A) = \int_{A_0}^A p(\xi) \, d\xi.$$

- **Term 4:** For the viscous term in the source term we obtain:

$$T_4 = \int_0^L K_R \frac{Q}{A} v \, dz = \int_0^L K_R v^2 \, dz.$$

Summing the terms T_1 to T_4, we obtain:

$$\frac{1}{2}\rho\frac{d}{dt}\int_0^L A v^2 \, dz + \frac{d}{dt}\int_0^L \Psi(A) \, dz + \rho K_R \int_0^L v^2 \, dz + \left(\frac{1}{2}\rho A v^3 + Qp\right)\Big|_0^L = 0.$$

Integrating this equation on both sides wrt. the time variable t from 0 to T yields for the total energy

$$\mathcal{E}(t) = \frac{1}{2}\rho\int_0^L A(z,t)\, v^2(z,t) \, dz + \int_0^L \Psi(A(z,t)) \, dz \tag{3.83}$$

of the vessel the following equation:

$$\mathcal{E}(T) + \rho K_R \int_0^T \int_0^L v^2 \, dz \, dt + \int_0^T Q \left(\frac{1}{2} \rho v^2 + p \right) \Big|_0^L dt = \mathcal{E}(0) .$$

Considering the energy term in (3.83), two different components can be identified. While the first term quantifies the kinetic energy of flow, the second term determines the potential energy contained in the vessel wall. The energy balance from above can be derived for each vessel $i \in \{m, b_1, b_2\}$. Summing the corresponding balance equations, one obtains:

$$\sum_{i \in \{m, b_1, b_2\}} \mathcal{E}_i(T) + \rho K_R \int_0^T \int_0^{L_i} v_i^2 \, dz \, dt + \int_0^T Q_i \left(\frac{1}{2} \rho v_i^2 + p_i \right) \Big|_0^{L_i} dt = \sum_{i \in \{m, b_1, b_2\}} \mathcal{E}_i(0) .$$

Rearranging the boundary terms yields:

$$\begin{aligned} &\sum_{i \in \{m, b_1, b_2\}} \mathcal{E}_i(T) + \rho K_R \int_0^T \int_0^{L_i} v_i^2 \, dz \, dt + \\ &\int_0^T Q_m(L_m, t)\, p_{t,m}(L_m, t) - Q_{b_1}(0, t)\, p_{t,b_1}(0, t) - Q_{b_2}(0, t)\, p_{t,b_2}(0, t)\, dt = \\ &\sum_{i \in \{m, b_1, b_2\}} \mathcal{E}_i(0) + \\ &\underbrace{\int_0^T Q_m(0, t)\, p_{t,m}(0, t) - Q_{b_1}(L_{b_1}, t)\, p_{t,b_1}(L_{b_1}, t) - Q_{b_2}(L_{b_2}, t)\, p_{t,b_2}(L_{b_2}, t)\, dt}_{=:b(T)}, \end{aligned}$$

where $p_{t,i} = \frac{1}{2} \rho v_i^2 + p_i$ is the total pressure. Imposing the continuity of the total pressure at the boundaries $z_m = L_m$, $z_{b_1} = 0$ und $z_{b_2} = 0$:

$$p_{t,m}(L_m, t) = p_{t,b_1}(0, t) = p_{t,b_2}(0, t) , \ \forall t \in [0, T] , \tag{3.84}$$

and mass conservation

$$Q_m(L_m, t) = Q_{b_1}(0, t) + Q_{b_2}(0, t)$$

an inequality results, in which the energy terms $\mathcal{E}_i(T)$, $i \in \{m, b_1, b_2\}$ are bounded from above by the initial conditions and the boundary conditions $b(T)$:

$$\sum_{i \in \{m, b_1, b_2\}} \mathcal{E}_i(T) \leq \sum_{i \in \{m, b_1, b_2\}} \mathcal{E}_i(0) + b(T) .$$

It turns out that the continuity of the total pressure (3.84) implies boundedness of the energy terms by given data. Thus we use in the following (3.84) as a second condition besides the mass conservation (3.82) to couple the flow problems at a bifurcation. Since at each bifurcation there are six unknowns (two boundary values i.e. one for A and Q) to determine, we require besides the physical based coupling conditions (3.82) and (3.84) three further equations to compute all the boundary values. A possible way how to derive the remaining equations is described in Chap. 4.

The coupling condition based on the continuity of the total pressure (3.84) can be enhanced to account for a possible energy loss by the geometry of the bifurcation, which is governed by the branching angles α_1 and α_2. In [25], further terms are added to (3.84) based on [56]. This results in the following equations:

$$\begin{aligned} p_{t,m}(L_m,t) - \operatorname{sign}(v_m)\, f_m(v_m) &= p_{t,b_1}(0,t) + \operatorname{sign}\left(v_{b_1}\right) f_1\left(v_{b_1},\alpha_1\right), \\ p_{t,m}(L_m,t) - \operatorname{sign}(v_m)\, f_m(v_m) &= p_{t,b_2}(0,t) + \operatorname{sign}\left(v_{b_2}\right) f_2\left(v_{b_2},\alpha_2\right). \end{aligned} \tag{3.85}$$

The functions $f_i,\ i \in \{m,\ b_1,\ b_2\}$ are positive and equal to zero, if the corresponding velocity v_i is zero. A possible choice of these functions is given by:

$$f_m = \gamma_m v_m^2,\ \ f_i = \gamma_i v_{b_i}^2 \sqrt{2(1-\cos\alpha_i)},\ \ i \in \{1,2\},$$

where γ_m and γ_i are positive coefficients. However, the authors of [25] noted that the calibration of suitable coefficients is difficult. For this reason, we neglect in the upcoming chapters the functions f_i and set them to zero.

Transport processes through a bifurcation

After describing, how the 1D flow equations can be coupled at a bifurcation, we derive further balance equations for transport equations such that besides blood volume a solute can be transported through the bifurcation without any loss. Computing the concentration variables $\Gamma_i,\ i \in \{m, b_1, b_2\}$, we perform a distinction of cases as in [37]. This distinction of cases is based on the orientation of the coordinate axes and the sign of the flow rates Q_i. For this reason, we define two kinds of boundary conditions: Inflow and outflow interfaces. The index indicating an inflow interface is labelled by $in_j \in \{m, b_1, b_2\}$, whereas $out_j \in \{m, b_1, b_2\},\ j \in \{1,2\}$ is assigned to an outflow interface. A boundary at a bifurcation is defined as inflow interface, if it holds:

$$Q_m(L_m) \geq 0 \text{ or } Q_i(0) < 0,\ \ i \in \{b_1, b_2\}, \tag{3.86}$$

while an outflow interface is charaterized by:

$$Q_i(0) \geq 0,\ \ i \in \{b_1, b_2\} \text{ or } Q_m(L_m) < 0. \tag{3.87}$$

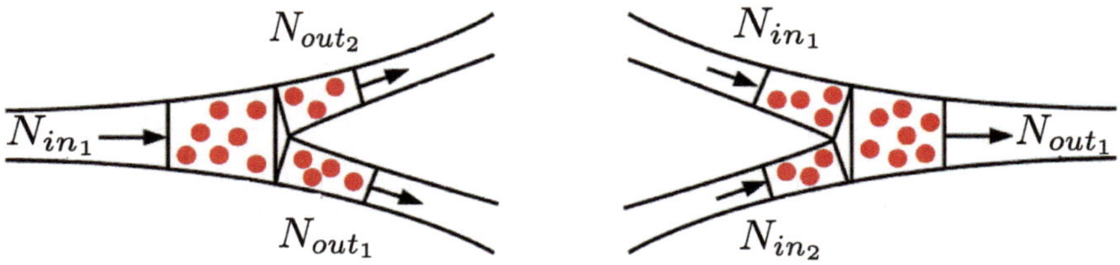

Fig. 3.16 Transport of a substance (red particles) through a bifurcation. Modeling this process, we distinguish between two cases: In the first case a substance is transported through a single inflow interface into the bifurcation and removed by two outflow interfaces out_1 and out_2. Here, we obtain the balance equation wrt. the number of particles: $N_{in_1} = N_{out_1} + N_{out_2}$ (left). In the second case transport occurs via two inflow interfaces in_1 and in_2. The incoming number of particles $N_{out_1} = N_{in_1} + N_{in_2}$ is removed by a single outflow interface (right)

Thereby, only the cases are relevant, in which there is a a single inflow interface or two different inflow interfaces (see Fig. 3.16). All the other cases can be excluded, due to the fact that they are in contradiction with mass conservation (3.82).

Case 1: Single inflow interface

In this case it holds for the index sets:

$$\{in_1,\ out_1,\ out_2\} = \{m,\ b_1,\ b_2\}\,.$$

At the boundary in_1, the concentration variable Γ_{in_1} can be determined by means of upwinding (see Chap. 4). The remaining equations result from the continuity of the volumetric concentration c [mmol/cm³]:

$$\frac{\Gamma_{in_1}}{A_{in_1}} = c = \frac{\Gamma_{out_1}}{A_{out_1}},\ \frac{\Gamma_{in_1}}{A_{in_1}} = c = \frac{\Gamma_{out_2}}{A_{out_2}}. \tag{3.88}$$

By means of (3.88), the concentration variable Γ_{in_1} as well as (3.82) and (3.84) all the boundary conditions can be determined for this case. Continuity of the volumetric concentration ensures the mass conservation of the transported solute. This can be shown as follows: We denote by N_i [mmol], $i \in \{m, b_1, b_2\}$ the number of particles passing the interface of vessel i within a time interval $[t, t + \Delta t]$. This yields the following balance equation for the transport process:

$$N_{in_1} = N_{out_1} + N_{out_2}.$$

The number of particles N_i can be computed as follows:

$$N_i = \int_t^{t+\Delta t} \frac{\Gamma_i \cdot |Q_i|}{A_i}\, dt.$$

By this, we have:

$$\int_t^{t+\Delta t} \frac{\Gamma_{in_1} \cdot \left|Q_{in_1}\right|}{A_{in_1}} \, dt = \int_t^{t+\Delta t} \frac{\Gamma_{out_1} \cdot \left|Q_{out_1}\right|}{A_{out_1}} \, dt + \int_t^{t+\Delta t} \frac{\Gamma_{out_2} \cdot \left|Q_{out_2}\right|}{A_{out_2}} \, dt.$$

Using the continuity of the volumetric concentration, one obtains:

$$\int_t^{t+\Delta t} c\left|Q_{in_1}\right| \, dt = \int_t^{t+\Delta t} c\left|Q_{out_1}\right| \, dt + \int_t^{t+\Delta t} c\left|Q_{out_2}\right| \, dt$$

and

$$\int_t^{t+\Delta t} \left|Q_{in_1}\right| - \left|Q_{out_1}\right| - \left|Q_{out_2}\right| \, dt = 0.$$

Due to the fact that the choice of $\Delta t > 0$ is arbitrary, it follows together with Sect. 7.5 (Fundamental theorem of variation calculus):

$$\left|Q_{in_1}\right| = \left|Q_{out_1}\right| + \left|Q_{out_2}\right|.$$

Case 2: Two inflow interfaces

In this case, it holds for the index sets:

$$\{in_1, \; in_2, \; out_1\} = \{m, \; b_1, \; b_2\}.$$

At the interfaces in_1 and in_2 the concentration variables Γ_{in_1} and Γ_{in_2} can be determined as in Case 1 by upwinding (see Chap. 4). In order to compute the third concentration variable Γ_{out_1}, we use a balance equation corresponding to the mass conservation for the transport process:

$$\frac{\Gamma_{out_1} Q_{out_1}}{A_{out_1}} = \frac{\Gamma_{in_1} Q_{in_1}}{A_{in_1}} + \frac{\Gamma_{in_2} Q_{in_2}}{A_{in_2}}. \tag{3.89}$$

Generalization for More than Two Branches

In the following Chaps. 4 and 5 we consider only networks that exhibit vessels with only two branches. However, one can be faced with networks exhibiting more than two branches at a branching point. For this reason it is necessary to generalize the coupling condition, we have derived so far. We assume that at a branching point n vessels are linked with each other. The single vessels are numbered by the following index set:

$$I = \{1, \ldots, n\}.$$

The set I is divided in a set I_{in} and a set I_{out}, where in the set I_{in} all the vessels are contained exhibiting an inflow interface at a branching point, while I_{out} contains all vessels removing blood from a branching point. Mathematically, an assignement of an index $i \in I$ can be formulated as follows:

$$\begin{aligned} i \in I_{out} &\Leftrightarrow (z_i = 0 \wedge Q_i > 0) \vee (z_i = L_i \wedge Q_i \leq 0)\,, \\ i \in I_{in} &\Leftrightarrow (z_i = 0 \wedge Q_i \leq 0) \vee (z_i = L_i \wedge Q_i > 0)\,, \end{aligned} \tag{3.90}$$

where z_i is the space coordinate of vessel $i \in I$ that is adjacent to a branching point. Using these definitions the coupling equations read as follows:

- **Continuity of total pressure:**

$$p_{t,1}\left(z_1, t\right) = p_{t,i}\left(z_i, t\right), \;\; I \setminus \{1\}$$

- **Mass conservation (blood flow):**

$$\sum_{i \in I} Q_i\left(z_i\right) \sigma\left(z_i\right) = 0, \;\; \sigma\left(z_i\right) = \begin{cases} 1, & \text{falls } z_i = L_i, \\ -1, & \text{falls } z_i = 0 \end{cases}$$

- **Mass conservation (transport):**

$$\frac{|Q_i|}{\sum_{j \in I_{out}} |Q_j|} = \frac{\frac{\Gamma_i |Q_i|}{A_i}}{\sum_{j \in I_{in}} \frac{\Gamma_j |Q_j|}{A_j}}, \;\; i \in I_{out}$$

- **Upwinding:** If $i \in I_{in}$ the value at the interface is determined by means of the solution in the interior of the domain $(0, L_i)$. The numerical implementation is described in Chap. 4.

3.6.4 Modeling of a Stenosis (SB)

In the final subsection of this chapter, we are concerned with modeling of stenoses, i.e., narrowings of larger arteries by plaque formed in terms of arteriosclerosis (see Sect. 1.2.1). For this purpose, we use two approaches that have already been applied in terms of modeling outflow boundaries (OB) (see Sect. 3.6.2) and bifurcations (see Sect. 3.6.3). These are on the one hand side the usage of 1D-0D coupled differential systems of equations and on the other hand side a domain decomposition approach. As a domain that is decomposed, we consider a vessel Ω containing a stenosis, where the section covered by the stenosis is denoted by $\Omega_s \subset \Omega$. The stenosis splits the vessel in a proximal part $\Omega_p \subset \Omega$ and a distal part $\Omega_d \subset \Omega$, such that it holds:

$$\Omega = \Omega_p \cup \Omega_s \cup \Omega_d.$$

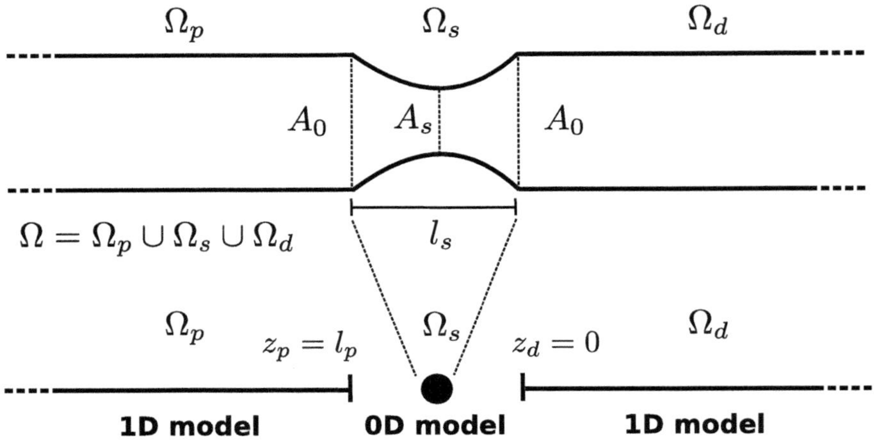

Fig. 3.17 Modeling of a stenosis Ω_s by a domain decomposition approach and a 1D-0D coupled differential equation system. The vessel Ω, is divided by the stenosis into a proximal part Ω_p, the stenosis Ω_s and a distal part Ω_d

In this context, we assume that the length of Ω_s is small compared to the lengths of Ω_p and Ω_d. Based on this assumption, we assign to each of the subvessels Ω_p and Ω_d a 1D blood flow model given by (3.23) (see Fig. 3.17).

The influence of the stenosis is described by an ODE, i.e., the spatial extension of the stenosis is neglected (lumped parameter model). A derivation of this ODE can be found in [57, 58]. It has the following form:

$$\frac{K_u \rho l_s}{A_0} \frac{dQ_s}{dt} = \Delta P_s - \frac{K_v \mu}{A_0 D_0} Q_s - \frac{K_t \rho}{2A_0^2} \left(\frac{A_0}{A_s} - 1 \right)^2 Q_s \left| Q_s \right| . \tag{3.91}$$

The solution variable of the differential equation is given by the flow variable Q_s, which can be used to impose a boundary condition at the boundaries $z_p = l_p$ and $z_d = 0$. To enforce the mass conservation, one requires:

$$Q_p \left(l_p, t \right) = Q_d \left(0, t \right) = Q_s \left(t \right), \; t > 0.$$

Here l_p denotes the length of the proximal vessel Ω_p, while Q_p and Q_d represent the flow rates in Ω_p and Ω_d. The right hand side of the ODE (3.91) consists of three different terms. The first term quantifies the pressure drop ΔP_s along the stenosis and is defined as follows:

$$\Delta P_s \left(t \right) = p_p \left(l_p, t \right) - p_d \left(0, t \right), \; t > 0,$$

where p_p and p_d denote the pressures in the subvessels Ω_p and Ω_d. The viscous flow resistance is taken into account by the second term. μ stands for blood viscosity, A_0

is the cross-sectional area of Ω if there is no stress in the vessel wall, and D_0 is the associated diameter. K_v is an empirical parameter given by the following expression [59]:

$$K_v = 32.0 \cdot (0.83\, l_s + 1.64\, D_s) \cdot \left(\frac{A_0}{A_s}\right)^2 \cdot \frac{1}{D_0}. \tag{3.92}$$

l_s is the length of the stenosis, while A_s denotes the smallest cross-sectional area in the stenosis. D_s is the diameter of the cross-sectional area A_s. The last term in the right hand side of (3.91) includes the influence of the narrowing. In particular, according to [57, 58], this term accounts for the energy loss due to turbulence. The parameter ρ stands for the density of blood. $K_u = 1.2$ and $K_t = 1.52$ are additional empirical parameters weightening the inertia term and the turbulence term. Another important quantity that we require in the upcoming chapters of this book is the degree of a stenosis S_d, which is defined as follows:

$$S_d = 1 - \frac{A_s}{A_0}. \tag{3.93}$$

In order to model transport processes through the stenosis, we enforce the continuity of the volumetric concentration:

$$\frac{\Gamma_p}{A_p}\left(l_p, t\right) = \frac{\Gamma_d}{A_d}(0, t)\,,\ t \geq 0. \tag{3.94}$$

Depending on the sign of Q_s, the missing coupling condition is determined by an upwinding procedure, i.e., if $Q_s > 0$ holds, $\Gamma_p\left(l_p, t\right)$ is given by the solution of the 1D transport equation in Ω_p. If $Q_s \leq 0$, $\Gamma_d\,(0, t)$ is determined by the solution of the 1D transport equation in Ω_d. The numerical implementation can be found in Chap. 4.

3.7 Validation of the 1D Model

Concluding this chapter, we would like to point out several references from literature that deal with the validation of the dimension reduced blood flow model. Considering the corresponding references, we can identify two ways to evaluate the simulation results of the 1D blood flow model.

One way is to compare the simulation results of the 1D model with simulation results of a more accurate 3D model. Here, the 3D model is given by the incompressible Navier-Stokes Eqs. (3.1a) and (3.1b). In [6, 7], one can find a comparison between the 3D model, which does not take into account the deformation of the vessel walls, and the 1D model. We denote the 3D model with rigid walls by 3D-SW. The results of the comparisons between the 3D-SW and the 1D model can be summarized as follows:

(SW 1)
Pressures and velocities reported at a certain position of a vessel and averaged with respect to time are in accordance with the models 3D-SW and 1D-DW. The 3D quantities are averaged over the cross-sectional areas at the positions of the corresponding vessel.

(SW 2)
There are significant differences between the shape of the pressure and velocity curves reported during a heart beat. Both the shape and the extrema in the diastolic and systolic phases differ from each other. This can be explained by the fact that the 3D-SW model is not damping the pressure or velocity peaks, significantly.

(SW 3)
Similar observations as in (SW 2) can be made with respect to the wall shear stresses (WSS).

In particular, observations (SW 2) and (SW 3) demonstrate that for a more accurate comparison, the 3D flow Eqs. (3.1a) and (3.1b) must be combined with an FSI model. Therefore, in [7], the incompressible Navier-Stokes equations were enhanced by an FSI model. Similar considerations can be found in [62]. We will use the abbreviation 3D-DW for this model class in the following. Based on various geometries generated both artificially and by image reconstruction, the following observations can be listed:

(DW 1)
The 3D pressure and velocity curves averaged over the cross-sectional area show good agreement with the shape of the corresponding 1D-DW curves.

(DW 2)
Regarding the pressure and velocity values, there is a very good agreement between the two models during diastole.

(DW 3)
Differences during systole can be explained by the simplifications in the derivation of the 1D-DW model. These are e.g. neglecting the vessel curvatures and using a fixed velocity profile.

Certainly, the results of the 1D-DW model can be directly compared with medical data. This has been done in a large number of publications. As an example, we would like to refer to [60, 61]. The comparisons with medical measurements allow similar observations as in the comparisons with 3D-DW:

(E1)
The measured pressure and velocity curves have essentially the same shape as the curves of the 1D-DW model.

(E2)
Smaller deviations occur in the pressure curves, especially in the systolic phase.

(E3)
More significant differences can be observed in the flow rates and velocities of some vessels.

Summarizing, it can be noted that the 1D-DW model can reproduce the characteristic shape of the blood pressure and velocity curves despite the many simplifications. Furthermore, realistic values for the pressure can be obtained except of minor local deviations.

References

1. John L, Pustějovská P, Steinbach O (2017) On the influence of the wall shear stress vector form on hemodynamic indicators. Comput Vis Sci 18(4–5):113–122. Springer
2. Helmig R (1997) Multiphase flow and transport processes in the subsurface. Springer, Berlin
3. Galdi G, Rannacher R, Robertson A, Turek S (2008) Hemodynamical flows: modeling and analysis and simulation, vol 37. Birkhäuser, Basel
4. Crosetto P, Reymond P, Deparis S, Kontaxakis D, Stergiopulos N, Quarteroni A (2011) Fluid–structure interaction simulation of aortic blood flow. Comput Fluids 43(1):46–57. Elsevier
5. Formaggia L, Quarteroni A, Veneziani A (eds) (2010) Cardiovascular mathematics: modeling and simulation of the circulatory system, vol 1. Springer Science & Business Media
6. Grinberg L, Cheever E, Anor T, Madsen J, Karniadakis G (2011) Modeling blood flow circulation in intracranial arterial networks: a comparative 3D/1D simulation study. Ann Biomed Eng 39(1):297–309. Springer
7. Reymond P, Crosetto P, Deparis S, Quarteroni A, Stergiopulos N (2013) Physiological simulation of blood flow in the aorta: comparison of hemodynamic indices as predicted by 3-D FSI, 3-D rigid wall and 1-D models. Med Eng Phys 35(6):784–791. Elsevier
8. Torii R, Oshima M, Kobayashi T, Takagi K, Tezduyar T (2009) Fluid–structure interaction modeling of blood flow and cerebral aneurysm: significance of artery and aneurysm shapes. Comput Methods Appl Mech Eng 198(45):3613–3621. Elsevier
9. Quarteroni A, Veneziani A, Zunino P (2002) Mathematical and numerical modeling of solute dynamics in blood flow and arterial walls. SIAM J Numer Anal 39(5):1488–1511. SIAM
10. Barnard A, Hunt W, Timlake W, Varley E (1966) A theory of fluid flow in compliant tubes. Biophys J 6(6):717. The Biophysical Society
11. Čanić S, Kim E (2003) Mathematical analysis of the quasilinear effects in a hyperbolic model blood flow through compliant axi-symmetric vessels. Math Methods Appl Sci 26(14):1161–1186. Wiley Online Library
12. D'Angelo C (2007) Multiscale modelling of metabolism and transport phenomena in living tissues. ÉCOLE POLYTECHNIQUE FÉDÉRALE DE LAUSANNE
13. Ottesen J, Olufsen M, Larsen J (2004) Applied mathematical models in human physiology. Siam
14. Hughes T (1974) A study of the one-dimensional theory of arterial pulse propagation. University of California, Struct Eng Lab
15. Hughes T, Lubliner J (1973) On the one-dimensional theory of blood flow in the larger vessels. Math Biosci 18(1):161–170. Elsevier
16. Smith N, Pullan A, Hunter P (2002) An anatomically based model of transient coronary blood flow in the heart. SIAM J Appl Math 62(3):990–1018. SIAM

17. Ghigo A, Lagrée P, Fullana J (2018) A time-dependent non-Newtonian extension of a 1D blood flow model. J NonNewton Fluid Mech 253:36–49. Elsevier
18. Puelz C, Čanić S, Rivière B, Rusin C (2017) Comparison of reduced models for blood flow using Runge–Kutta discontinuous Galerkin methods. Appl Numer Math 115:114–141. Elsevier
19. Raines J, Jaffrin M, Shapiro A (197) A computer simulation of arterial dynamics in the human leg. J Biomech 7(1):77–91. Elsevier
20. Porenta G, Young D, Rogge T (1986) A finite-element model of blood flow in arteries including taper, branches, and obstructions. J Biomech Eng 108(2):161–167. American Society of Mechanical Engineers
21. Olufsen M (1999) Structured tree outflow condition for blood flow in larger systemic arteries. Am J Physiol-Heart Circulatory Physiol 276(1):H257–H268. Am Physiological Soc
22. Sherwin S, Formaggia L, Peiro J, Franke V (2003) Computational modelling of 1D blood flow with variable mechanical properties and its application to the simulation of wave propagation in the human arterial system. Int J Numer Methods Fluids 43(6–7):673–700. Wiley Online Library
23. Battista C (2015) Parameter estimation of viscoelastic wall models in a one-dimensional circulatory network. North Carolina State University
24. Quarteroni A, Sacco R, Saleri F (2010) Numerical mathematics, vol 37. Springer Science & Business Media
25. Formaggia L, Lamponi D, Quarteroni A (2003) One-dimensional models for blood flow in arteries. J Eng Math 47(3–4):251–276. Springer
26. Hooke R (1678) De potentia restitutiva, or of spring explaining the power of springing bodies, vol 1678. John Martyn, London, UK, pp 23
27. Westerhof N, Bosman F, De Vries C, Noordergraaf A (1969) Analog studies of the human systemic arterial tre. J Biomech 2(2):121IN1135IN3137IN5139–134136138143. Elsevier
28. Toro E (2016) Brain venous haemodynamics, neurological diseases and mathematical modelling. A review. Appl Math Comput 272:542–579. Elsevier
29. Amadori D, Ferrari S, Formaggia L (2007) Derivation and analysis of a fluid-dynamical model in thin and long elastic vessels. Networks Heterogen Media 2(1):99. American Institute of Mathematical Sciences
30. Quarteroni A, Formaggia L (2004) Mathematical modelling and numerical simulation of the cardiovascular system. Handbook Numer Anal 12:3–127. Elsevier
31. Valdez-Jasso D, Haider M, Banks H, Santana D, Germán Y, Armentano R, Olufsen M (2009) Analysis of viscoelastic wall properties in ovine arteries. IEEE Trans Biomed Eng 5(2):210–219. IEEE
32. Fung Y (2013) Biomechanics: mechanical properties of living tissues. Springer Science & Business Media
33. DeVault K, Gremaud P, Novak V, Olufsen M, Vernieres G, Zhao P (2008) Blood flow in the circle of Willis: modeling and calibration. Multiscale Model Simul 7(2):888–909. SIAM
34. Čanić S, Tambača J, Guidoboni G, Mikelić A, Hartley C, Rosenstrauch D (2006) Modeling viscoelastic behavior of arterial walls and their interaction with pulsatile blood flow. SIAM J Appl Math 67(1):164–193. SIAM
35. Raghu R, Vignon-Clementel I, Figueroa C, Taylor C (2011) Comparative study of viscoelastic arterial wall models in nonlinear one-dimensional finite element simulations of blood flow. J Biomech Eng 133(8):081003. American Society of Mechanical Engineers
36. Acosta S, Puelz C, Rivière B, Penny D, Rusin C (2015) Numerical method of characteristics for one-dimensional blood flow. J Comput Phys 294:96–109. Elsevier
37. Köppl T, Wohlmuth B, Helmig R (2013) Reduced one-dimensional modelling and numerical simulation for mass transport in fluids. Int J Numer Methods Fluids 72(2):135–156. Wiley Online Library
38. Alastruey J, Parker K, Sherwin S et al (2012) Arterial pulse wave haemodynamics. 11th Int Confer Press Surges 30:401–443. Virtual PiE Led t/a BHR Group Lisbon, Portugal
39. Formaggia L, Nobile F, Quarteroni A (2002) A one dimensional model for blood flow: application to vascular prosthesis. Lect Notes Comput Sci Eng 19:137–154. Springer

40. Liang F, Takagi S, Himeno R, Liu H (2009) Multi-scale modeling of the human cardiovascular system with applications to aortic valvular and arterial stenoses. Med Biol Eng Comput 47(7):743–755. Springer
41. Hackbusch W (1986) Theorie und Numerik elliptischer Differentialgleichungen. Springer
42. Odehnal B, Stachel H, Glaeser G (2020) The universe of quadrics. Springer Nature
43. Li D, Li T (1994) Global classical solutions for quasilinear hyperbolic systems, vol 32. John Wiley & Sons Incorporated
44. Li T, Zhao K (2011) Global existence and long-time behavior of entropy weak solutions to a quasilinear hyperbolic blood flow model. Networks Heterogen Media 6(4):625. American Institute of Mathematical Sciences
45. Fernández M, Milisic V, Quarteroni A (2005) Analysis of a geometrical multiscale blood flow model based on the coupling of ODEs and hyperbolic PDEs. Multiscale Model Simul 4(1):215–236. SIAM
46. Milisic V, Quarteroni A (2004) Analysis of lumped parameter models for blood flow simulations and their relation with 1D models. Math Model Num Anal 38(4):613–632
47. Hoppensteadt F, Peskin C (2012) Modeling and simulation in medicine and the life sciences, vol 10. Springer Science & Business Media
48. Spencer M, Denison A (1963) Pulsatile blood flow in the vascular system. Handbook Physiol 2:839–864. Amer. Physiol, Soc
49. Formaggia L, Lamponi D, Tuveri M, Veneziani A (2006) Numerical modeling of 1D arterial networks coupled with a lumped parameters description of the heart. Comput Methods Biomech Biomed Eng 9(5):273–288. Taylor & Francis
50. Epstein S, Willemet M, Chowienczyk P, Alastruey J (2015) Reducing the number of parameters in 1D arterial blood flow modeling: less is more for patient-specific simulations. Am J Physiol-Heart Circulatory Physiol 309(1):H222–H234. American Physiological Society Bethesda, MD
51. Hambley A, Kumar N, Kulkarni A (2008) Electrical engineering: principles and applications. Pearson Prentice Hall Upper Saddle River, NJ
52. Chen P, Quarteroni A, Rozza G (2013) Simulation-based uncertainty quantification of human arterial network hemodynamics. Int J Numer Methods Biomed Eng 29(6):698–721. Wiley Online Library
53. Ismail M, Wall W, Gee M (2013) Adjoint-based inverse analysis of windkessel parameters for patient-specific vascular models. J Comput Phys 244:113–130. Elsevier
54. Kissas G, Yang Y, Hwuang E, Witschey W, Detre J, Perdikaris P (2020) Machine learning in cardiovascular flows modeling: predicting arterial blood pressure from non-invasive 4D flow MRI data using physics-informed neural networks. Comput Methods Appl Mech Eng 358:112623. Elsevier
55. Formaggia L, Gerbeau J, Nobile F, Quarteroni A (2001) On the coupling of 3D and 1D Navier–Stokes equations for flow problems in compliant vessels. Comput Methods Appl Mech Eng 191(6–7):561–582. Elsevier
56. Holden H, Risebro N (1999) Riemann problems with a kink. SIAM J Math Anal 30(3):497–515. SIAM
57. Young D, Tsai F (1973) Flow characteristics in models of arterial stenoses–I. Steady flow. J Biomech 6(4):395–410. Elsevier
58. Young D, Tsai F (1973) Flow characteristics in models of arterial stenoses–II. Unsteady flow. J Biomech 6(5):547–559. Elsevier
59. Seeley B, Young D (1976) Effect of geometry on pressure losses across models of arterial stenoses. J Biomech 9(7):439–448. Elsevier
60. Boileau E, Nithiarasu P, Blanco P, Müller L, Fossan F, Hellevik L, Donders W, Huberts W, Willemet M, Alastruey J (2015) A benchmark study of numerical schemes for one-dimensional arterial blood flow modelling. Int J Numer Methods Biomed Eng 31(10):e02732. Wiley Online Library
61. Olufsen, MS, Peskin CS, Kim, WY, Pedersen EM, Nadim A, Larsen J (2000) Numerical simulation and experimental validation of blood flow in arteries with structured-tree outflow conditions. Ann Biomed Eng 28(11):1281–1299. Springer

62. Xiao N, Alastruey J, Figueroa A (2014) A systematic comparison between 1-D and 3-D hemodynamics in compliant arterial models. Int J Numer Methods Biomed Eng 30(2):204–231. Wiley Online Library
63. Rodero GC (2017) Analysis of blood flow in one dimensional elastic artery using Navier-Stokes conservation laws

Chapter 4
Numerical Solution Methods

Abstract In the previous chapter, we have derived 1D models for blood flow and transport processes using Reynold's transport theorem and averaging operations. These models have the form of PDEs, which can be assigned to the class of first order hyperbolic equations. Moreover, it turns out that these systems consist of non-linear equations that are strongly coupled to each other. Therefore, these systems of equations are generally not solvable by analytical methods, but only by the use of numerical methods. In order to compute the numerical solution of the 1D blood flow model and the transport equation, a variety of numerical solution methods have been developed in the past decades. These range from simple finite difference methods [1, 2] to neural networks [3]. In the following sections of this chapter, we will present some simple solution schemes. These are, on the one hand side the Lax-Wendroff method for the blood flow model and on the other hand side an implicit upwinding method for the transport equation. The choice of these methods is motivated as follows: The Lax-Wendroff method is a method that does not require an analytical expression of the characteristic variables. Therefore, the method can easily be applied to both flat velocity profiles ($\alpha = 1$) and other velocity profiles ($\alpha \neq 1$). This is a major difference compared to other numerical methods, which require an explicit expression for the characteristic variables to incorporate boundary conditions. In order to decouple the simulation of flow and transport, we use an upwinding method combined with the implicit Euler method for time integration. The chapter contains the following sections: In the first section, we describe the construction of the two methods for a single vessel. We show that both methods can be considered as finite volume methods, where the fluxes at the respective cell interfaces are approximated by finite differences. Afterwards we consider the numerical treatment of the various boundary conditions. In the following section, the convergence behavior of the two methods is investigated. Section 4.4 contains a series of numerical experiments that confirm the theoretical results from the previous sections and chapters. We also discuss the advantages and disadvantages of the proposed discretization methods and provide an overview on further discretization methods that have been applied to the blood flow model.

T. Köppl and R. Helmig, *Dimension Reduced Modeling of Blood Flow in Large Arteries*, Mathematical Engineering, https://doi.org/10.1007/978-3-031-33087-2_4

4.1 Construction of the Numerical Methods

We first consider a single vessel of length L in a simulation period $(0, T) \subset \mathbb{R}$:

$$\Omega_T = \left\{ (z, t) \in \mathbb{R}^2 \middle| 0 < z < L,\ 0 < t < T \right\}$$

and introduce a discretization of Ω_T by a grid $\Omega_{\Delta z, \Delta t}$ with grid widths Δz and Δt:

$$\Omega_{\Delta z, \Delta t} = \{ (z_i, t_n) \in \Omega_T | z_i = i \cdot \Delta z,\ t_n = n \cdot \Delta t,\ i = 0, \ldots, N_z,\ n = 0, \ldots, N_T \}.$$

$N_z + 1$ and $N_T + 1$ are the number of grid points in space and time. The grid width Δt is also called the time stepsize, since the points t_n discretize the time interval $(0, T)$. This means that $\Omega_{\Delta z, \Delta t}$ decomposes the space-time domain Ω_T into cells C_i^n, given by:

$$C_i^n = (z_i, z_{i+1}) \times (t_n, t_{n+1}) \text{ und } \overline{\Omega}_T = \bigcup_{i=0}^{N_z-1} \bigcup_{n=0}^{N_T-1} \overline{C}_i^n.$$

Wrt. $\Omega_{\Delta z, \Delta t}$ we define a dual grid $\tilde{\Omega}_{\Delta z, \Delta t}$:

$$\tilde{\Omega}_{\Delta z, \Delta t} = \left\{ \left(\tilde{z}_{i+\frac{1}{2}}, t_n \right) \in \Omega_T \middle| \tilde{z}_{i+\frac{1}{2}} = \frac{z_i + z_{i+1}}{2} \right\},$$

composed of the cells

$$\tilde{C}_i^n = \left(\tilde{z}_{i-\frac{1}{2}}, \tilde{z}_{i+\frac{1}{2}} \right) \times (t_n, t_{n+1}).$$

The solution $\mathbf{U_f}$ of the blood flow model is approximated at the grid nodes (z_i, t_n), the associated vector is denoted by $\mathbf{U}_{f,i}^n$ and the solution variables by A_i^n and Q_i^n, respectively, see Fig. 4.1. Using the definitions that have just been introduced, we now derive a discretization of the blood flow model. Thereby, we integrate the PDE-system (3.21) or (3.29a) component-wise on both sides wrt. to an inner cell $\tilde{C}_i^n \subset \Omega_T$:

$$\int_{t_n}^{t_{n+1}} \int_{\tilde{z}_{i-\frac{1}{2}}}^{\tilde{z}_{i+\frac{1}{2}}} \frac{\partial \mathbf{U_f}}{\partial t}\, dz\, dt + \int_{t_n}^{t_{n+1}} \int_{\tilde{z}_{i-\frac{1}{2}}}^{\tilde{z}_{i+\frac{1}{2}}} \frac{\partial \mathbf{F_f}}{\partial z} (\mathbf{U_f})\, dz\, dt = \int_{t_n}^{t_{n+1}} \int_{\tilde{z}_{i-\frac{1}{2}}}^{\tilde{z}_{i+\frac{1}{2}}} \mathbf{S_f} (\mathbf{U_f})\, dz\, dt.$$

Integration of the first term with respect to time and the second term with respect to the space variable yields:

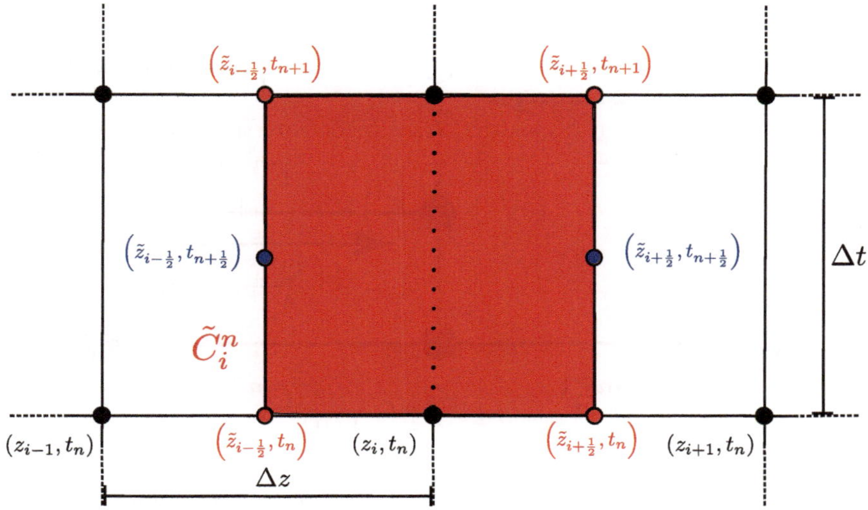

Fig. 4.1 Notation associated with a cell $\tilde{C}_i^n$

$$\int_{\tilde{z}_{i-\frac{1}{2}}}^{\tilde{z}_{i+\frac{1}{2}}} \mathbf{U_f}\left(z, t_{n+1}\right) - \mathbf{U_f}\left(z, t_n\right)\, dz +$$

$$\int_{t_n}^{t_{n+1}} \mathbf{F_f}\left(\mathbf{U_f}\left(\tilde{z}_{i+\frac{1}{2}}, t\right)\right) - \mathbf{F_f}\left(\mathbf{U_f}\left(\tilde{z}_{i-\frac{1}{2}}, t\right)\right)\, dt = \int_{t_n}^{t_{n+1}} \int_{\tilde{z}_{i-\frac{1}{2}}}^{\tilde{z}_{i+\frac{1}{2}}} \mathbf{S_f}\left(\mathbf{U_f}\right)\, dz\, dt.$$

Approximating $\mathbf{U_f}$ on $\tilde{C}_i^n$ by $\mathbf{U}_{f,i}^n$, the previous equation can be rewritten using $\tilde{z}_{i+\frac{1}{2}} - \tilde{z}_{i-\frac{1}{2}} = \Delta z$ and $t_{n+1} - t_n = \Delta t$:

$$\Delta z \left(\mathbf{U}_{f,i}^{n+1} - \mathbf{U}_{f,i}^{n}\right) + \Delta t \left[\mathbf{F_f}\left(\mathbf{U_f}\left(\tilde{z}_{i+\frac{1}{2}}, t_{n+\frac{1}{2}}\right)\right) - \mathbf{F_f}\left(\mathbf{U_f}\left(\tilde{z}_{i-\frac{1}{2}}, t_{n+\frac{1}{2}}\right)\right)\right]$$
$$= \Delta z \Delta t \cdot \mathbf{S_f}\left(\mathbf{U_f}\left(z_i, t_{n+\frac{1}{2}}\right)\right).$$

Approximating the first two integrals numerically, we use the midpoint rule, where the time point $t_{n+\frac{1}{2}}$ is defined as follows:

$$t_{n+\frac{1}{2}} = \frac{t_{n+1} + t_n}{2}.$$

The integral wrt. the source term is approximated in space and time by the midpoint rule. Now the issue arises, in which way the fluxes at the left and right edges of $\tilde{C}_i^n$ can be approximated. In order to write the equation in a compact way, we introduce the following notation for the fluxes at the respective edges (Fig. 4.2):

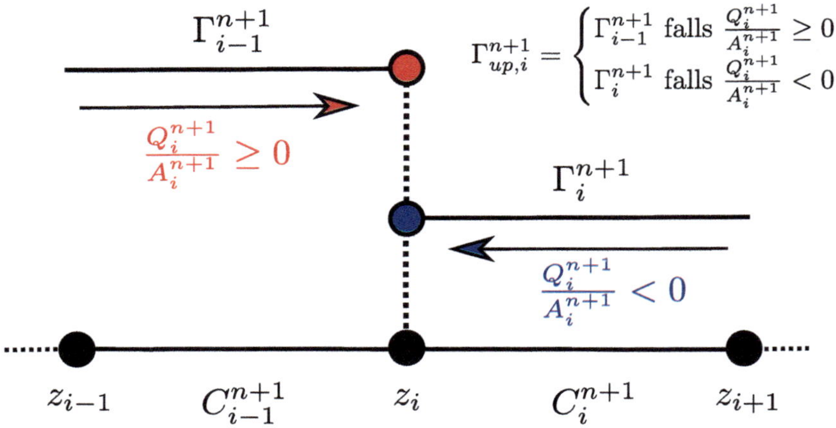

Fig. 4.2 Upwinding process wrt. a node (z_i, t_{n+1}). If the velocity field points in the positive direction, $\Gamma(z_i, t_{n+1}) = \Gamma_{up}(z_i, t_{n+1})$ is given by Γ_{i-1}^{n+1}. Otherwise it holds that: $\Gamma(z_i, t_{n+1}) = \Gamma_{up}(z_i, t_{n+1}) = \Gamma_i^{n+1}$

$$\mathbf{F}_{f,i+\frac{1}{2}}^{n+\frac{1}{2}} = \mathbf{F_f}\left(\mathbf{U_f}\left(\tilde{z}_{i+\frac{1}{2}}, t_{n+\frac{1}{2}}\right)\right) \text{ and } \mathbf{S}_{f,i+\frac{1}{2}}^{n+\frac{1}{2}} = \mathbf{S_f}\left(\mathbf{U_f}\left(\tilde{z}_{i+\frac{1}{2}}, t_{n+\frac{1}{2}}\right)\right).$$

By this, we obtain:

$$\mathbf{U}_{f,i}^{n+1} = \mathbf{U}_{f,i}^{n} - \frac{\Delta t}{\Delta z}\left(\mathbf{F}_{f,i+\frac{1}{2}}^{n+\frac{1}{2}} - \mathbf{F}_{f,i-\frac{1}{2}}^{n+\frac{1}{2}}\right) + \frac{\Delta t}{2}\left(\mathbf{S}_{f,i+\frac{1}{2}}^{n+\frac{1}{2}} + \mathbf{S}_{f,i-\frac{1}{2}}^{n+\frac{1}{2}}\right). \tag{4.1}$$

The unknowns

$$\mathbf{U}_{f,i+\frac{1}{2}}^{n+\frac{1}{2}} = \mathbf{U_f}\left(\tilde{z}_{i+\frac{1}{2}}, t_{n+\frac{1}{2}}\right)$$

and the fluxes at the cell edges can be determined by a Finite Difference method. One appropriate method which can be used for this purpose is the Lax-Friedrichs method [4]. Using the half time stepsize, the following approximations for $\mathbf{U}_{f,i+\frac{1}{2}}^{n+\frac{1}{2}}$ and $\mathbf{U}_{f,i-\frac{1}{2}}^{n+\frac{1}{2}}$ are obtained:

$$\begin{aligned}
\mathbf{U}_{f,i+\frac{1}{2}}^{n+\frac{1}{2}} &= \frac{\mathbf{U}_{f,i+1}^{n} + \mathbf{U}_{f,i}^{n}}{2} + \frac{\Delta t}{2}\left(-\frac{\mathbf{F}_{f,i+1}^{n} - \mathbf{F}_{f,i}^{n}}{\Delta z} + \frac{\mathbf{S}_{f,i+1}^{n} + \mathbf{S}_{f,i}^{n}}{2}\right), \\
\mathbf{U}_{f,i-\frac{1}{2}}^{n+\frac{1}{2}} &= \frac{\mathbf{U}_{f,i}^{n} + \mathbf{U}_{f,i-1}^{n}}{2} + \frac{\Delta t}{2}\left(-\frac{\mathbf{F}_{f,i}^{n} - \mathbf{F}_{f,i-1}^{n}}{\Delta z} + \frac{\mathbf{S}_{f,i}^{n} + \mathbf{S}_{f,i-1}^{n}}{2}\right).
\end{aligned} \tag{4.2}$$

These approximations yield a second order of convergence in space and time (see Sect. 4.3). In a next step, we derive the discretization of the transport Eq. (3.24). Contrary to the derivation of the Lax-Wendroff method, we now integrate (3.24) wrt. C_i^n:

$$\int_{t_n}^{t_{n+1}} \int_{z_i}^{z_{i+1}} \frac{\partial \Gamma}{\partial t} \, dz \, dt + \int_{t_n}^{t_{n+1}} \int_{z_i}^{z_{i+1}} \frac{\partial}{\partial z} \left(\omega \frac{Q}{A} \Gamma \right) \, dz \, dt = 0.$$

Further computations yield:

$$\int_{z_i}^{z_{i+1}} \Gamma(z, t_{n+1}) - \Gamma(z, t_n) \, dz + \int_{t_n}^{t_{n+1}} \left(\omega \frac{Q}{A} \Gamma \right) (z_{i+1}, t) - \left(\omega \frac{Q}{A} \Gamma \right) (z_i, t) \, dt = 0.$$

Approximating Γ on C_i^n by $\Gamma_i^n = \Gamma\left(\tilde{z}_{i+\frac{1}{2}}, t_n\right)$, we obtain:

$$\Delta z \left(\Gamma_i^{n+1} - \Gamma_i^n \right) + \int_{t_n}^{t_{n+1}} \left(\omega \frac{Q}{A} \Gamma \right) (z_{i+1}, t) - \left(\omega \frac{Q}{A} \Gamma \right) (z_i, t) \, dt = 0.$$

We replace the time integral over the convective part of the transport equation by the rectangular rule. Evaluating the integrand at the time $t = t_{n+1}$, an implicit method results:

$$\Gamma_i^{n+1} = \Gamma_i^n - \frac{\Delta t \omega}{\Delta z} \left[\left(\frac{Q}{A} \right) (z_{i+1}, t_{n+1}) \Gamma(z_{i+1}, t_{n+1}) - \left(\frac{Q}{A} \right) (z_i, t_{n+1}) \Gamma(z_i, t_{n+1}) \right].$$

Since we approximate the solution vector $\mathbf{U_f}$ in terms of the Lax-Wendroff method at the grid nodes (z_i, t_n) by $\mathbf{U}_{f,i}^n = \left(A_i^n, Q_i^n \right)^T$, it holds:

$$\Gamma_i^{n+1} = \Gamma_i^n - \frac{\Delta t}{\Delta z} \left(\omega \frac{Q_{i+1}^{n+1}}{A_{i+1}^{n+1}} \Gamma(z_{i+1}, t_{n+1}) - \omega \frac{Q_i^{n+1}}{A_i^{n+1}} \Gamma(z_i, t_{n+1}) \right).$$

Now the issue arises, how $\Gamma(z_{i+1}, t_{n+1})$ and $\Gamma(z_i, t_{n+1})$ can be determined. This is not obvious, since the solution variable Γ is not approximated on the primal space-time grid, but on the dual space-time grid. One way to do this is to use an upwinding procedure. Here, we check the sign of the velocities Q_i^{n+1}/A_i^{n+1} and $Q_{i+1}^{n+1}/A_{i+1}^{n+1}$ and approximate Γ by the value, which is opposite to the flow direction. By means of these considerations, we obtain:

$$\Gamma(z_{i+1}, t_{n+1}) = \Gamma_{up}(z_{i+1}, t_{n+1}) = \begin{cases} \Gamma_i^{n+1}, & \text{if } \frac{Q_{i+1}^{n+1}}{A_{i+1}^{n+1}} \geq 0, \\ \Gamma_{i+1}^{n+1}, & \text{if } \frac{Q_{i+1}^{n+1}}{A_{i+1}^{n+1}} < 0. \end{cases}$$

and

$$\Gamma\left(z_i, t_{n+1}\right) = \Gamma_{up}\left(z_i, t_{n+1}\right) = \begin{cases} \Gamma_{i-1}^{n+1}, \text{ if } \frac{Q_i^{n+1}}{A_i^{n+1}} \geq 0, \\ \Gamma_i^{n+1}, \text{ if } \frac{Q_i^{n+1}}{A_i^{n+1}} < 0. \end{cases}$$

All in all, we obtain the following scheme:

$$\Gamma_i^{n+1} = \Gamma_i^n - \frac{\Delta t}{\Delta z}\left(\omega \frac{Q_{i+1}^{n+1}}{A_{i+1}^{n+1}} \Gamma_{up}\left(z_{i+1}, t_{n+1}\right) - \omega \frac{Q_i^{n+1}}{A_i^{n+1}} \Gamma_{up}\left(z_i, t_{n+1}\right)\right). \tag{4.3}$$

Contrary to the Lax-Wendroff method, this method is not subject to any restriction regarding the time stepsize (see Sect. 4.3). Thus, for a flow with a periodic pattern, the simulation of the transport can be decoupled from the flow if the pressure and velocity field has been stored for a heart beat. If one wants to simulate transport processes during a longer period of time with periodic flow conditions, it is advantageous not to be restricted in the choice of the time step. As one can observe, (4.3) results in a linear system of equations:

$$A_\Gamma^{n+1} \Gamma^{n+1} = \mathbf{b}_\Gamma^n, \tag{4.4}$$

with the unknowns:

$$\Gamma^{n+1} = \left(\Gamma_0^{n+1}, \ldots, \Gamma_{N_z-1}^{n+1}\right)^T.$$

The matrix $A_\Gamma^{n+1} \in \mathbb{R}^{N_z \times N_z}$ is sparse for large N_z and has the structure of a tridiagonal matrix. The vector $\mathbf{b}_\Gamma^n \in \mathbb{R}^{N_z}$ contains the solution values from the previous time $t = t_n$. Extending these considerations to a network with N vessels, we obtain a block matrix $A_\Gamma^{n+1} \in \mathbb{R}^{\sum_i N_{z,i} \times \sum_i N_{z,i}}$, where $N_{z,i} + 1$ is the number of grid nodes in space for vessel i. The individual block matrices are denoted by $A_\Gamma^{ij} \in \mathbb{R}^{N_{z,i} \times N_{z,j}}$:

$$A_\Gamma^{n+1} = \begin{pmatrix} A_\Gamma^{11} & \ldots & A_\Gamma^{1N} \\ \vdots & \ddots & \vdots \\ A_\Gamma^{N1} & \ldots & A_\Gamma^{NN} \end{pmatrix}. \tag{4.5}$$

According to this structure, the vector of unknowns and the right hand side are structured as follows:

$$\Gamma^{n+1} = \left(\Gamma_1^{n+1}, \ldots, \Gamma_N^{n+1}\right)^T, \; \Gamma_i^{n+1} = \left(\Gamma_{0,i}^{n+1}, \ldots, \Gamma_{N_{z,i}-1,i}^{n+1}\right)^T$$

and

$$\mathbf{b}_\Gamma^n = \left(\mathbf{b}_1^n, \ldots, \mathbf{b}_N^n\right)^T, \; \mathbf{b}_i^n = \left(\mathbf{b}_{0,i}^n, \ldots, \mathbf{b}_{N_{z,i}-1,i}^n\right)^T. \tag{4.6}$$

The diagonal matrices A_Γ^{ii}, $i \in \{1, \ldots, N\}$ arise from (4.3), while the remaining matrices are either zero blocks, represent the coupling of the vessels at the bifurcations, or describe the influence of the outflow and inflow boundaries. The assembly of these matrices and the right-hand side $\mathbf{b}_\Gamma^n$ is discussed in the following sections.

4.2 Implementation of Boundary Conditions

After describing the discretization of the blood flow model and the transport equation for a single blood vessel, we consider in the upcoming section the numerical treatment of the boundary conditions presented in Sect. 3.6.

4.2.1 Numerical Treatment of Inflow Boundaries

For simplicity, we assume that a blood vessel with an inflow boundary is oriented such that the inflow boundary is located at $z = 0$ or the grid node $z_0 = 0 = 0 \cdot \Delta z$. Considering the results from Sect. 3.3, it can be noted that at least one boundary condition for one of the solution variables is needed.

Furthermore, the boundary condition may be given by the volumetric flow rate e.g., by Q_{AV} from (3.69). Considering the numerical scheme in (4.1) one can see that for $i = 0$ the solution vector $\mathbf{U}_{f,-1/2}^{n+1/2}$ has to be determined for $t = t_{n+1/2}$. Equation (4.2) suggests that for the computation of $\mathbf{U}_{f,-1/2}^{n+1/2}$ the vector $\mathbf{U}_{f,-1}^{n+1/2}$ is required. Since this vector is not approximated on the grid $\Omega_{\Delta z,\Delta t}$, it is necessary to derive additional equations besides the boundary condition Q_{AV} such that the missing boundary condition A_0^{n+1} can be determined. For this purpose, we consider the first equation in (3.21) and the corresponding discretization in (4.1). In case of the grid node $z_0 = 0$ and time $t = t_{n+1}$ we obtain:

$$A_0^{n+1} = A_0^n - \frac{\Delta t}{\Delta z}\left(Q_{\frac{1}{2}}^{n+\frac{1}{2}} - Q_{-\frac{1}{2}}^{n+\frac{1}{2}}\right) \text{ und } Q_{AV}\left(t_{n+1}\right) = Q_0^{n+1}. \tag{4.7}$$

$Q_{1/2}^{n+1/2}$ results from the first equation in (4.2). Computing $Q_{-1/2}^{n+1/2}$ we approximate the flow rate at the grid nodes $z_0 = 0$ and $t = t_{n+1/2}$ by the arithmetic mean, which is corresponding to a linear interpolation between $Q_{1/2}^{n+1/2}$ and $Q_{-1/2}^{n+1/2}$ evaluated at the center of the interval $(-\Delta z/2, \Delta z/2)$ i.e., $z = 0$:

$$Q_{AV}\left(t_{n+\frac{1}{2}}\right) = Q_0^{n+\frac{1}{2}} \approx \frac{Q_{\frac{1}{2}}^{n+\frac{1}{2}} + Q_{-\frac{1}{2}}^{n+\frac{1}{2}}}{2}. \tag{4.8}$$

This means that the unknown values $Q_{-1/2}^{n+1/2}$ and A_0^{n+1} can be determined from the Eqs. (4.7) and (4.8). The whole procedure is summarized in Algorithm 1. Instead of directly prescribing Q_{AV} by (3.69), Q_{AV} can also be determined by (3.74) modeling the left ventricle.

In a next step, we try to incorporate the boundary condition (3.76) into the discretization of the transport equation. Thereby, we consider the discretization of the transport Eq. (4.3) for $i = 0$:

$$\Gamma_0^{n+1} = \Gamma_0^n - \frac{\Delta t}{\Delta z}\left(\omega\frac{Q_1^{n+1}}{A_1^{n+1}}\Gamma_{up}\left(z_1, t_{n+1}\right) - \omega\frac{Q_0^{n+1}}{A_0^{n+1}}\Gamma_{up}\left(z_0, t_{n+1}\right)\right).$$

Algorithm 1: Boundary conditions at an inflow boundary (1)

INPUT: Grid $\Omega_{\Delta z,\Delta t}$ with an inflow boundary, time stepsize Δt, flow rate: Q_{AV}, model parameter: E-modulus, wall thickness etc., time interval: $\left[t_n, t_{n+1}\right]$;
OUTPUT: Boundary condition: A_0^{n+1};

1) Compute Q_0^{n+1} and $Q_0^{n+1/2}$ by evaluating $Q_{AV}(t)$ at $t = t_{n+1}$ and $t = t_{n+1/2}$;

2) Compute $Q_{\frac{1}{2}}^{n+\frac{1}{2}}$ by means of (4.2);

3) Compute $Q_{-\frac{1}{2}}^{n+\frac{1}{2}}$ by means of (4.8);

4) Compute A_0^{n+1} by means of (4.7) ;

In the following Algorithm 2 we describe how to incorporate (3.76) and (4.3) into the system matrix A_Γ^{n+1} (4.5) and the right-hand side $\mathbf{b}_\Gamma^n$ (4.6).

4.2.2 *Numerical Treatment of Outflow Boundaries*

As in the case of the inflow boundaries, we first fix the orientation the coordinate axes. Thereby, we assume that the outflow boundary of a vessel coincides with $z = L$, where L denotes the length of the vessel. Furthermore, let $N_z + 1$ be the number of grid points in space and G_0 is the elasticity parameter of the corresponding vessel wall. A_0 is the cross-sectional area for the static pressure $p = 0$. To determine the missing boundary condition, we consider the three-element Windkessel model governed by the ODE (3.79):

$$p - p_v + CR_2\frac{dp}{dt} = \left(R_1 + R_2\right)Q + CR_1R_2\frac{dQ}{dt}.$$

Algorithm 2: Boundary conditions at an inflow boundary (2)

INPUT: Grid $\Omega_{\Delta z,\Delta t}$ containing an inflow boundary, A_i^{n+1} and Q_i^{n+1} at the grid nodes of $\Omega_{\Delta z,\Delta t}$ for $t = t_{n+1}$, time stepsize Δt, model parameter: Volumetric concentration C_{in}, E-modulus, wall thickness etc., time period: $[t_n, t_{n+1}]$;
OUTPUT: Matrix A_Γ^{n+1} and right-hand side $\mathbf{b}_\Gamma^n$;

Determine the index j of a vessel containing an inflow boundary;
Determine the index k of the variable $\Gamma_{0,j}^{n+1}$ in the vector Γ^{n+1}: $k = \sum_{l=1}^{j-1} N_{z,l} + 1$;
Determine Q_0^{n+1} and A_0^{n+1};
if $Q_0^{n+1} \geq 0$ **then**
  Compute the k-th component of $\mathbf{b}_\Gamma^n$ using (3.76) and (4.3):
  $\mathbf{b}_\Gamma^n(k) = \mathbf{b}_\Gamma^n(k) + \omega \frac{\Delta t Q_0^{n+1}}{\Delta z} C_{in}$;
end
else
  Adapt A_Γ^{jj} as follows: $A_\Gamma^{jj}(1,1) = A_\Gamma^{jj}(1,1) - \omega \frac{\Delta t Q_0^{n+1}}{\Delta z A_0^{n+1}}$;
end

The parameters R_1 and R_2 denote the two resistances of the Windkessel model, while C stands for the capacity. Discretizing this differential equation by means of the explicit Euler method wrt. a time step $[t_n, t_{n+1}]$, $\Delta t = t_{n+1} - t_n$ [5, 6] yields:

$$\frac{Q_{N_z}^{n+1} - Q_{N_z}^n}{\Delta t} = \frac{p_{N_z}^n - p_v}{CR_1R_2} + \frac{1}{R_1}\frac{p_{N_z}^{n+1} - p_{N_z}^n}{\Delta t} - \frac{R_1 + R_2}{CR_1R_2} Q_{N_z}^n. \tag{4.9}$$

p is replaced by the pressure $p_{N_z}^{n+1}$ computed at the time $t = t_{n+1}$ at the grid node z_{N_z}:

$$p_{N_z}^{n+1} = G_0\left(\sqrt{\frac{A_{N_z}^{n+1}}{A_0}} - 1\right). \tag{4.10}$$

Accordingly, $Q_{N_z}^n$ is the flow rate approximated at time $t = t_n$ and the grid node z_{N_z}. To compute $A_{N_z}^{n+1}$, we discretize the first equation in (3.21) by the implicit Euler method and a backward Finite difference in space:

$$A_{N_z}^{n+1} = A_{N_z}^n - \frac{\Delta t}{\Delta z}\left(Q_{N_z}^{n+1} - Q_{N_z-1}^{n+1}\right). \tag{4.11}$$

Using (4.9)–(4.11), a fixed point method for determining $p_{N_z}^{n+1}$ can be formulated. A summary of the whole algorithm can be found in Algorithm 3. Concluding this subsection, the boundary condition for the transport problem at an outflow boundary is considered. For this purpose, (4.3) at $i + 1 = N_z$ is used:

Algorithm 3: Boundary values at an outflow boundary (1)

INPUT: Grid $\Omega_{\Delta z,\Delta t}$, time stepsize Δt, Windkesselparameters: R_1, R_2 and C, maximal iteration number: $k_{\max}$, tolerance: TOL, model parameter: E-modulus, wall thickness etc.;
OUTPUT: Boundary values: $A_{N_z}^{n+1}$, $Q_{N_z}^{n+1}$ and $p_{N_z}^{n+1}$;

Consider the time period: $\left[t_n, t_{n+1}\right]$ with $\Delta t = t_{n+1} - t_n$;
$k = 0$;
$p_0 = p_{N_z}^n$;
$e_k = p_0$;
while $k \leq k_{\max}$ *AND* $e_k > TOL$ **do**
  Replace $p_{N_z}^{n+1}$ in (4.9) by p_k and solve for $Q_{N_z,k}^{n+1}$, in a previous step one has to replace $Q_{N_z}^{n+1}$ by $Q_{N_z,k}^{n+1}$ in (4.9) ;
  Compute $A_{N_z,k}^{n+1}$ by means of (4.11), in a previous step one has to replace $A_{N_z}^{n+1}$ by $A_{N_z,k}^{n+1}$;
  Compute p_{k+1} by means of (4.10), in a previous step one has to replace $p_{N_z}^{n+1}$ by p_{k+1} ;
  $e_k = |p_{k+1} - p_k|$;
  $p_k = p_{k+1}$;
  $k = k + 1$;
end
$p_{N_z}^{n+1} = p_k$;
$A_{N_z}^{n+1} = A_{N_z,k}^{n+1}$;
$Q_{N_z}^{n+1} = Q_{N_z,k}^{n+1}$;

$$\Gamma_{N_z-1}^{n+1} = \Gamma_{N_z-1}^{n} - \frac{\Delta t}{\Delta z}\left(\omega \frac{Q_{N_z}^{n+1}}{A_{N_z}^{n+1}} \Gamma_{up}\left(z_{N_z}, t_{n+1}\right) - \omega \frac{Q_{N_z-1}^{n+1}}{A_{N_z-1}^{n+1}} \Gamma_{up}\left(z_{N_z-1}, t_{n+1}\right)\right). \tag{4.12}$$

In order to determine Γ_{up} at the grid node $z = z_{N_z-1}$, we use the usual upwinding method:

$$\Gamma_{up}\left(z_{N_z-1}, t_{n+1}\right) = \begin{cases} \Gamma_{N_z-2}^{n+1}, & \text{if } \omega \frac{Q_{N_z-1}^{n+1}}{A_{N_z-1}^{n+1}} \geq 0, \\ \\ \Gamma_{N_z-1}^{n+1}, & \text{if } \omega \frac{Q_{N_z-1}^{n+1}}{A_{N_z-1}^{n+1}} < 0. \end{cases}$$

At $z = z_{N_z} = L$, we couple the 1D transport equation according to (3.80) with an ODE describing the mean concentration c of the substance in the omitted vessels and tissue. Solving this ODE by means of the implicit Euler method yields:

$$\begin{aligned} V_c\left(t_{n+1}\right) c\left(t_{n+1}\right) = V_c\left(t_n\right) c\left(t_n\right) + \Delta t \Bigg(& \frac{Q_{N_z}^{n+1}}{A_{N_z}^{n+1}} \Gamma_{up}\left(z_{N_z}, t_{n+1}\right) \\ & - \frac{p_c\left(t_{n+1}\right) - p_v}{R_2} \cdot c\left(t_{n+1}\right) - r\left(c\left(t_{n+1}\right), t_{n+1}\right)\Bigg). \end{aligned} \tag{4.13}$$

Moreover, for $Q_{N_z}^{n+1} \geq 0$ one obtains:

$$\Gamma_{up}\left(z_{N_z}, t_{n+1}\right) = \Gamma_{N_z-1}^{n+1}. \tag{4.14}$$

$V_c\left(t_{n+1}\right)$ and $p_c\left(t_{n+1}\right)$ can be computed by the following equations

$$V_c\left(t_{n+1}\right) = V_c\left(t_n\right) + \Delta t\left(Q_{N_z}^{n+1} - \frac{p_c\left(t_{n+1}\right) - p_v}{R_2}\right) \tag{4.15}$$

and

$$p_c\left(t_{n+1}\right) = p_{N_z}^{n+1} - R_1 Q_{N_z}^{n+1}. \tag{4.16}$$

These equations result from an implicit Euler discretization of (3.77) or (3.78). If r is non-linear, $c\left(t_{n+1}\right)$ must be determined by a suitable solver, e.g., by a Newton method. We assume that $\Gamma_{N_z-1}^{n+1}$ is already known, i.e., the value of $c\left(t_{n+1}\right)$ has to be determined by solving the non-linear Eq. (4.13).

Moreover, (4.13) does not have to be integrated into the system of Eq. (4.4), which keeps the system of Eq. (4.4) linear. Thus, if $Q_{N_z}^{n+1} \geq 0$ holds at an outflow boundary, then Γ^{n+1} in (4.4) is first determined. In a next step, the concentration values $c\left(t_{n+1}\right)$ are computed using (4.13) (see Algorithm 4). If $Q_{N_z}^{n+1} < 0$ holds, then (3.80) yields:

Algorithm 4: Boundary values at an outflow boundary (2)

INPUT: Grid $\Omega_{\Delta z, \Delta t}$ containing an outflow boundary, time period: $\left[t_n, t_{n+1}\right]$, time stepsize Δt, approximations of A and Q at the grid nodes of $\Omega_{\Delta z, \Delta t}$ for $t = t_{n+1}$, model parameters: E-modulus, wall thickness, resistances at the outflow boundaries, etc.;
OUTPUT: New concentration value $c\left(t_{n+1}\right)$;

- -

Determine the index j of a vessel with an outflow boundary;
Determine $A_{N_{z,j},j}^{n+1}$, $Q_{N_{z,j},j}^{n+1}$ and $\Gamma_{N_{z,j}-1,j}^{n+1}$;
Determine $V_c\left(t_{n+1}\right)$ and $p_c\left(t_{n+1}\right)$ by means of (4.15) and (4.16);
if $Q_{N_{z,j},j}^{n+1} \geq 0$ **then**
  Solve (4.13) by means of a non-linear solver to obtain $c\left(t_{n+1}\right)$. As an initial solution one can choose $c\left(t_n\right)$;
end

$$V_c\left(t_{n+1}\right) c\left(t_{n+1}\right) = V_c\left(t_n\right) c\left(t_n\right) + \Delta t\left(Q_{N_z}^{n+1} c\left(t_{n+1}\right) - \frac{p_c\left(t_{n+1}\right) - p_v}{R_2} \cdot c\left(t_{n+1}\right) - r\left(c\left(t_{n+1}\right), t_{n+1}\right)\right). \tag{4.17}$$

All in all, one obtains from (4.15) and (4.17):

$$\Gamma_{up}\left(z_{N_z}, t_{n+1}\right) = c\left(t_{n+1}\right) \cdot A_{N_z}^{n+1}. \tag{4.18}$$

In Algorithm 5 we describe how the previous considerations affect the vector $\mathbf{b}_\Gamma^n$.

Algorithm 5: Boundary values at an outflow boundary (3)

INPUT: Grid $\Omega_{\Delta z,\Delta t}$ containing an outflow boundary, time period: $\left[t_n, t_{n+1}\right]$, time stepsize Δt, approximations of A and Q at the grid nodes of $\Omega_{\Delta z,\Delta t}$ for $t = t_{n+1}$, model parameters: E-modulus, wall thickness, resistance at the outflow boundaries etc.;
OUTPUT: New right-hand side $\mathbf{b}_\Gamma^n$;

- -

Determine the index j of the vessel ;
Determine the index k of the variable $\Gamma_{N_{z,j}-1,j}^{n+1}$ in the vector Γ^{n+1}: $k = \sum_{l=1}^{j} N_{z,l}$;
Determine $A_{N_{z,j},j}^{n+1}$, $Q_{N_{z,j},j}^{n+1}$ and $\Gamma_{N_{z,j}-1,j}^{n+1}$;
Determine $V_c\left(t_{n+1}\right)$ and $p_c\left(t_{n+1}\right)$ by (4.15) and (4.16);
if $Q_{N_{z,j},j}^{n+1} < 0$ **then**
  Solve (4.17) by means of a non-linear solver to obtain $c\left(t_{n+1}\right)$. As initial solution choose $c\left(t_n\right)$;
  $\mathbf{b}_\Gamma^n(k) = \mathbf{b}_\Gamma^n(k) - \omega\frac{\Delta t Q_{N_{z,j},j}^{n+1}}{\Delta z_j} c\left(t_{n+1}\right)$;
end

4.2.3 *Numerical Treatment of Bifurcations*

After discussing in detail the numerical treatment of the boundary conditions at inlets and outlets, we still need to consider the coupling conditions at a bifurcation. This is necessary to fully describe blood flow and transport processes in a network. Here we treat not only the case of a bifurcation, but the more general case of a branching with more than two branches. Thereby, we number the vessels adjacent to a branching point using the index set $I = \{1, \ldots, m\}$. Based on the considerations from Sect. 3.6.3, the coupling conditions for the blood flow problem and a time step $\left[t_n, t_{n+1}\right]$ can be formulated as follows:

$$p_{t,1}\left(y_1, t_{n+1}\right) = p_{t,i}\left(y_i, t_{n+1}\right),\ I \setminus \{1\}, \tag{4.19}$$

$$\sum_{i \in I} Q_i\left(y_i, t_{n+1}\right)\sigma\left(y_i\right) = 0,\ \sigma\left(y_i\right) = \begin{cases} 1, & \text{if } y_i = L_i, \\ -1, & \text{if } y_i = 0. \end{cases} \tag{4.20}$$

By means of (4.19) we enforce the continuity of the total pressure at a bifurcation, while (4.20) ensures the mass conservation. $y_i \in \{0, L_i\}$ represents the spatial coor-

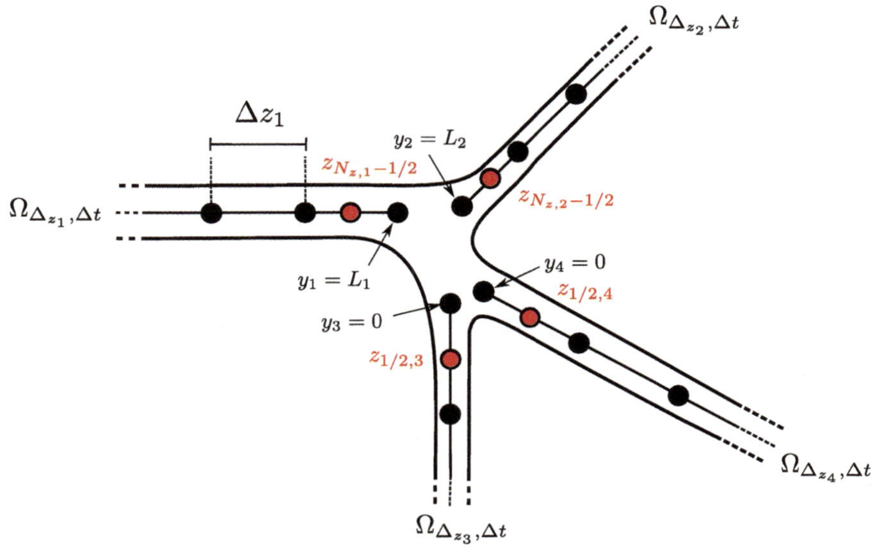

Fig. 4.3 Different orientations of discretization grids $\Omega_{\Delta z_i,\Delta t}$, $i \in \{1, 2, 3, 4\}$ at a bifurcation

dinate of the vessels adjacent to the bifurcation (see Fig. 4.3), where L_i is the length of a vessel $i \in I$. According to Sect. 3.6.3 the total pressure $p_{t,i}(y_i, t_{n+1})$ is defined as follows:

$$p_{t,i}(y_i, t_{n+1}) = \frac{\rho}{2}\left(\frac{Q_i(y_i, t_{n+1})}{A_i(y_i, t_{n+1})}\right)^2 + G_{0,i}\left(\sqrt{\frac{A_i(y_i, t_{n+1})}{A_{0,i}}} - 1\right).$$

$A_{0,i}$ is the section area for $p_i = 0$ and $G_{0,i}$ is the corresponding elasticity parameter. It can be seen at a first glance that from (4.19) and (4.20) m equations can be obtained. However, since $2m$ unknowns have to be determined, we still require m further equations to obtain a well-posed problem. In order to derive m further equations, we use the so-called *ghost-point method*. Deriving this method, we assign in a first step to each vessel i its discretization grid $\Omega_{\Delta z_i,\Delta t}$. Wrt. to $\Omega_{\Delta z_i,\Delta t}$, $N_{z,i}$ denotes the index of the last node in space. The spatial grid width of $\Omega_{\Delta z_i,\Delta t}$ is given by Δz_i.

Next, we place inbetween the grid nodes 0 and Δz_i or $z_{N_{z,i}-1} = \left(N_{z,i} - 1\right) \cdot \Delta z_i$ and $z_{N_{z,i}} = N_{z,i} \cdot \Delta z_i = L_i$ a further grid node that is not contained in the respective grid $\Omega_{\Delta z_i,\Delta t}$ (siehe Fig. 4.3). Since neither $z_{1/2,i}$ nor $z_{N_{z,i}-1/2}$ are contained in $\Omega_{\Delta z_i,\Delta t}$ this method is referred to as *ghost-point Methode*. At the newly created grid nodes $z_{1/2,i}$ or $z_{N_{z,i}-1/2}$ the mass conservation equation:

$$\frac{\partial A}{\partial t} + \frac{\partial Q}{\partial z} = 0,$$

is discretized by means of a finite difference approach [7, 9]. Thereby, the spatial derivatives wrt. Q are replaced by central differences. In order to approximate the section area A at the nodes $z_{1/2,i}$ and $z_{N_{z,i}-1/2}$ the arithmetic mean is used. For the discretization wrt. time the implict Euler method is used. All in all, one obtains for

$$z_{1/2,i} = \Delta z_i/2 \text{ or } z_{N_{z,i}-1/2} = L_i - \Delta z_i/2:$$

$$\frac{A\left(y_i - \sigma\left(y_i\right)\Delta z_i, t_{n+1}\right) + A\left(y_i, t_{n+1}\right)}{2\Delta t} - \frac{A\left(y_i - \sigma\left(y_i\right)\Delta z_i, t_n\right) + A\left(y_i, t_n\right)}{2\Delta t} + \frac{\sigma\left(y_i\right)\left(Q\left(y_i, t_{n+1}\right) - Q\left(y_i - \sigma\left(y_i\right)\Delta z_i, t_{n+1}\right)\right)}{\Delta z_i} = 0. \tag{4.21}$$

Summarizing (4.19)–(4.21), one obtains a non-linear system of equations with the

---

Algorithm 6: Simulation of blood flow through a bifurcation

INPUT: m vessels, adjacent to a bifurcation, discretization grids $\Omega_{\Delta z_i,\Delta t}, \ldots, \Omega_{\Delta z_m,\Delta t}$, model parameters: lengths of the vessels, elasticity parameters etc., time interval $\left[t_n, t_{n+1}\right]$
OUTPUT: $2m$ boundary values for A and Q

- -

1) Consider the boundaries $y_i \in \{0, L_i\}$ for each vessel $i \in \{1, \ldots, m\}$;

2) Determine the factors $\sigma\left(y_i\right)$ (see (4.20));

3) Insert the ghost nodes $y_i - \sigma\left(y_i\right)\Delta z_i/2$ between y_i and $y_i - \sigma\left(y_i\right)\Delta z_i$ (see Fig. 4.3);

4) Discretize the mass balance equation using (4.21);

5) Establish the coupling Eqs. (4.19) and (4.21);

6) Assemble the non-linear system of Eq. (4.22);

7) Solve the system of Eq. (4.22) e.g., by a damped Newton method, as an initial solution use the values from the previous time step;

8) Assign the solutions of (4.22) to the boundaries of the respective vessels;

---

vector of unknowns $\mathbf{x} \in \mathbb{R}^{2m}$:

$$F\left(\mathbf{x}\right) = \left(F_1\left(\mathbf{x}\right), F_2\left(\mathbf{x}\right), \ldots, F_{2m}\left(\mathbf{x}\right)\right)^T = \mathbf{0}. \tag{4.22}$$

$F: \mathbb{R}^{2m} \rightarrow \mathbb{R}^{2m}$ is a non-linear vector valued function and the vector $\mathbf{x}$ of unknows, is defined as follows:

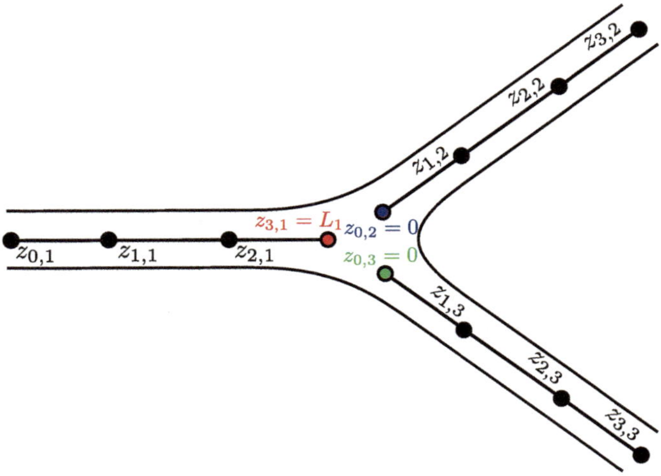

Fig. 4.4 Bifurcation with two branches, i.e., it holds: $m = 3$. The axes are oriented in such a way that $y_1 = L_1$, $y_2 = 0$ and $y_3 = 0$ are adjacent to the bifurcation. Each 1D grid contains $N_{z,1} + 1 = N_{z,2} + 1 = N_{z,3} + 1 = 4$ nodes

$$\mathbf{x} = \left(A\left(y_1, t_{n+1}\right), \ldots, A\left(y_m, t_{n+1}\right), Q\left(y_1, t_{n+1}\right), \ldots, Q\left(y_m, t_{n+1}\right)\right)^T .$$

The resulting non-linear system of Eq. (4.22) can be solved e.g., by a damped Newton method [8]. In order to demonstrate the assembly of the non-linear system of equations, we consider a simple example. Therefore, we set $m = 3$ i.e., we consider a bifurcation (see Fig. 4.4). At the bifurcation there is the boundary $y_1 = L_1$ of the main vessel and the inlets $y_2 = 0$ and $y_3 = 0$ of the two branches. For this configuration the coupling equations read as follows:

$$\frac{\rho}{2}\left(\frac{Q_1\left(L_1, t_{n+1}\right)}{A_1\left(L_1, t_{n+1}\right)}\right)^2 + G_{0,1}\left(\sqrt{\frac{A_1\left(L_1, t_{n+1}\right)}{A_{0,1}}} - 1\right) = \frac{\rho}{2}\left(\frac{Q_2\left(0, t_{n+1}\right)}{A_2\left(0, t_{n+1}\right)}\right)^2 + G_{0,2}\left(\sqrt{\frac{A_2\left(0, t_{n+1}\right)}{A_{0,2}}} - 1\right) \tag{4.23}$$

$$\frac{\rho}{2}\left(\frac{Q_1\left(L_1, t_{n+1}\right)}{A_1\left(L_1, t_{n+1}\right)}\right)^2 + G_{0,1}\left(\sqrt{\frac{A_1\left(L_1, t_{n+1}\right)}{A_{0,1}}} - 1\right) = \frac{\rho}{2}\left(\frac{Q_3\left(0, t_{n+1}\right)}{A_3\left(0, t_{n+1}\right)}\right)^2 + G_{0,3}\left(\sqrt{\frac{A_3\left(0, t_{n+1}\right)}{A_{0,3}}} - 1\right) \tag{4.24}$$

$$Q_1(L_1, t_{n+1}) - Q_2(0, t_{n+1}) - Q_3(0, t_{n+1}) = 0 \tag{4.25}$$

$$\frac{A_1\left(z_{N_{z,1}-1}, t_{n+1}\right) + A_1(L_1, t_{n+1})}{2\Delta t} - \frac{A_1\left(z_{N_{z,1}-1}, t_n\right) + A_1(L_1, t_n)}{2\Delta t} + \frac{Q_1(L_1, t_{n+1}) - Q_1\left(z_{N_{z,1}-1}, t_{n+1}\right)}{\Delta z_1} = 0. \tag{4.26}$$

$$\frac{A_2(\Delta z_2, t_{n+1}) + A_2(0, t_{n+1})}{2\Delta t} - \frac{A_2(\Delta z_2, t_n) + A(0, t_n)}{2\Delta t} + \frac{Q_2(\Delta z_2, t_{n+1}) - Q_2(0, t_{n+1})}{\Delta z_2} = 0 \tag{4.27}$$

$$\frac{A_3(\Delta z_3, t_{n+1}) + A_3(0, t_{n+1})}{2\Delta t} - \frac{A_3(\Delta z_3, t_n) + A(0, t_n)}{2\Delta t} + \frac{Q_3(\Delta z_3, t_{n+1}) - Q_3(0, t_{n+1})}{\Delta z_3} = 0. \tag{4.28}$$

The vector $\mathbf{x} \in \mathbb{R}^6$ containing the unknowns, is given by:

$$\begin{aligned}\mathbf{x} = (&A_1(L_1, t_{n+1}), A_2(0, t_{n+1}), A_3(0, t_{n+1}), \ldots\\ &Q_1(L_1, t_{n+1}), Q_2(0, t_{n+1}), Q_3(0, t_{n+1}))^T = (x_1, x_2, x_3, x_4, x_5, x_6)^T,\end{aligned}$$

and the function $F : \mathbb{R}^6 \to \mathbb{R}^6$ is governed by (4.29)–(4.34):

$$F_1(\mathbf{x}) = \frac{\rho}{2}\left(\frac{x_4}{x_1}\right)^2 + G_{0,1}\left(\sqrt{\frac{x_1}{A_{0,1}}} - 1\right) - \frac{\rho}{2}\left(\frac{x_5}{x_2}\right)^2 - G_{0,2}\left(\sqrt{\frac{x_2}{A_{0,2}}} - 1\right) \tag{4.29}$$

$$F_2(\mathbf{x}) = \frac{\rho}{2}\left(\frac{x_4}{x_1}\right)^2 + G_{0,1}\left(\sqrt{\frac{x_1}{A_{0,1}}} - 1\right) - \frac{\rho}{2}\left(\frac{x_6}{x_3}\right)^2 - G_{0,3}\left(\sqrt{\frac{x_3}{A_{0,3}}} - 1\right) \tag{4.30}$$

$$F_3(\mathbf{x}) = x_4 - x_5 - x_6 \tag{4.31}$$

$$F_4(\mathbf{x}) = \frac{A_1\left(z_{N_{z,1}-1}, t_{n+1}\right) + x_1}{2\Delta t} - \frac{A_1\left(z_{N_{z,1}-1}, t_n\right) + A_1\left(L_1, t_n\right)}{2\Delta t} + \frac{x_4 - Q_1\left(z_{N_{z,1}-1}, t_{n+1}\right)}{\Delta z_1} \tag{4.32}$$

$$F_5(\mathbf{x}) = \frac{A_2\left(\Delta z_2, t_{n+1}\right) + x_2}{2\Delta t} - \frac{A_2\left(\Delta z_2, t_n\right) + A\left(0, t_n\right)}{2\Delta t} + \frac{Q_2\left(\Delta z_2, t_{n+1}\right) - x_5}{\Delta z_2} \tag{4.33}$$

$$F_6(\mathbf{x}) = \frac{A_3\left(\Delta z_3, t_{n+1}\right) + x_3}{2\Delta t} - \frac{A_3\left(\Delta z_3, t_n\right) + A\left(0, t_n\right)}{2\Delta t} + \frac{Q_3\left(\Delta z_3, t_{n+1}\right) - x_6}{\Delta z_3} \tag{4.34}$$

The dimension of this system of equations can be further reduced, by solving e.g., the Eqs. (4.26)–(4.28) wrt. to the unknowns x_4, x_5 and x_6 and replacing them in the first three Eqs. (4.23)–(4.25).

A further way to couple the discrete systems at a bifurcation is described in [5]. While the system of Eqs. (4.23)–(4.28) is based on a simple discretization of the mass conservation equation to obtain a sufficient number of equations, in [5] the intermediate steps of the Lax-Wendroff method are incorporated into the system of equations. This means that the Eqs. (4.26)–(4.28) are replaced by the Lax-Wendroff scheme. By this the approximation order of the discretization is improved, since the discretizations based on the *ghost nodes* have a lower approximation order compared to the Lax-Wendroff method (see Sect. 4.3).

On the other hand the number of unknowns is increased, which implies that the computational effort for assembling the system of equations and computing Jacobians is increased. Moreover, the size of the time step may have to be restricted in a more severe way, such that a locally convergent solver as the Newton method can converge. Finally, our simulation results in Sect. 4.4 reveal that the simpler system of equations, which is based on the ghost point method, yields satisfactory simulation results. Due to these observations, we restict ourselves to the simpler system of Eq. (4.22).

Next, we turn our attention to the transport of some component through a bifurcation. Thereby, we decompose the index set I according to (3.90) into two further index set I_{in} und I_{out}. The vessels from which blood volume is transported into the bifurcation are assigned to the index set I_{in}, while the remaining vessels are assigned to the index set I_{out}. Following Sect. 3.6 the coupling conditions for $i \in I_{out}$ are given by:

$$\frac{\left|Q_i\left(y_i, t_{n+1}\right)\right|}{\sum_{j\in I_{out}}\left|Q_j\left(y_j, t_{n+1}\right)\right|} = \frac{\frac{\Gamma_i(y_i,t_{n+1})|Q_i(y_i,t_{n+1})|}{A_i(y_i,t_{n+1})}}{\sum_{j\in I_{in}}\frac{\Gamma_j(y_j,t_{n+1})|Q_j(y_j,t_{n+1})|}{A_j(y_j,t_{n+1})}}, \quad i \in I_{out}.$$

By this we obtain for the boundary values $\Gamma_{up,i}(y_i, t_{n+1})$, $i \in I_{out}$, the following equations:

$$|Q_i(y_i, t_{n+1})| \sum_{j \in I_{in}} \frac{\Gamma_{up,j}\left(y_j, t_{n+1}\right)\left|Q_j\left(y_j, t_{n+1}\right)\right|}{A_j\left(y_j, t_{n+1}\right)} - \frac{\Gamma_{up,i}(y_i, t_{n+1})\,|Q_i(y_i, t_{n+1})|}{A_i(y_i, t_{n+1})} \sum_{j \in I_{out}} \left|Q_j\left(y_j, t_{n+1}\right)\right| = 0.$$

The boundary values $\Gamma_{up,j}\left(y_j, t_{n+1}\right)$ for $j \in I_{in}$ are given by:

$$\Gamma_{up,j}\left(y_j, t_{n+1}\right) = \Gamma_j\left(y_j - \sigma\left(y_j\right)\frac{\Delta z_j}{2}, t_{n+1}\right). \tag{4.35}$$

and the scheme (4.3) i.e.:

$$\begin{aligned}\Gamma_j\left(y_j - \sigma\left(y_j\right)\frac{\Delta z_j}{2}, t_{n+1}\right) &= \Gamma_j\left(y_j - \sigma\left(y_j\right)\frac{\Delta z_j}{2}, t_n\right) \\ &- \frac{\Delta t \omega}{\Delta z_j}\left(\sigma\left(y_i\right)\left(\frac{Q_j\left(y_j, t_{n+1}\right)}{A_j\left(y_j, t_{n+1}\right)}\Gamma_j\left(y_j - \sigma\left(y_j\right)\frac{\Delta z_j}{2}, t_{n+1}\right)\right.\right. \\ &\left.\left. - \frac{Q_j\left(y_j - \sigma\left(y_j\right)\Delta z_j, t_{n+1}\right)}{A_j\left(y_j - \sigma\left(y_j\right)\Delta z_j, t_{n+1}\right)}\Gamma_{up,j}\left(y_j - \sigma\left(y_j\right)\Delta z_j, t_{n+1}\right)\right)\right).\end{aligned}$$

All in all one obtains for $i \in I_{out}$:

$$\Gamma_{up,i}(y_i, t_{n+1}) = \frac{A_i(y_i, t_{n+1}) \sum_{j \in I_{in}} \frac{\Gamma_j\left(y_j - \sigma\left(y_j\right)\frac{\Delta z_j}{2}, t_{n+1}\right)\left|Q_j\left(y_j, t_{n+1}\right)\right|}{A_j\left(y_j, t_{n+1}\right)}}{\sum_{j \in I_{out}}\left|Q_j\left(y_j, t_{n+1}\right)\right|}. \tag{4.36}$$

Illustrating the coupling of the discretized transport models, we show how to assemble the system of equations that is to be solved in an implicit Euler step. Thereby, we consider a single bifurcation. To do this, we use the orientation of the coordinate axes from Fig. 4.4 and the velocity field from Fig. 4.5. The different 1D grids exhibit 4 nodes, respectively. From Fig. 4.4 one can see, that it holds:

$$\frac{Q_i\left(z_{j,i}, t_n\right)}{A_i\left(z_{j,i}, t_n\right)} = \frac{Q_{j,i}^n}{A_{j,i}^n} > 0,\ i \in I,\ j \in \{0, 1, 2, 3\},\ \forall t_n \geq 0.$$

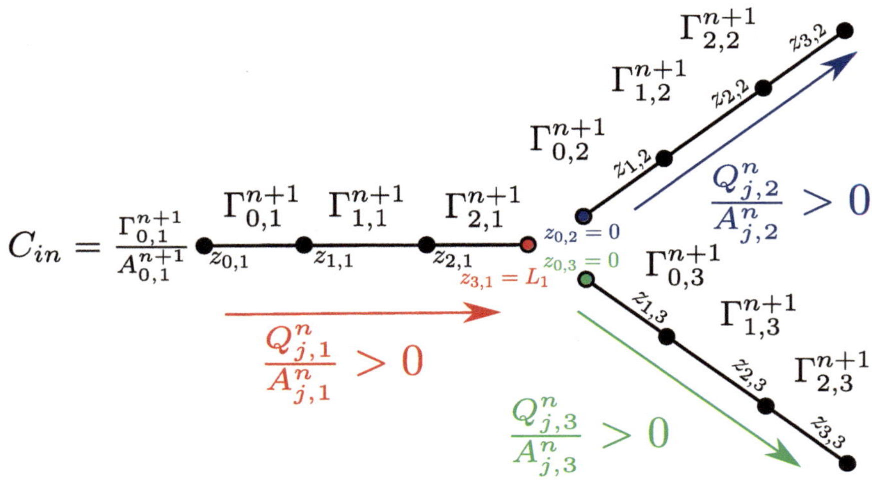

Fig. 4.5 Bifurcation with two branches i.e., it holds $m = 3$. The axes are orientated in such a way that the coordinates $y_1 = L_1$, $y_2 = 0$ and $y_3 = 0$ are adjacent to the bifurcation. The velocity fields in the different vessels have a positive sign. At the inlet of Vessel 1 a constant concentration C_{in} is prescribed

At the inlet of Vessel 1 a constant concentration

$$C_{in} = \frac{\Gamma_{0,1}^n}{A_{0,1}^n}, \quad \forall t_n \geq 0$$

is prescribed. For the given model problem, the index sets read as follows:

$$I = \{1, 2, 3\} \text{ and } I_{in} = \{1\}, \quad I_{out} = \{2, 3\}.$$

Since the concentration variables Γ are determined for each time step $[t_n, t_{n+1}]$ at the cell centers, a linear system of equations must be solved to determine the concentration values for $t = t_{n+1}$. In the following, we assemble for the transport problem in Fig. 4.4, the corresponding system of equations. In a similar way, transport of a substance can be determined for all other flow conditions at a bifurcation. The vector of unknows for Vessel i at $t = t_{n+1}$ is denoted by $\Gamma_i^{n+1} \in \mathbb{R}^3$:

$$\Gamma_i^{n+1} = \left(\Gamma_{0,i}^{n+1}, \Gamma_{1,i}^{n+1}, \Gamma_{2,i}^{n+1}\right)^T.$$

The whole vector $\Gamma^{n+1} \in \mathbb{R}^9$ has the following form:

$$\Gamma^{n+1} = \left(\Gamma_1^{n+1}, \Gamma_2^{n+1}, \Gamma_3^{n+1}\right)^T.$$

Due to the structure of Γ^{n+1}, the matrix $A_\Gamma^{n+1} \in \mathbb{R}^{9\times 9}$ of the system of equations is divided into nine blocks and the right-hand side $\mathbf{b}_\Gamma^n \in \mathbb{R}^9$ into three subvectors $\mathbf{b}_i^n \in \mathbb{R}^3$, $i \in I$:

Algorithm 7: Transport process through a bifurcation (2)

INPUT: m vessels that are adjacent to a bifurcation, grids $\Omega_{\Delta z_i,\Delta t}, \ldots, \Omega_{\Delta z_m,\Delta t}$, model parameter: lengths of the vessels, elasticity parameters etc., time step $[t_n, t_{n+1}]$, approximations of A and Q at the grid nodes of $\Omega_{\Delta z_i,\Delta t}$ at $t = t_{n+1}$
OUTPUT: A new matrix A_Γ^{n+1} and a new right-hand side $\mathbf{b}_\Gamma^n$;

- -

Split the index set $I = \{1, \ldots, m\}$ into the stets I_{in} and I_{out} according to (3.90)
for $i \in I_{in}$ **do**
  Consider the boundary $y_i \in \{0, L_i\}$;
  Determine the global index k_i of vessel i;
  if $y_i == 0$ **then**
    $A_\Gamma^{k_i k_i}(1,1) = A_\Gamma^{k_i k_i}(1,1) - \frac{\Delta t\omega}{\Delta z_i}\frac{Q_i(0,t_{n+1})}{A_i(0,t_{n+1})}$;
  end
  else
    $A_\Gamma^{k_i k_i}(N_{z,k_i}, N_{z,k_i}) = A_\Gamma^{k_i k_i}(N_{z,k_i}, N_{z,k_i}) + \frac{\Delta t\omega}{\Delta z_i}\frac{Q_i(L_i,t_{n+1})}{A_i(L_i,t_{n+1})}$;
  end
end
Compute $Q_{out} = \sum_{i\in I_{in}} |Q_i(y_i, t_{n+1})|$;
for $i \in I_{out}$ **do**
  Consider the boundary $y_i \in \{0, L_i\}$;
  Determine the global index k_i of vessel i;
  for $j \in I_{in}$ **do**
    Consider the boundary $y_j \in \{0, L_j\}$;
    Determine the global index k_j of vessel j;
    if $y_i == 0$ **then**
      if $y_j == 0$ **then**
        $A_\Gamma^{k_i k_j}(1,1) -= \frac{\Delta t\omega}{\Delta z_i}\frac{Q_i(0,t_{n+1})}{A_i(0,t_{n+1})}\left(\frac{A_i(0,t_{n+1})}{Q_{out}}\frac{|Q_j(0,t_{n+1})|}{A_j(0,t_{n+1})}\right)$;
      end
      else
        $A_\Gamma^{k_i k_j}(1, N_{z,k_j}) -= \frac{\Delta t\omega}{\Delta z_i}\frac{Q_i(0,t_{n+1})}{A_i(0,t_{n+1})}\left(\frac{A_i(0,t_{n+1})}{Q_{out}}\frac{|Q_j(L_j,t_{n+1})|}{A_j(L_j,t_{n+1})}\right)$;
      end
    end
    else
      if $y_j == 0$ **then**
        $A_\Gamma^{k_i k_j}(N_{z,k_i}, 1) += \frac{\Delta t\omega}{\Delta z_i}\frac{Q_i(L_i,t_{n+1})}{A_i(L_i,t_{n+1})}\left(\frac{A_i(L_i,t_{n+1})}{Q_{out}}\frac{|Q_j(0,t_{n+1})|}{A_j(0,t_{n+1})}\right)$;
      end
      else
        $A_\Gamma^{k_i k_j}(N_{z,k_i}, N_{z,k_j}) += \frac{\Delta t\omega}{\Delta z_i}\frac{Q_i(L_i,t_{n+1})}{A_i(L_i,t_{n+1})}\left(\frac{A_i(L_i,t_{n+1})}{Q_{out}}\frac{|Q_j(L_j,t_{n+1})|}{A_j(L_j,t_{n+1})}\right)$;
      end
    end
  end
end

$$\underbrace{\begin{pmatrix} A_\Gamma^{11} & A_\Gamma^{12} & A_\Gamma^{13} \\ A_\Gamma^{21} & A_\Gamma^{22} & A_\Gamma^{23} \\ A_\Gamma^{31} & A_\Gamma^{32} & A_\Gamma^{33} \end{pmatrix}}_{=A_\Gamma^{n+1}} \underbrace{\begin{pmatrix} \Gamma_1^{n+1} \\ \Gamma_2^{n+1} \\ \Gamma_3^{n+1} \end{pmatrix}}_{=\Gamma_{n+1}} = \underbrace{\begin{pmatrix} \mathbf{b}_1^n \\ \mathbf{b}_2^n \\ \mathbf{b}_3^n \end{pmatrix}}_{=\mathbf{b}_\Gamma^n}.$$

The matrices on the diagonal emerge from the discretization (4.3), while the remaining matrices consider the coupling at the bifurcation.Taking into account the boundary condition $\Gamma_{0,1}^{n+1} = C_{in} A_{0,1}^{n+1}$ and the sign of the velocity fields $Q_{j,i}^{n+1}/A_{j,i}^{n+1}$, we obtain for the matrices on the diagonal:

$$A_\Gamma^{11} = \begin{pmatrix} 1+\omega\frac{\Delta t\, Q_{0,1}^{n+1}}{\Delta z_1 A_{0,1}^{n+1}} & 0 & 0 \\ -\omega\frac{\Delta t\, Q_{0,1}^{n+1}}{\Delta z_1 A_{0,1}^{n+1}} & 1+\omega\frac{\Delta t\, Q_{1,1}^{n+1}}{\Delta z_1 A_{1,1}^{n+1}} & 0 \\ 0 & -\omega\frac{\Delta t\, Q_{1,1}^{n+1}}{\Delta z_1 A_{1,1}^{n+1}} & 1+\omega\frac{\Delta t\, Q_{2,1}^{n+1}}{\Delta z_1 A_{2,1}^{n+1}} \end{pmatrix}$$

$$A_\Gamma^{22} = \begin{pmatrix} 1+\omega\frac{\Delta t\, Q_{1,2}^{n+1}}{\Delta z_2 A_{1,2}^{n+1}} & 0 & 0 \\ -\omega\frac{\Delta t\, Q_{1,2}^{n+1}}{\Delta z_2 A_{1,2}^{n+1}} & 1+\omega\frac{\Delta t\, Q_{2,2}^{n+1}}{\Delta z_2 A_{2,2}^{n+1}} & 0 \\ 0 & -\omega\frac{\Delta t\, Q_{2,2}^{n+1}}{\Delta z_2 A_{2,2}^{n+1}} & 1+\omega\frac{\Delta t\, Q_{3,2}^{n+1}}{\Delta z_2 A_{3,2}^{n+1}} \end{pmatrix}$$

$$A_\Gamma^{33} = \begin{pmatrix} 1+\omega\frac{\Delta t\, Q_{1,3}^{n+1}}{\Delta z_3 A_{1,3}^{n+1}} & 0 & 0 \\ -\omega\frac{\Delta t\, Q_{1,3}^{n+1}}{\Delta z_3 A_{1,3}^{n+1}} & 1+\omega\frac{\Delta t\, Q_{2,3}^{n+1}}{\Delta z_3 A_{2,3}^{n+1}} & 0 \\ 0 & -\omega\frac{\Delta t\, Q_{2,3}^{n+1}}{\Delta z_3 A_{2,3}^{n+1}} & 1+\omega\frac{\Delta t\, Q_{3,3}^{n+1}}{\Delta z_3 A_{3,3}^{n+1}} \end{pmatrix}$$

Because of $1 \notin I_{out}$, the coupling Eq. (4.36) does not have to be considered for the first vessel. From this it follows:

$$A_\Gamma^{12} = A_\Gamma^{13} = \begin{pmatrix} 0 & 0 & 0 \\ 0 & 0 & 0 \\ 0 & 0 & 0 \end{pmatrix}.$$

In order to be able to determine the block matrices

$$A_\Gamma^{21},\ A_\Gamma^{23},\ A_\Gamma^{31} \text{ und } A_\Gamma^{32}$$

the coupling conditions for $i \in I_{out}$ are considered:

$$\left|Q_{0,2}^{n+1}\right| \frac{\Gamma_{2,1}^{n+1}\left|Q_{3,1}^{n+1}\right|}{A_{3,1}^{n+1}} - \frac{\Gamma_{0,2}^{n+1}\left|Q_{0,2}^{n+1}\right|}{A_{0,2}^{n+1}}\left(\left|Q_{0,2}^{n+1}\right| + \left|Q_{0,3}^{n+1}\right|\right) = 0,$$

$$\left|Q_{0,3}^{n+1}\right| \frac{\Gamma_{2,1}^{n+1}\left|Q_{3,1}^{n+1}\right|}{A_{3,1}^{n+1}} - \frac{\Gamma_{0,3}^{n+1}\left|Q_{0,3}^{n+1}\right|}{A_{0,3}^{n+1}}\left(\left|Q_{0,2}^{n+1}\right| + \left|Q_{0,3}^{n+1}\right|\right) = 0.$$

Due to $\left|Q_{0,2}^{n+1}\right| > 0,\ \left|Q_{0,3}^{n+1}\right| > 0$ and the mass conservation

$$\left|Q_{3,1}^{n+1}\right| = \left|Q_{0,2}^{n+1}\right| + \left|Q_{0,3}^{n+1}\right|$$

it follows:

$$\Gamma_{0,2}^{n+1} = \frac{A_{0,2}^{n+1}}{A_{3,1}^{n+1}}\Gamma_{2,1}^{n+1} \text{ und } \Gamma_{0,3}^{n+1} = \frac{A_{0,3}^{n+1}}{A_{3,1}^{n+1}}\Gamma_{2,1}^{n+1}.$$

This is corresponding to the coupling Eq. (3.88) from Sect. 3.6.3. Therefore it holds for the coupling matices:

$$A_\Gamma^{23} = A_\Gamma^{32} = \begin{pmatrix} 0 & 0 & 0 \\ 0 & 0 & 0 \\ 0 & 0 & 0 \end{pmatrix}$$

as well as

$$A_\Gamma^{21} = \begin{pmatrix} 0 & 0 & -\omega\frac{\Delta t}{\Delta z_2}\cdot\frac{Q_{0,2}^{n+1}}{A_{0,2}^{n+1}}\frac{A_{0,2}^{n+1}}{A_{3,1}^{n+1}} \\ 0 & 0 & 0 \\ 0 & 0 & 0 \end{pmatrix} \text{ and } A_\Gamma^{31} = \begin{pmatrix} 0 & 0 & -\omega\frac{\Delta t}{\Delta z_3}\cdot\frac{Q_{0,3}^{n+1}}{A_{0,3}^{n+1}}\frac{A_{0,3}^{n+1}}{A_{3,1}^{n+1}} \\ 0 & 0 & 0 \\ 0 & 0 & 0 \end{pmatrix}.$$

The vectors of the right-hand side are given by:

$$\mathbf{b}_1^n = \begin{pmatrix} \omega\frac{\Delta t\, Q_{0,1}^{n+1}}{\Delta z_1}C_{in} \\ \Gamma_{1,1}^n \\ \Gamma_{2,1}^n \end{pmatrix},\ \mathbf{b}_2^n = \begin{pmatrix} \Gamma_{0,2}^n \\ \Gamma_{1,2}^n \\ \Gamma_{2,2}^n \end{pmatrix} \text{ und } \mathbf{b}_3^n = \begin{pmatrix} \Gamma_{0,3}^n \\ \Gamma_{1,3}^n \\ \Gamma_{2,3}^n \end{pmatrix}.$$

4.2.4 Numerical Simulation of a Stenosis

Now, we consider the numerical coupling of two 1D models adjacent to a stenosis. As it has been mentioned in Sect. 3.6.4, the volume flow Q_s through a stenosis can be computed using the ordinary differential Eq. (3.91):

$$\frac{K_u \rho l_s}{A_0}\frac{dQ_s}{dt} = \Delta P_s - \frac{K_v \mu}{A_0 D_0}Q_s - \frac{K_t \rho}{2A_0^2}\left(\frac{A_0}{A_s} - 1\right)^2 Q_s\left|Q_s\right|,$$

where A_s is the minimal section area of the stenosis and A_0 is the section area of the healthy vessel at $p = 0$. Thus the parameter

$$S_d = 1 - \frac{A_s}{A_0}$$

denotes the narrowing degree of a stenosis. The remaining parameters are discussed below (3.91). To the 1D models, that are adjacent to the stenosis, we assign the indices p and d, where p stands for proximal (near the heart) and d for distal (far away from the heart). Wlog. we assume that the boundaries $z_p = L_p$ and $z_d = 0$ are adjacent to the stenosis. L_p is the length of the proximal vessel. The corresponding grids are denoted by $\Omega_{\Delta z_d, \Delta t}$ und $\Omega_{\Delta z_p, \Delta t}$ (see Fig. 4.6) and the solution variables are given by $A_d,\ Q_d$ and $A_p,\ Q_p$.

If one intends to perform a Lax-Wendroff step for both models according to (4.1), we use the *ghost-point* method to incorporate the boundary conditions. For this purpose, we insert between the boundary nodes and their neighbors further nodes:

$$z_{N_{z,p}-1/2} = \Delta z_p \cdot \left(N_{z,p} - 1/2\right) \text{ und } z_{1/2,d} = \Delta z_d/2.$$

An approximation of the solution variables at the new nodes yields just as in (4.21):

$$\frac{A_p\left(z_{N_{z,p}-1}, t_{n+1}\right) + A_p\left(z_{N_{z,p}}, t_{n+1}\right)}{2\Delta t} - \frac{A_p\left(z_{N_{z,p}-1}, t_n\right) + A_p\left(z_{N_{z,p}}, t_n\right)}{2\Delta t} + \frac{Q_p\left(z_{N_{z,p}}, t_{n+1}\right) - Q_p\left(z_{N_{z,p}-1}, t_{n+1}\right)}{\Delta z_p} = 0. \tag{4.37}$$

$$\frac{A_d\left(0, t_{n+1}\right) + A_d\left(\Delta z_d, t_{n+1}\right)}{2\Delta t} - \frac{A_d\left(0, t_n\right) + A_d\left(\Delta z_d, t_n\right)}{2\Delta t} + \frac{Q_d\left(\Delta z_d, t_{n+1}\right) - Q_d\left(0, t_{n+1}\right)}{\Delta z_p} = 0. \tag{4.38}$$

The missing boundary conditions are governed by Q_s:

$$Q_s\left(t_{n+1}\right) = Q_d\left(0, t_{n+1}\right) = Q_p\left(z_{N_{z,p}}, t_{n+1}\right). \tag{4.39}$$

Computing $Q_s\left(t_{n+1}\right)$ the differential equation from above is discretized by an explicit Euler method wrt. to a time interval $\left[t_n, t_{n+1}\right]$. This yields:

$$Q_s^{n+1} = Q_s^n + \frac{\Delta t A_0}{K_u \rho l_s}\left(\Delta P_s^n - \frac{K_v \mu}{A_0 D_0} Q_s^n - \frac{K_t \rho}{2A_0^2}\left(\frac{A_0}{A_s} - 1\right)^2 Q_s^n \left|Q_s^n\right|\right). \tag{4.40}$$

Thereby the following notation is used: $Q_s^{n+1} = Q_s\left(t_{n+1}\right)$. The pressure difference ΔP_s^n is determined as follows:

$$\Delta P_s^n = p_p\left(z_{N_{z,p}}, t_n\right) - p_d\left(0, t_n\right)$$
$$= G_{0,p}\left(\sqrt{\frac{A_p\left(z_{N_{z,p}}, t_n\right)}{A_{0,p}}} - 1\right) - G_{0,d}\left(\sqrt{\frac{A_d\left(0, t_n\right)}{A_{0,d}}} - 1\right). \tag{4.41}$$

Please note that for the extreme cases $A_s \approx 0$ i.e., $S_d \approx 1.0$ (4.40) is not used. Otherwise there would be a divison by zero within the third term of the right-hand side. If $A_s \approx 0$ holds, we just set $Q_s = 0$. All in all one can determine the four missing boundary values for the fluid variables by means of (4.37)–(4.41) (see Algorithm 8). Simulating a transport process through the stenosis, we consider the solutions variables Γ_p and Γ_d for the proximal and distal vessel. Moreover, we use the notation in Fig. 4.6.

Algorithm 8: Coupling of two 1D problems at a stenosis (1)

INPUT: Grid $\Omega_{\Delta z_d,\Delta t}$ and $\Omega_{\Delta z_p,\Delta t}$, lengths: L_p und L_d, elasticity parameters $G_{0,p}$ and $G_{0,d}$, stenosis parameters in (3.91), time step $\left[t_n, t_{n+1}\right]$, $\Delta t = t_{n+1} - t_n$, A_p, A_d, Q_p and Q_d for $t = t_n$ at all the grid nodes, A_p, A_d, Q_p and Q_d for $t = t_{n+1}$ at the inner grid nodes.
OUTPUT: Boundary values: $A_{N_{z,p}}^{n+1}$, $Q_{N_{z,p}}^{n+1}$, $A_{0,d}^{n+1}$ and $Q_{0,d}^{n+1}$;

Solve the ODE (3.91) by means of the explict Euler method (4.40) to obtain Q_s^{n+1} ;
Determine the boundary values $A_{N_{z,p}}^{n+1}$, $Q_{N_{z,p}}^{n+1}$, $A_{0,d}^{n+1}$ and $Q_{0,d}^{n+1}$ by means of Eqs. (4.37)–(4.39);

$$\Gamma_p\left(z_{N_{z,p}-1/2}, t_{n+1}\right) = \Gamma_{N_{z,p}-1}^{n+1},\ \Gamma_d\left(\frac{\Delta z_d}{2}, t_{n+1}\right) = \Gamma_{0,d}^{n+1},$$
$$\Gamma_p\left(z_{N_{z,p}-3/2}, t_{n+1}\right) = \Gamma_{N_{z,p}-2}^{n+1},\ \Gamma_d\left(\frac{3\Delta z_d}{2}, t_{n+1}\right) = \Gamma_{1,d}^{n+1}.$$

According to (4.3) the averaged concentrations $\Gamma_{N_{z,p}-1}^{n+1}$ and $\Gamma_{0,d}^{n+1}$ at the boundaries are computed as follows:

$$\Gamma_{N_{z,p}-1}^{n+1} = \Gamma_{N_{z,p}-1}^{n} - \frac{\Delta t\omega}{\Delta z_p}\left(\frac{Q_s^{n+1}}{A_{N_{z,p}}^{n+1}}\Gamma_p\left(z_{N_{z,p}}, t_{n+1}\right) - \frac{Q_{N_{z,p}-1}^{n+1}}{A_{N_{z,p}-1}^{n+1}}\Gamma_p\left(z_{N_{z,p}-1}, t_{n+1}\right)\right) \tag{4.42}$$

$$\Gamma_{0,d}^{n+1} = \Gamma_{0,d}^{n} - \frac{\Delta t\omega}{\Delta z_d}\left(\frac{Q_{1,d}^{n+1}}{A_{1,d}^{n+1}}\Gamma_d\left(z_{1,d}, t_{n+1}\right) - \frac{Q_s^{n+1}}{A_{0,d}^{n+1}}\Gamma_d\left(0, t_{n+1}\right)\right). \tag{4.43}$$

To determine the concentrations at the grid nodes one has to distinguish between two cases wrt. to the sign of Q_s^{n+1}. If we take into account the continuity of the volumetric concentration (3.94):

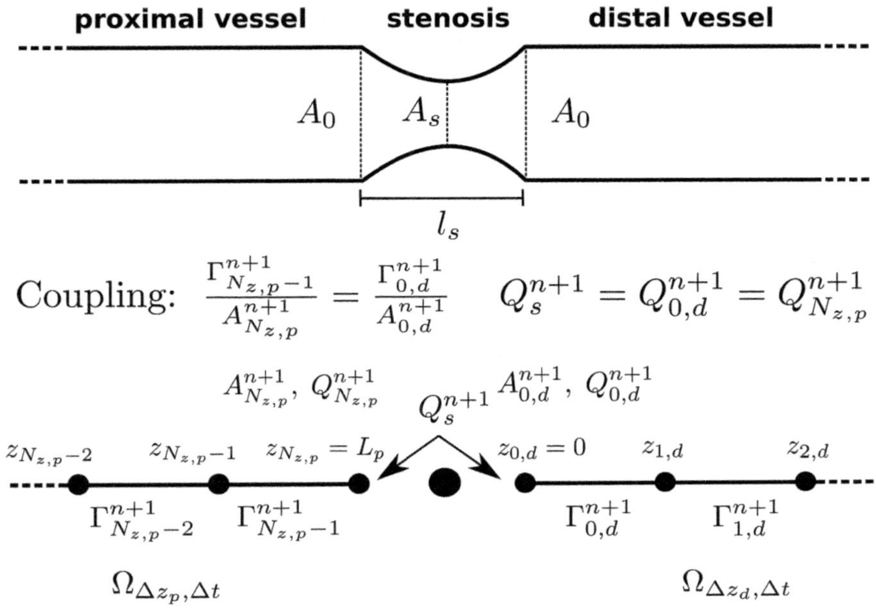

Fig. 4.6 Coupling of two 1D models at a stenosis: The volumetric flow rate Q_s^{n+1} through the stenosis serves as a boundary condition for the proximal and distal vessel, which are adjacent to the stenosis. To simulate the transport of a substance, we enforce the continuity of its volumetric concentration

$$\frac{\Gamma^{n+1}_{N_{z,p}-1}}{A^{n+1}_{N_{z,p}}} = \frac{\Gamma^{n+1}_{0,d}}{A^{n+1}_{0,d}}, \tag{4.44}$$

it follows:

- **Case 1:** $Q_s^{n+1} \geq 0$. In this case, the concentration $\Gamma_d(0, t_{n+1})$ is replaced by the coupling condition (4.44) as follows:

$$\Gamma_d(0, t_{n+1}) = \frac{A^{n+1}_{0,d}}{A^{n+1}_{N_{z,p}}} \Gamma^{n+1}_{N_{z,p}-1}.$$

Furthermore, it holds:

$$\Gamma_p\left(z_{N_{z,p}}, t_{n+1}\right) = \Gamma^{n+1}_{N_{z,p}-1}.$$

The remaining concentrations depend on the signs of the volumetric flow rates $Q^{n+1}_{N_{z,p}-1}$ and $Q^{n+1}_{1,d}$:

$$\Gamma_p\left(z_{N_{z,p}-1}, t_{n+1}\right) = \begin{cases} \Gamma^{n+1}_{N_{z,p}-2}, & \text{if } Q^{n+1}_{N_{z,p}-1} \geq 0, \\ \Gamma^{n+1}_{N_{z,p}-1}, & \text{if } Q^{n+1}_{N_{z,p}-1} < 0. \end{cases}$$

and

$$\Gamma_d\left(z_{1,d}, t_{n+1}\right) = \begin{cases} \Gamma^{n+1}_{0,d}, & \text{if } Q^{n+1}_{1,d} \geq 0, \\ \Gamma^{n+1}_{1,d}, & \text{if } Q^{n+1}_{1,d} < 0. \end{cases}$$

- **Case 2:** $Q_s^{n+1} < 0$. In this case the averaged concentration $\Gamma_p\left(z_{N_{z,p}-1}, t_{n+1}\right)$ results from the coupling condition (4.44):

$$\Gamma_p\left(z_{N_{z,p}-1}, t_{n+1}\right) = \frac{A^{n+1}_{N_{z,p}}}{A^{n+1}_{0,d}} \Gamma^{n+1}_{0,d}.$$

  In addition to that, we have:

$$\Gamma_d\left(z_{0,d}, t_{n+1}\right) = \Gamma^{n+1}_{0,d}.$$

  The remaining averaged concentrations are determined as in **Case 1**.

Incorporation of (4.42) and (4.43) into the system of Eq. (4.4) is described in Algorithm 9.

Algorithm 9: Coupling of two 1D models at a stenosis (2).

INPUT: Grid $\Omega_{\Delta z_d, \Delta t}$ and $\Omega_{\Delta z_p, \Delta t}$, lengths: L_p and L_d, elasticity parameters $G_{0,p}$ and $G_{0,d}$, stenosis parameters in (3.91), time step $\left[t_n, t_{n+1}\right]$, $\Delta t = t_{n+1} - t_n$, A_p, A_d, Q_p and Q_d at the grid nodes of $\Omega_{\Delta z_d, \Delta t}$ and $\Omega_{\Delta z_p, \Delta t}$ for $t = t_{n+1}$, volumetric flow rate Q_s through a stenosis for $t = t_{n+1}$.
OUTPUT: New system matrix A_Γ^{n+1} and right-hand side $\mathbf{b}_\Gamma^n$;

- -

Find the index j_p of the proximal vessel;
Find the index j_d of the distal vessel;
if $Q_s^{n+1} \geq 0$ **then**

$$A_\Gamma^{j_p j_p}\left(N_{z,j_p}, N_{z,j_p}\right) = A_\Gamma^{j_p j_p}\left(N_{z,j_p}, N_{z,j_p}\right) + \frac{\omega \Delta t}{\Delta z_p} \frac{Q_s^{n+1}}{A^{n+1}_{N_{z,p}}};$$

$$A_\Gamma^{j_d j_p}\left(1, N_{z,j_p}\right) = A_\Gamma^{j_d j_p}\left(1, N_{z,j_p}\right) - \frac{\omega \Delta t}{\Delta z_d} \frac{Q_s^{n+1}}{A^{n+1}_{0,d}} \frac{A^{n+1}_{0,d}}{A^{n+1}_{N_{z,p}}};$$

end
else

$$A_\Gamma^{j_d j_d}(1, 1) = A_\Gamma^{j_d j_d}(1, 1) - \frac{\omega \Delta t}{\Delta z_d} \frac{Q_s^{n+1}}{A^{n+1}_{0,d}};$$

$$A_\Gamma^{j_p j_d}\left(N_{z,j_p}, 1\right) = A_\Gamma^{j_p j_d}\left(N_{z,j_p}, 1\right) + \frac{\omega \Delta t}{\Delta z_p} \frac{Q_s^{n+1}}{A^{n+1}_{N_{z,p}}} \frac{A^{n+1}_{N_{z,p}}}{A^{n+1}_{0,d}};$$

end

Summary of the Sub-Algorithms

Concluding this section, we summarize the algorithms used to simulate blood flow and transport processes through an entire network. The computation of the flow field is given by Algorithm 10, while the transport process is described in Algorithm 11. The two algorithms are related to a single time step. The whole procedure is summarized in Algorithm 12.

Algorithm 10: Simulation of blood flow in a single time step

INPUT: Network composed of N vessels, N_E inlets, N_A outlets, N_B Bifurcations, N_S stenoses, model parameters: lengths of the vessels, elasticity parameters, resistances etc., time step $\left[t_n, t_{n+1}\right]$, time stepsize Δt.

- -

for *Vessel* $j \in \{1, \ldots, N\}$ **do**

Consider the grid $\Omega_{\Delta z_j, \Delta t}$ with $N_{z,j}+1$ grid nodes $z_{i,j}$;

Consider the approximations at the grid nodes for $t = t_n$:

$$\mathbf{A}_j^n = \left(A_{0,j}^n, \ldots, A_{N_{z,j},j}^n\right)^T, \quad \mathbf{Q}_j^n = \left(Q_{0,j}^n, \ldots, Q_{N_{z,j},j}^n\right)^T$$

Compute the approximations at the inner grid nodes for the time point $t = t_{n+1}$:

$$A_{1,j}^{n+1}, \ldots, A_{N_{z,j}-1,j}^{n+1}, Q_{1,j}^{n+1}, \ldots, Q_{N_{z,j}-1,j}^{n+1}$$

by a Lax-Wendroff step (see Eq. (4.1));

end

for $i = 1, \ldots, N_E$ **do**

Determine the index j of the vessel adjacent to the inlet i;

Compute $A_{0,j}^{n+1}$ and $Q_{0,j}^{n+1}$ at the respective boundary nodes using Algorithm 1;

end

for $i = 1, \ldots, N_A$ **do**

Determine the index j of the vessel adjacent to the outlet i;

Compute $A_{N_{z,j},j}^{n+1}$ and $Q_{N_{z,j},j}^{n+1}$ at the respective boundary nodes using Algorithm 3;

end

for $i = 1, \ldots, N_B$ **do**

Determine the indices $j_1, \ldots, j_m$ of the vessels adjacent to the bifurcation i;

Compute the boundary conditions at the respective boundary nodes of the Vessels $j_1, \ldots, j_m$ by the help of Algorithm 6;

end

for $i = 1, \ldots, N_S$ **do**

Determine the indices p and d of the vessels containing the stenosis i, p represents the proximal vessel and d the distal vessel;

Compute $A_{N_{z,p},p}^{n+1}$, $Q_{N_{z,p},p}^{n+1}$, $A_{0,d}^{n+1}$ and $Q_{0,d}^{n+1}$ at the respective boundary nodes by means of Algorithm 8;

end

Algorithm 11: Simulation of transport processes for a single time step

INPUT: Network with N vessels, N_E inflow boundaries, N_A outflow boundaries, N_B bifurcations, N_S stenoses, model parameters: lengths of the vessels, elasticity parameters, outflow boundaries etc., time step $[t_n, t_{n+1}]$, time stepsize Δt.

---

for *Vessel* $j \in \{1, \ldots, N\}$ **do**
  Compute the matrices A_Γ^{jj} in (4.5) according to (4.3);
  Set the right hand sides $\mathbf{b}_j^n$ in (4.6) to: $\mathbf{b}_j^n = \Gamma^n$;
end
for $i = 1, \ldots, N_E$ **do**
  Incorporate the boundary condition at the inflow boundary i by means of Algorithm 2 into the system of Eq. (4.4);
end
for $i = 1, \ldots, N_A$ **do**
  Incorporate the boundary condition at the outflow boundary i by means of Algorithm 5 or 5 into the right hand side $\mathbf{b}_\Gamma^n$;
end
for $i = 1, \ldots, N_B$ **do**
  Incorporate the boundary conditions at bifurcation i by means of Algorithm 7 into the system of Eq. (4.4);
end
for $i = 1, \ldots, N_S$ **do**
  Incorporate the boundary conditions at stenosis i by means of Algorithm 9 into the system of Eq. (4.4);
end
Solve the system of Eq. (4.4): $A_\Gamma^{n+1}\Gamma^{n+1} = \mathbf{b}_\Gamma^n$;
for $i = 1, \ldots, N_A$ **do**
  Update of the concentrations at outflow boundary i according to Algorithm 4;
end

4.3 Stability and Convergence Analysis

In the first two sections of this chapter, we have been concerned with constructing the Lax-Wendroff method and an implicit Finite Volume method. Furthermore the numerical treatment of boundary conditions from Sect. 3.6 has been described. An important aspect, that has been omitted in these sections, are the approximation properties of both methods. In the following subsections, we will investigate some numerical features of the Lax-Wendroff method and the implicit Finite Volume method.

Algorithm 12: Simulation of blood flow and transport processes in a network

---

INPUT: Network consisting of N vessels, N_E inflow boundaries, N_A outflow boundaries, N_B bifurcations, N_S stenoses, model parameters: Lengths of the vessels, elasticity parameters, resistances at outlets etc., end of the simulation time: T_{end}, step size Δt.

- -

Index time step: $n = 0$, starting time: $t_n = 0$;
while $t_n \leq T_{end}$ **do**
  $t_{n+1} = t_n + \Delta t$;
  Consider the time step: $[t_n, t_{n+1}]$;
  Compute the flow field by means of Algorithm 10;
  Compute the new concentrations by means of Algorithm 11;
  for *Vessel* $j \in \{1, \ldots, N\}$ **do**
    Replace the old solution vectors by the new ones:
    $\mathbf{A}_j^n = \mathbf{A}_j^{n+1},\ \mathbf{Q}_j^n = \mathbf{Q}_j^{n+1},\ \Gamma_j^n = \Gamma_j^{n+1}$;
  end
  $t_n = t_{n+1},\ n = n + 1$;
end

4.3.1 Important Mathematical Concepts

In order to be able to investigate the approximation properties of the Lax-Wendroff method and the implicit Finite Volume method several concepts and notations have to be introduced. In a first step, we restrict ourselves to a Cauchyproblem (see Eqs. (3.30) and (3.31)) i.e., the influence of boundary conditions is neglected and the model equations are defined on $\mathbb{R}$:

$$\frac{\partial \mathbf{U}}{\partial t} + \frac{\partial \mathbf{F}}{\partial z}(\mathbf{U}) = \mathbf{S}(\mathbf{U}), \text{ in } \mathbb{R} \times (0, \infty), \tag{4.45}$$
$$\mathbf{U}(z, 0) = \mathbf{U_0}(z), \text{ auf } \mathbb{R}.$$

The exact definitions of $\mathbf{U}$, $\mathbf{F}$ and $\mathbf{S}$ can be found at the beginning of Sect. 3.3. As in case of the derivation of both methods the simulation domain $\mathbb{R} \times (0, \infty)$ is discretized by an uniform grid consisting of the following nodes:

$$(z_i, t_j) = (i \cdot \Delta z, j \cdot \Delta t) \text{ mit } i \in \mathbb{Z},\ j \in \mathbb{N}_0$$

$\Delta z > 0$ and $\Delta t > 0$ denote as in the previous sections the grid size in space and time. In the following, it will be revealed that the ratio between Δz and Δt plays a crucial role analyzing Finite Difference methods. For this reason, we introduce a symbol for this ratio:

$$\lambda = \frac{\Delta t}{\Delta z}.$$

At each grid node (z_i, t_j) we approximate the solution of (4.45), where the corresponding numerical solutions are denoted by $\mathbf{U}_i^j$. Computing $\mathbf{U}_i^j$ we use a numerical method that can be represented as follows:

$$\mathbf{U}_i^{j+1} = H\left(\mathbf{U}_{i-m}^{j+1}, \ldots, \mathbf{U}_{i+m}^{j+1}, \mathbf{U}_{i-m}^{j}, \ldots, \mathbf{U}_{i+m}^{j}\right). \tag{4.46}$$

Thereby, $m \in \mathbb{N}$ is fixed and the grid transfer function

$$H : \mathbb{R}^{n\cdot(4m+2)} \to \mathbb{R}^n$$

is continuous componentwise. In context with (4.46) different concepts can be found in literature. They are summarized in Definition 4.1.

Definition 4.1 The method given by (4.46) is called *Onestep-method*, since for the computation of $\mathbf{U}_i^{j+1}$ only solution values from the time step $[t_j, t_{j+1}]$ are used. A Onestep-method can be assigned to the following types of methods:

- *Explicit methods:* The method (4.46) is denoted as explicit, if H is depending only on solution values at the time t_j. This implies:

$$H\left(\mathbf{U}_{i-m}^{j+1}, \ldots, \mathbf{U}_{i+m}^{j+1}, \mathbf{U}_{i-m}^{j}, \ldots, \mathbf{U}_{i+m}^{j}\right) = H\left(\mathbf{U}_{i-m}^{j}, \ldots, \mathbf{U}_{i+m}^{j}\right).$$

- *Implicit method:* The method (4.46) is called implicit, if H is depending both on the solution values for t_j and t_{j+1}.
- *Linear methods:* If H is a linear function, (4.46) is referred to as linear method.
- *Finite Difference method:* If the grid transfer function H is based on approximations of the derivatives occuring in (4.45) by means of difference quotients, (4.46) is denoted as Finite Difference method.

Next, we try to provide some examples for the concepts in Definition 4.1. For this purpose, we provide three standard examples for Finite Difference methods. The first method is known as *explicit central Euler method*. The idea of this method is to replace the time derivative by a forward difference quotient:

$$\frac{\partial \mathbf{U}}{\partial t}(z_i, t_j) \approx \frac{\mathbf{U}_i^{j+1} - \mathbf{U}_i^j}{\Delta t}.$$

As the name of the method suggests, the spatial derivative of the flux function $\mathbf{F}$ is replaced by a central difference quotient:

$$\frac{\partial \mathbf{F}}{\partial z}\left(\mathbf{U}_i^j\right) \approx \frac{\mathbf{F}\left(\mathbf{U}_{i+1}^j\right) - \mathbf{F}\left(\mathbf{U}_{i-1}^j\right)}{2\Delta z}$$

Setting $\mathbf{S}(\mathbf{U}) = 0$ in (4.45), Eq. (4.45) is known as *conservation equation*. Applying the explicit central Euler method to the conservation equation, we obtain:

$$\frac{\mathbf{U}_i^{j+1}-\mathbf{U}_i^j}{\Delta t}+\frac{\mathbf{F}\left(\mathbf{U}_{i+1}^j\right)-\mathbf{F}\left(\mathbf{U}_{i-1}^j\right)}{2\Delta z}=0$$

and

$$\mathbf{U}_i^{j+1}=H\left(\mathbf{U}_{i-1}^j,\mathbf{U}_i^j,\mathbf{U}_{i+1}^j\right),$$

where

$$H\left(\mathbf{U}_{i-1}^j,\mathbf{U}_i^j,\mathbf{U}_{i+1}^j\right)=\mathbf{U}_i^j-\frac{\lambda}{2}\left(\mathbf{F}\left(\mathbf{U}_{i+1}^j\right)-\mathbf{F}\left(\mathbf{U}_{i-1}^j\right)\right) \tag{4.47}$$

i.e., one sets $m=1$. If the flux function $\mathbf{F}$ is linear, H is a linear function and thus the explicit central Euler method is linear. The *implicit central Euler method* is obtained by evaluating the derivative of the flux function $\mathbf{F}$ not at $\left(z_i,t_j\right)$ but at $\left(z_i,t_{j+1}\right)$. By this, it follows:

$$\mathbf{U}_i^{j+1}=H\left(\mathbf{U}_{i-1}^{j+1},\mathbf{U}_i^j,\mathbf{U}_{i+1}^{j+1}\right),$$

with

$$H\left(\mathbf{U}_{i-1}^{j+1},\mathbf{U}_i^j,\mathbf{U}_{i+1}^{j+1}\right)=\mathbf{U}_i^j-\frac{\lambda}{2}\left(\mathbf{F}\left(\mathbf{U}_{i+1}^{j+1}\right)-\mathbf{F}\left(\mathbf{U}_{i-1}^{j+1}\right)\right). \tag{4.48}$$

Replacing in (4.47) the vector $\mathbf{U}_i^j$ by the mean value between $\mathbf{U}_{i+1}^j$ and $\mathbf{U}_{i-1}^j$, one can design a further explicit method that is known as *Lax-Friedrichs method*. The Lax-Friedrichs method has the following grid transfer function:

$$H\left(\mathbf{U}_{i-1}^{j+1},\mathbf{U}_{i+1}^{j+1}\right)=\frac{1}{2}\left(\mathbf{U}_{i-1}^j-\lambda\mathbf{F}\left(\mathbf{U}_{i-1}^j\right)\right)+\frac{1}{2}\left(\mathbf{U}_{i+1}^j-\lambda\mathbf{F}\left(\mathbf{U}_{i+1}^j\right)\right). \tag{4.49}$$

In the following subsection, we will show that the Lax-Wendroff method and the implicit Finite Volume method can be derived in a similar way. In Sect. 4.1 both methods are obtained by means of integration over cells arounde the grid nodes $\left(z_i,t_j\right)$ and appropriate quadrature rules. Application of the grid transfer functions (4.47)–(4.49) can be represented by means of stencil (see Fig. 4.7). A node with a filling means that the corresponding value is used to compute $\mathbf{U}_i^{j+1}$.

Conservative and Consistent Finite-Difference Methods

An important principle, which is considered in terms of numerical methods, is to transfer certain properties of the continuous problem to the respective numerical solution method. In case of scalar conservation equations ($n=1$)

$$\frac{\partial \mathbf{U}}{\partial t}+\frac{\partial \mathbf{F}}{\partial z}(\mathbf{U})=0, \tag{4.50}$$

the corresponding solutions have the following property: If $\mathbf{U}_0\in L^1(\mathbb{R})$, it holds:

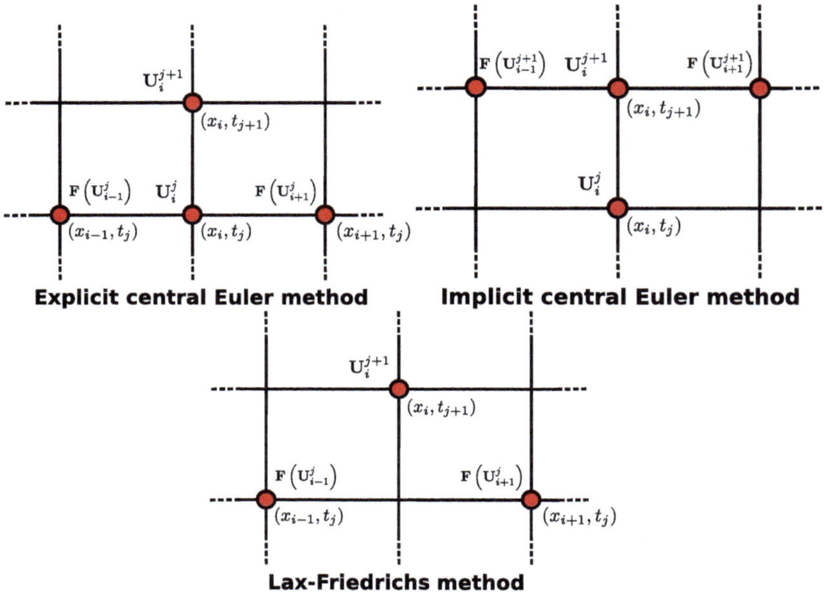

Fig. 4.7 Stencils for the explicit/implicit central Euler method and the Lax-Friedrichs method

$$\int_{\mathbb{R}} \mathbf{U}(z, t)\ dz = \int_{\mathbb{R}} \mathbf{U}_0(z)\ dz.$$

Integration of (4.50) wrt. a cell

$$\left[z_{i-1/2}, z_{i+1/2}\right) \times \left[t_j, t_{j+1}\right),$$

yields as in Sect. 4.1:

$$\frac{1}{\Delta z}\int_{z_{i-1/2}}^{z_{i+1/2}} \mathbf{U}\left(z, t_{j+1}\right)\ dz = \frac{1}{\Delta z}\int_{z_{i-1/2}}^{z_{i+1/2}} \mathbf{U}\left(z, t_j\right)\ dz$$
$$-\lambda\left(\frac{1}{\Delta t}\int_{t_j}^{t_{j+1}} \mathbf{F}\left(z_{i+\frac{1}{2}}, t\right)\ dt - \frac{1}{\Delta t}\int_{t_j}^{t_{j+1}} \mathbf{F}\left(z_{i-\frac{1}{2}}, t\right)\ dt\right).$$

On the left side of the above equation one can find the mean value of $\mathbf{U}$ at time $t = t_{j+1}$ around $z = z_i$. It is given by the mean value of $\mathbf{U}$ at $t = t_j$ around $z = z_i$ and time integrals of the flux function $\mathbf{F}$ wrt. $z = z_{i-1/2}$ and $z = z_{i+1/2}$ multiplied with λ. This structure motivates the concept of conservative numerical schemes.

Definition 4.2 An explicit Onestep-method is denoted as conservative, if the grid transfer function H can be represented as follows (conservation form):

$$
\begin{aligned}
&H\left(\mathbf{U}_{i-m}^{j},\ldots,\mathbf{U}_{i+m}^{j}\right)\\
=&\mathbf{U}_{i}^{j}-\lambda\left(\tilde{\mathbf{F}}\left(\mathbf{U}_{i-m+1}^{j},\ldots,\mathbf{U}_{i+m}^{j}\right)-\tilde{\mathbf{F}}\left(\mathbf{U}_{i-m}^{j},\ldots,\mathbf{U}_{i+m-1}^{j}\right)\right)\\
=&\mathbf{U}_{i}^{j}-\lambda\left(\tilde{\mathbf{F}}_{i+\frac{1}{2}}^{j}-\tilde{\mathbf{F}}_{i-\frac{1}{2}}^{j}\right)
\end{aligned}
$$

where $\tilde{\mathbf{F}}:\mathbb{R}^{2\cdot nm}\rightarrow\mathbb{R}^{n}$ is a continuous function. The function $\tilde{\mathbf{F}}$ is called *numerical flux function*. By this, the method is given by:

$$
\mathbf{U}_{i}^{j+1}=H\left(\mathbf{U}_{i-m}^{j},\ldots,\mathbf{U}_{i+m}^{j}\right).
$$

If a method is conservative, it follows for the scalar case ($n=1$):

$$
\begin{aligned}
\int_{\mathbb{R}}\mathbf{U}^{j}(z)\ dz=\Delta z\sum_{i}\mathbf{U}_{i}^{j}&=\Delta z\sum_{i}H\left(\mathbf{U}_{i-m}^{j-1},\ldots,\mathbf{U}_{i+m}^{j-1}\right)\\
&=\int_{\mathbb{R}}H\left(\mathbf{U}^{j-1}\right)(z)\ dz=\int_{\mathbb{R}}\mathbf{U}^{j-1}(z)\ dz.
\end{aligned}
$$

Thereby it holds:

$$
\mathbf{U}^{j}(z)=\mathbf{U}_{i}^{j}\ \text{und}\ H\left(\mathbf{U}^{j}\right)(z)=H\left(\mathbf{U}_{i-m}^{j},\ldots,\mathbf{U}_{i+m}^{j}\right)(z),\ \ z\in\left(z_{i-1/2},z_{i+1/2}\right).
$$

Continuing this procedure, we obtain:

$$
\int_{\mathbb{R}}\mathbf{U}_{0}(z)\ dz=\int_{\mathbb{R}}\mathbf{U}^{j}(z)\ dz.
$$

This means that a conservative method incorporates the main feature of conservation laws. Representing a conservative method in the following form:

$$
\underbrace{\frac{\mathbf{U}_{i}^{j+1}-\mathbf{U}_{i}^{j}}{\Delta t}}_{\approx\frac{\partial\mathbf{U}}{\partial t}}+\underbrace{\frac{\tilde{\mathbf{F}}_{i+\frac{1}{2}}^{j}-\tilde{\mathbf{F}}_{i-\frac{1}{2}}^{j}}{\Delta z}}_{\approx\frac{\partial\mathbf{F}}{\partial z}}=0,
$$

one can note that a central difference quotient is used to approximate the derivative of the flux function. If for t_j there is a constant solution $\mathbf{U}^{j}$ in space, the central difference quotient has to vanish. This can be guaranteed, if the numerical flux $\tilde{\mathbf{F}}$ and the continuous flux function $\mathbf{F}$ fulfill the following relationship:

$$
\tilde{\mathbf{F}}(\mathbf{U},\ldots,\mathbf{U})=\mathbf{F}(\mathbf{U}).
$$

From this observation, we can define the term “consistent numerical flux”.

Definition 4.3 $\tilde{F}$ is called *consistent numerical flux*, if it exhibits the following features:

- **Lipschitz continuity:** Let $\mathbf{U}_{-m}, \ldots, \mathbf{U}_m \in I_n = [a, b]^n \subset \mathbb{R}^n$, $a < b$ and $\mathbf{V}_{-m}, \ldots, \mathbf{V}_m \in I_n$. $\tilde{F}$ is referred to as Lipschitz continuous, if there is a constant $L_g > 0$ such that the following inequality is fulfilled:

$$\left\| \tilde{\mathbf{F}} \left(\mathbf{U}_{-m}, \ldots, \mathbf{U}_m \right) - \tilde{\mathbf{F}} \left(\mathbf{V}_{-m}, \ldots, \mathbf{V}_m \right) \right\| \leq L_g \sum_{i=-m}^{m} \| \mathbf{U}_i - \mathbf{V}_i \| .$$

- **Consistency:** $\tilde{\mathbf{F}}$ and $\mathbf{F}$ are consistent to each over, if it holds:

$$\tilde{\mathbf{F}} \left(\mathbf{U}, \ldots, \mathbf{U} \right) = \mathbf{F} \left(\mathbf{U} \right) .$$

A Onestep-method is called **consistent**, if it has a consistent numerical flux function.

If the flux function $\mathbf{F}$ is Lipschitz continuous, one can show in a straight forward way that both the central Euler method and the Lax-Friedrichs method are conservative and consistent. For the central Euler method, one obtains:

$$\begin{aligned}
\mathbf{U}_i^{j+1} &= \mathbf{U}_i^j - \frac{\lambda}{2} \left(\mathbf{F} \left(\mathbf{U}_{i+1}^j \right) - \mathbf{F} \left(\mathbf{U}_{i-1}^j \right) \right) \\
&= \mathbf{U}_i^j - \lambda \left(\frac{1}{2} \left(\mathbf{F} \left(\mathbf{U}_{i+1}^j \right) + \mathbf{F} \left(\mathbf{U}_i^j \right) \right) - \frac{1}{2} \left(\mathbf{F} \left(\mathbf{U}_i^j \right) + \mathbf{F} \left(\mathbf{U}_{i-1}^j \right) \right) \right) .
\end{aligned}$$

From this, we obtain, the numerical flux function:

$$\tilde{\mathbf{F}} \left(\mathbf{U}, \mathbf{V} \right) = \frac{1}{2} \left(\mathbf{F} \left(\mathbf{U} \right) + \mathbf{F} \left(\mathbf{V} \right) \right) .$$

Setting $\mathbf{U} = \mathbf{V}$, the consistency of the method follows directly. The Lax-Friedrichs method can be represented in the following form:

$$\begin{aligned}
\mathbf{U}_i^{j+1} &= \frac{1}{2} \left(\mathbf{U}_{i-1}^j + \mathbf{U}_{i+1}^j \right) - \frac{\lambda}{2} \left(\mathbf{F} \left(\mathbf{U}_{i+1}^j \right) - \mathbf{F} \left(\mathbf{U}_{i-1}^j \right) \right) \\
&= \mathbf{U}_i^j - \lambda \left[\frac{1}{\lambda} \mathbf{U}_i^j - \frac{1}{2\lambda} \left(\mathbf{U}_{i-1}^j + \mathbf{U}_{i+1}^j \right) + \frac{1}{2} \left(\mathbf{F} \left(\mathbf{U}_{i+1}^j \right) - \mathbf{F} \left(\mathbf{U}_{i-1}^j \right) \right) \right] \\
&= \mathbf{U}_i^j - \lambda \left[\frac{1}{2\lambda} \left(\mathbf{U}_i^j - \mathbf{U}_{i+1}^j \right) + \frac{1}{2} \left(\mathbf{F} \left(\mathbf{U}_{i+1}^j \right) + \mathbf{F} \left(\mathbf{U}_i^j \right) \right) \right. \\
&\qquad \left. - \frac{1}{2\lambda} \left(\mathbf{U}_i^j - \mathbf{U}_{i-1}^j \right) - \frac{1}{2} \left(\mathbf{F} \left(\mathbf{U}_{i-1}^j \right) + \mathbf{F} \left(\mathbf{U}_i^j \right) \right) \right] .
\end{aligned}$$

This means that the Lax-Friedrichs method has the following numerical flux function:

$$\tilde{\mathbf{F}}(\mathbf{U}, \mathbf{V}) = \frac{1}{2\lambda}(\mathbf{U} - \mathbf{V}) + \frac{1}{2}(\mathbf{F}(\mathbf{U}) + \mathbf{F}(\mathbf{V})).$$

Obviously, this method is conservative, too. For a conservative and consistent Finite Difference method, one can formulate a convergence theorem, which is known as *Theorem of Lax-Wendroff* in literature.

Theorem 4.1 *Let H be the grid transfer function of a conservative and consistent Finite Difference method. Furthermore, let us consider the limit $\Delta t \to 0$ with a fixed grid ratio $\lambda = \Delta t/\Delta z$. The solution for a certain time stepsize Δt, is denoted by $\mathbf{U}_{\Delta t}$. If it holds that:*

$$\|\mathbf{U}_{\Delta t}\| \leq C \text{ and } \mathbf{U}_{\Delta t} \to \mathbf{U},$$

$\mathbf{U}$ *is called weak solution of* (4.45). $\mathbf{U}$ *is a weak solution of* (4.45), *if for all* $\phi \in C_0^\infty(\mathbb{R} \times [0, \infty); \mathbb{R}^n)$ *it holds (see also (3.55)):*

$$\int_0^\infty \int_{\mathbb{R}} \left[\frac{\partial \phi}{\partial t}\mathbf{U} + \frac{\partial \phi}{\partial z}\mathbf{F}(\mathbf{U})\right] dzdt + \int_{\mathbb{R}} \phi(z, 0)\, \mathbf{U}_0(z)\ dz = 0.$$

Proof A proof of this statement can be found e.g., in [1] and [4, Chap. 12].

One drawback of this theorem is that convergence of the numerical method has to be assumed. Furthermore, it is not guaranteed that the method is converging to a unique weak solution. Another drawback of the Theorem of Lax and Wendroff is that it contains no statement on the speed of convergence or error estimates. This makes the comparison of different methods difficult.

Convergence Order, Consistency Order and Stability

Before we can introduce a measure for the approximation properties of a Finite Difference methods, we require a definition of discrete norms to estimate approximation errors on a grid. In particular, discrete norms approximating the continuous norms of $L^p(\mathbb{R})$ spaces are of interest, $p \in [1, \infty]$. For a solution vector

$$\mathbf{U}^j = \left(\ldots, \mathbf{U}_{i-1}^j, \mathbf{U}_i^j, \mathbf{U}_{i+1}^j, \ldots\right)^T, \ \mathbf{U}_i^j \in \mathbb{R}^n$$

at a time $t = t_j$ these norms read as follows:

$$\left\|\mathbf{U}^j\right\|_{\Delta,p} = \left(\Delta z \sum_{i=-\infty}^{\infty} \left\|\mathbf{U}_i^j\right\|_2^p\right)^{\frac{1}{p}}, \ p \in [1, \infty) \tag{4.51}$$

and

$$\left\|\mathbf{U}^j\right\|_{\Delta,\infty} = \sup_{k\in\{1,\ldots,n\}} \sup_i \left|\mathbf{U}_i^j(k)\right|.$$

$\mathbf{U}_i^j(k)$ denotes the k-th component of the vector $\mathbf{U}_i^j$. By this, we can define the convergence order of a Onestep-method.

Definition 4.4 Let $\mathbf{V}$ a sufficiently smooth solution of (4.45). As for the numerical approximation $\mathbf{U}_i^j$ the evaluation of $\mathbf{V}$ at a grid node (z_i, t_j), $i \in \mathbb{Z}$, $j \in \mathbb{N}$ is denoted by $\mathbf{V}_i^j$. A Finite Difference method is called convergent of order l in space and q in time, if for all components $k \in \{1, \ldots, m\}$ of the solution vector it holds:

$$\left\|\mathbf{V}^j - \mathbf{U}^j\right\|_{\Delta,p} = \left\|\mathbf{E}^j\right\|_{\Delta,p} \in \mathcal{O}\left(\Delta z^l + \Delta t^q\right).$$

By this it follows:

$$\left\|\mathbf{E}^j\right\|_{\Delta,p} \to 0 \text{ for } \Delta z,\ \Delta t \to 0.$$

Definition 4.5 The local truncation error L_i^j in a grid point (z_i, t_j) is defined as follows:

$$L_i^j = \frac{\mathbf{V}_i^j - H\left(\mathbf{V}^j, \mathbf{V}^{j-1}\right)}{\Delta t}.$$

The method is consistent of order p in space and q in time, if it holds:

$$L_i^j \in \mathcal{O}\left(\Delta z^p\right) + \mathcal{O}\left(\Delta t^q\right),$$

where this notation indicates that each component of L_i^j is in $\mathcal{O}\left(\Delta z^p\right) + \mathcal{O}\left(\Delta t^q\right)$.

The terms consistency and convergence, which have been introduced in Definition 4.4 and 4.5 govern the behavior of a numerical method for Δz, $\Delta t \to 0$. Furthermore, it is of interest, how the numerical method performs for fixed discretization parameters Δz and Δt. In particular, it is desirable that discretization errors are not accumulating as the number of time steps is increasing. A numerical method, keeping the numerical solutions bounded is denoted as numerical stable. This property can be formulated as follows:

Definition 4.6 A numerical method applied to a Cauchyproblem (4.45) is numerical stable, if for each time T there are two constants $C_T > 0$ and $\delta_0 > 0$, such that

$$\left\|\mathbf{U}^n\right\|_{\Delta,p} \leq C_T \left\|\mathbf{U}^0\right\|_{\Delta,p},$$

for all $n\Delta t \leq T$, $0 < \Delta z \leq \delta_0$ and $0 < \Delta t \leq \delta_0$.

One way to prove the $\Delta, 2$ stability of a method, is the von Neumann stability analysis. For this reason, we assume that the initial condition $\mathbf{U}_0 \in \mathbb{R}^n$ is 2π-periodic, such that it can be represented by a Fourier series (see Sect. 7.7):

$$\mathbf{U}_0(z) = \sum_{m=-\infty}^{\infty} \xi_m e^{imz}.$$

The vectors $\xi_m \in \mathbb{R}^n$ are called m-th Fourier coefficients and are defined as follows:

$$\xi_m = \frac{1}{2\pi} \int_{-\pi}^{\pi} \mathbf{U}_0(z)\, e^{-imz}\, dz$$

where the integration is performed componentwise. Thus is holds for the grid node (z_l, t_0):

$$\mathbf{U}_l^0 = \sum_{m=-\infty}^{\infty} \xi_m e^{iml\Delta z}.$$

Applying a method j-times to $\mathbf{U}_l^0$, one obtains:

$$\mathbf{U}_l^j = \sum_{m=-\infty}^{\infty} G^j \xi_m e^{iml\Delta z},$$

where the matrix $G \in \mathbb{C}^{n\times n}$ results from the numerical method. We also refer to the matrix G as *amplification matrix*. If the method is non-linear, G is based on the linearized numerical method. By means of this notation, von Neumann's stability condition can be formulated.

Definition 4.7 Let $G \in \mathbb{C}^{n\times n}$ be the amplification matrix of a numercial method. This method is stable according to von Neumann, if it holds that:

$$\|G\| \leq 1.$$

If a method is stable according to von Neumann, it is also stable with respect to the Δ, 2-norm. To prove this statement, we consider a finite Fourier series on the interval $[0, 2\pi]$ to represent the initial condition:

$$\mathbf{U}^0(z) = \sum_{m=-\frac{N}{2}}^{\frac{N}{2}-1} \xi_m e^{imz}.$$

N is a positive integer. Since we consider a grid with N equidistant grid nodes:

$$z_l = l\Delta z,\ l = 0, \ldots, N \text{ mit } \Delta z = \frac{2\pi}{N},$$

it is sufficient to consider a finite Fourier series. Fourier modes of higher order can not be represented on this grid (Aliasing effect [10]). Wrt. to this grid, we obtain:

$$\mathbf{U}_l^0 = \sum_{m=-\frac{N}{2}}^{\frac{N}{2}-1} \xi_m e^{iml\Delta z} \text{ and } \mathbf{U}_l^j = \sum_{m=-\frac{N}{2}}^{\frac{N}{2}-1} G^j \xi_m e^{iml\Delta z}.$$

From this representation, it follows for the individual components of the Δ, 2-norm:

$$\left\| \mathbf{U}_l^j \right\|_2^2 = \sum_{r,m=-\frac{N}{2}}^{\frac{N}{2}-1} \sum_{k=1}^{n} \left(G^j \xi_m\right)(k) \, \overline{\left(G^j \xi_r\right)}(k) \, e^{i(r-m)l\Delta z}.$$

Thus it holds:

$$\Delta z \sum_{l=0}^{N-1} \left\| \mathbf{U}_l^j \right\|_2^2 = \Delta z \sum_{l=0}^{N-1} \sum_{r,m=-\frac{N}{2}}^{\frac{N}{2}-1} \sum_{k=1}^{n} \left(G^j \xi_m\right)(k) \, \overline{\left(G^j \xi_r\right)}(k) \, e^{i(r-m)l\Delta z}.$$

Using Sect. 7.12, we obtain:

$$\Delta z \sum_{l=0}^{N-1} e^{i(r-m)l\Delta z} = 2\pi \delta_{rm}, \quad -\frac{N}{2} \leq r, m \leq \frac{N}{2} - 1.$$

From this, it follows:

$$\left\| \mathbf{U}^j \right\|_{\Delta,2}^2 = 2\pi \sum_{m=-\frac{N}{2}}^{\frac{N}{2}-1} \left\| G^j \xi_m \right\|_2^2 \leq 2\pi \sum_{m=-\frac{N}{2}}^{\frac{N}{2}-1} \|G\|^{2j} \, \|\xi_m\|^2 .$$

Since the method is stable according to von Neumann, Definition 4.7 implies that:

$$\|G\|^{2j} \leq 1.$$

All in all, we see that the Δ, 2-Norm of $\mathbf{U}^j$ is bounded by the constant $C_T = 1.0$ and the Δ, 2-norm of the initial condition:

$$\left\| \mathbf{U}^j \right\|_{\Delta,2}^2 \leq 2\pi \sum_{m=-\frac{N}{2}}^{\frac{N}{2}-1} \|\xi_m\|^2 = \left\| \mathbf{U}^0 \right\|_{\Delta,2}^2 \Rightarrow \left\| \mathbf{U}^j \right\|_{\Delta,2} \leq \left\| \mathbf{U}^0 \right\|_{\Delta,2} .$$

4.3.2 Analysis of the Lax-Wendroff Method

Having summarized the most important theoretical tools for the analysis of hyperbolic differential equations in the previous subsection, we would now like to apply them to the two-step Lax-Wendroff method. The grid transfer function H of the two-step Lax-Wendroff method has the following form according to (4.1) and (4.2):

$$H\left(\mathbf{U}_{i-1}^j, \mathbf{U}_i^j, \mathbf{U}_{i+1}^j\right) = \mathbf{U}_i^j - \frac{\Delta t}{\Delta z}\left(\mathbf{F}_{i+\frac{1}{2}}^{j+\frac{1}{2}} - \mathbf{F}_{i-\frac{1}{2}}^{j+\frac{1}{2}}\right) + \frac{\Delta t}{2}\left(\mathbf{S}_{i+\frac{1}{2}}^{j+\frac{1}{2}} + \mathbf{S}_{i-\frac{1}{2}}^{j+\frac{1}{2}}\right), \quad (4.52)$$

where the approximations at the intermediate steps are given by:

$$\mathbf{F}_{i+\frac{1}{2}}^{j+\frac{1}{2}} = \mathbf{F}\left(\mathbf{U}_{i+\frac{1}{2}}^{j+\frac{1}{2}}\right), \; \mathbf{F}_{i-\frac{1}{2}}^{j+\frac{1}{2}} = \mathbf{F}\left(\mathbf{U}_{i-\frac{1}{2}}^{j+\frac{1}{2}}\right), \; \mathbf{S}_{i+\frac{1}{2}}^{j+\frac{1}{2}} = \mathbf{S}\left(\mathbf{U}_{i+\frac{1}{2}}^{j+\frac{1}{2}}\right), \; \mathbf{S}_{i-\frac{1}{2}}^{j+\frac{1}{2}} = \mathbf{S}\left(\mathbf{U}_{i-\frac{1}{2}}^{j+\frac{1}{2}}\right),$$

$$\mathbf{U}_{i+\frac{1}{2}}^{j+\frac{1}{2}} = \frac{\mathbf{U}_{i+1}^{j} + \mathbf{U}_{i}^{j}}{2} + \frac{\Delta t}{2}\left(-\frac{\mathbf{F}_{i+1}^{j} - \mathbf{F}_{i}^{j}}{\Delta z} + \frac{\mathbf{S}_{i+1}^{j} + \mathbf{S}_{i}^{j}}{2}\right),$$

$$\mathbf{U}_{i-\frac{1}{2}}^{j+\frac{1}{2}} = \frac{\mathbf{U}_{i}^{j} + \mathbf{U}_{i-1}^{j}}{2} + \frac{\Delta t}{2}\left(-\frac{\mathbf{F}_{i}^{j} - \mathbf{F}_{i-1}^{j}}{\Delta z} + \frac{\mathbf{S}_{i}^{j} + \mathbf{S}_{i-1}^{j}}{2}\right).$$

Properties of the Lax-Wendroff Method

Despite of the fact that the method requires an intermediate step, the Lax-Wendroff method is an explicit one-step method, since for the computation of $\mathbf{U}_i^{j+1}$ only values at time $t = t_j$ are used. Depending on the form of the flux function $\mathbf{F}$ and the source term $\mathbf{S}$, a linear or non-linear method is obtained. Since we want to apply the Lax-Wendroff method in the following sections to the blood flow model from Sect. 3.2, which has a non-linear flux function and source term, we obtain a non-linear method. In a next step, we determine the numerical flux function of the method. It has the following form:

$$\tilde{\mathbf{F}}(\mathbf{U}, \mathbf{V}) = \mathbf{F}\left(\frac{\mathbf{U}+\mathbf{V}}{2} - \frac{\Delta t}{2\Delta z}\left(\mathbf{F}(\mathbf{U}) - \mathbf{F}(\mathbf{V})\right)\right),$$

$\mathbf{U}, \mathbf{V} \in \mathbb{R}^2$. Obviously, it holds for $\mathbf{U} = \mathbf{V}$:

$$\tilde{\mathbf{F}}(\mathbf{U}, \mathbf{U}) = \mathbf{F}\left(\frac{\mathbf{U}+\mathbf{U}}{2} - \frac{\Delta t}{2\Delta z}\left(\mathbf{F}(\mathbf{U}) - \mathbf{F}(\mathbf{U})\right)\right) = \mathbf{F}(\mathbf{U}).$$

This means that $\mathbf{F}$ and $\tilde{\mathbf{F}}$ are consistent to each other. If the flux function $\mathbf{F}$ is Lipschitz continuous with a Lipschitz constant L_g, it follows for suitable vectors $\mathbf{U}_k, \; \mathbf{V}_k \in [a, b]^n, \; k \in \{1, 2\}$:

$$\begin{aligned}
\left\|\tilde{\mathbf{F}}(\mathbf{U}_1, \mathbf{U}_2) - \tilde{\mathbf{F}}(\mathbf{V}_1, \mathbf{V}_2)\right\| = \\
& \left\|\mathbf{F}\left(\frac{\mathbf{U}_1 + \mathbf{U}_2}{2} - \frac{\Delta t}{2\Delta z}\left(\mathbf{F}(\mathbf{U}_1) - \mathbf{F}(\mathbf{U}_2)\right)\right)\right. \\
& \left. -\mathbf{F}\left(\frac{\mathbf{U}_1 + \mathbf{U}_2}{2} - \frac{\Delta t}{2\Delta z}\left(\mathbf{F}(\mathbf{U}_1) - \mathbf{F}(\mathbf{U}_2)\right)\right)\right\| \\
& \leq L_g \left\| \frac{\mathbf{U}_1 + \mathbf{U}_2}{2} - \frac{\Delta t}{2\Delta z}\left(\mathbf{F}(\mathbf{U}_1) - \mathbf{F}(\mathbf{U}_2)\right)\right. \\
& \qquad \left. -\frac{\mathbf{V}_1 + \mathbf{V}_2}{2} - \frac{\Delta t}{2\Delta z}\left(\mathbf{F}(\mathbf{V}_1) - \mathbf{F}(\mathbf{V}_2)\right)\right\|
\end{aligned}$$

Applying the triangle inequality and the Lipschitz condition several times, we obtain:

$$\left\|\tilde{\mathbf{F}}(\mathbf{U}_1, \mathbf{U}_2) - \tilde{\mathbf{F}}(\mathbf{V}_1, \mathbf{V}_2)\right\| \leq \left(\frac{L_g}{2} + \frac{\Delta t}{2\Delta z} L_g\right) \|\mathbf{U}_1 - \mathbf{V}_1\| + \left(\frac{L_g}{2} + \frac{\Delta t}{2\Delta z} L_g\right) \|\mathbf{U}_2 - \mathbf{V}_2\| .$$

If there is $\gamma > 0$ such that $\Delta t = \gamma \Delta z$, it holds:

$$\left\|\tilde{\mathbf{F}}(\mathbf{U}_1, \mathbf{U}_2) - \tilde{\mathbf{F}}(\mathbf{V}_1, \mathbf{V}_2)\right\| \leq \left(\frac{L_g}{2} + \frac{\gamma L_g}{2}\right) (\|\mathbf{U}_1 - \mathbf{V}_1\| + \|\mathbf{U}_2 - \mathbf{V}_2\|) .$$

Thus, the Lax-Wendroff method has a numerical consistent flux, if $\mathbf{F}$ is Lipschitz continuous. Moreover, the method is conservative for $\mathbf{S} \equiv 0$ by Definition 4.2.

Consistency Error of the Lax-Wendroff Method

After proving that the Lax-Wendroff method has a consistent numerical flux, we want to determine the consistency order of the Lax-Wendroff method. For this purpose, we consider a sufficiently smooth solution $\mathbf{U}$ of (4.45) at a grid node $z_i = i\Delta z$ and $t_{j+1} = (j+1)\Delta t$. In the following, we will denote it by $\mathbf{U}_i^{j+1}$. A Taylor expansion around (z_i, t_{j+1}) yields:

$$\mathbf{U}_i^{j+1} = \mathbf{U}(z_i, t_j + \Delta t) = \mathbf{U}_i^j + \Delta t \frac{\partial \mathbf{U}}{\partial t}(z_i, t_j) + \frac{\Delta t^2}{2} \frac{\partial^2 \mathbf{U}}{\partial t^2}(z_i, t_j) + \mathcal{O}(\Delta t^3) . \tag{4.53}$$

Using the differential Eq. (4.45), the second partial derivative with respect to time can be represented as follows:

$$\begin{aligned} \frac{\partial^2 \mathbf{U}}{\partial t^2} &= -\frac{\partial^2 \mathbf{F}}{\partial t \partial z} + \frac{\partial \mathbf{S}}{\partial t} = \frac{\partial}{\partial z}\left(-\frac{\partial \mathbf{F}}{\partial \mathbf{U}} \frac{\partial \mathbf{U}}{\partial t}\right) + \frac{\partial \mathbf{S}}{\partial \mathbf{U}} \frac{\partial \mathbf{U}}{\partial t} \\ &= \frac{\partial}{\partial z}\left(\frac{\partial \mathbf{F}}{\partial \mathbf{U}}\left(\frac{\partial \mathbf{F}}{\partial z} - \mathbf{S}(\mathbf{U})\right)\right) + \frac{\partial \mathbf{S}}{\partial \mathbf{U}}\left(\mathbf{S}(\mathbf{U}) - \frac{\partial \mathbf{F}}{\partial z}\right) \\ &= \frac{\partial \mathbf{S}}{\partial \mathbf{U}} \mathbf{S}(\mathbf{U}) - \frac{\partial \mathbf{S}}{\partial \mathbf{U}} \frac{\partial \mathbf{F}}{\partial z} - \frac{\partial}{\partial z}\left(\frac{\partial \mathbf{F}}{\partial \mathbf{U}} \mathbf{S}(\mathbf{U})\right) + \frac{\partial}{\partial z}\left(\frac{\partial \mathbf{F}}{\partial \mathbf{U}} \frac{\partial \mathbf{F}}{\partial z}\right) . \end{aligned} \tag{4.54}$$

In the next step, we perform a Taylor expansion of $\mathbf{F}_{i+\frac{1}{2}}^{j+\frac{1}{2}}$ wrt. to $\frac{\mathbf{U}_{i+1}^j + \mathbf{U}_i^j}{2}$. For more information on Taylor expansions of a vector-valued function, we refer to Sect. 7.11:

$$
\begin{aligned}
\mathbf{F}_{i+\frac{1}{2}}^{j+\frac{1}{2}} &= \mathbf{F}\left(\frac{\mathbf{U}_{i+1}^{j}+\mathbf{U}_{i}^{j}}{2}-\frac{\Delta t}{2\Delta z}\left(\mathbf{F}_{i+1}^{j}-\mathbf{F}_{i}^{j}\right)+\frac{\Delta t}{4}\left(\mathbf{S}_{i+1}^{j}+\mathbf{S}_{i}^{j}\right)\right)\\
&= \mathbf{F}\left(\frac{\mathbf{U}_{i+1}^{j}+\mathbf{U}_{i}^{j}}{2}+\Delta\mathbf{U}_{i+\frac{1}{2}}^{j}\right)\\
&= \mathbf{F}\left(\frac{\mathbf{U}_{i+1}^{j}+\mathbf{U}_{i}^{j}}{2}\right)+\frac{\partial\mathbf{F}}{\partial\mathbf{U}}\left(\frac{\mathbf{U}_{i+1}^{j}+\mathbf{U}_{i}^{j}}{2}\right)\Delta\mathbf{U}_{i+\frac{1}{2}}^{j}\\
&\quad+\frac{1}{2}\mathbf{H}_{\mathbf{F}}\left(\frac{\mathbf{U}_{i+1}^{j}+\mathbf{U}_{i}^{j}}{2}\right)\left(\Delta\mathbf{U}_{i+\frac{1}{2}}^{j}\right)^{2}+\mathcal{O}\left(\left(\Delta\mathbf{U}_{i+\frac{1}{2}}^{j}\right)^{3}\right).
\end{aligned}
$$

The expressions $\left(\Delta\mathbf{U}_{i+\frac{1}{2}}^{j}\right)^{2}$ and $\left(\Delta\mathbf{U}_{i+\frac{1}{2}}^{j}\right)^{3}$ indicate that the vector $\Delta\mathbf{U}_{i+\frac{1}{2}}^{j}$ is componentwise squared or cubed. $\mathbf{H}_{\mathbf{F}}$ is a second order tensor containing the second derivatives of the flux function $\mathbf{F}$. Now we derive a Taylor approximation of the arithmetic mean $\frac{\mathbf{U}_{i+1}^{j}+\mathbf{U}_{i}^{j}}{2}$:

$$
\begin{aligned}
\mathbf{U}_{i+1}^{j} &= \mathbf{U}\left(z_i+\frac{\Delta z}{2}+\frac{\Delta z}{2},t_j\right)\\
&= \mathbf{U}\left(z_i+\frac{\Delta z}{2},t_j\right)+\frac{\Delta z}{2}\frac{\partial\mathbf{U}}{\partial z}\left(z_i+\frac{\Delta z}{2},t_j\right)+\frac{\Delta z^2}{8}\frac{\partial^2\mathbf{U}}{\partial z^2}\left(z_i+\frac{\Delta z}{2},t_j\right)\\
&\quad+\frac{\Delta z^3}{48}\frac{\partial^3\mathbf{U}}{\partial z^3}\left(z_i+\frac{\Delta z}{2},t_j\right)+\mathcal{O}\left(\Delta z^4\right).
\end{aligned}
$$

Analogous to that, one obtains:

$$
\begin{aligned}
\mathbf{U}_{i}^{j} &= \mathbf{U}\left(z_i+\frac{\Delta z}{2}-\frac{\Delta z}{2},t_j\right)\\
&= \mathbf{U}\left(z_i+\frac{\Delta z}{2},t_j\right)-\frac{\Delta z}{2}\frac{\partial\mathbf{U}}{\partial z}\left(z_i+\frac{\Delta z}{2},t_j\right)+\frac{\Delta z^2}{8}\frac{\partial^2\mathbf{U}}{\partial z^2}\left(z_i+\frac{\Delta z}{2},t_j\right)\\
&\quad-\frac{\Delta z^3}{48}\frac{\partial^3\mathbf{U}}{\partial z^3}\left(z_i+\frac{\Delta z}{2},t_j\right)+\mathcal{O}\left(\Delta z^4\right).
\end{aligned}
$$

Addition of both equations yields:

$$
\frac{\mathbf{U}_{i+1}^{j}+\mathbf{U}_{i}^{j}}{2}=\mathbf{U}_{i+\frac{1}{2}}^{j}+\frac{\Delta z^2}{8}\frac{\partial^2\mathbf{U}}{\partial z^2}\left(z_i+\frac{\Delta z}{2},t_j\right)+\mathcal{O}\left(\Delta z^4\right). \tag{4.55}
$$

A Taylor expansion of $\mathbf{F}$ yields:

$$\mathbf{F}\left(\frac{\mathbf{U}_{i+1}^{j}+\mathbf{U}_{i}^{j}}{2}\right)=\mathbf{F}\left(\mathbf{U}_{i+\frac{1}{2}}^{j}\right)+\frac{\Delta z^{2}}{8}\frac{\partial \mathbf{F}}{\partial \mathbf{U}}\left(\mathbf{U}_{i+\frac{1}{2}}^{j}\right)\frac{\partial^{2}\mathbf{U}_{i+\frac{1}{2}}^{j}}{\partial z^{2}}+\mathcal{O}\left(\Delta z^{4}\right) \quad (4.56)$$

Next we compute a first order Taylor expansion of

$$\frac{\partial \mathbf{F}}{\partial \mathbf{U}}\left(\frac{\mathbf{U}_{i+1}^{j}+\mathbf{U}_{i}^{j}}{2}\right)\Delta\mathbf{U}_{i+\frac{1}{2}}^{j}$$

about $\mathbf{U}_{i+\frac{1}{2}}^{j}$. It is given by:

$$\frac{\partial \mathbf{F}}{\partial \mathbf{U}}\left(\frac{\mathbf{U}_{i+1}^{j}+\mathbf{U}_{i}^{j}}{2}\right)\Delta\mathbf{U}_{i+\frac{1}{2}}^{j}=\frac{\partial \mathbf{F}}{\partial \mathbf{U}}\left(\mathbf{U}_{i+\frac{1}{2}}^{j}\right)\Delta\mathbf{U}_{i+\frac{1}{2}}^{j}+\mathcal{O}\left(\Delta z^{2}\Delta t\right), \quad (4.57)$$

since it holds:

$$\Delta\mathbf{U}_{i+\frac{1}{2}}^{j}=-\frac{\Delta t}{2}\left(\frac{\partial \mathbf{F}}{\partial z}\left(z_{i+\frac{1}{2}},t_{j}\right)-\mathbf{S}\left(z_{i+\frac{1}{2}},t_{j}\right)+\mathcal{O}\left(\Delta z^{2}\right)\right)=\mathcal{O}\left(\Delta t\right). \quad (4.58)$$

In the following, we assume that $\Delta t=\gamma\Delta z$ ($\gamma>0$ fixed), from which follows:

$$\Delta\mathbf{U}_{i+\frac{1}{2}}^{j}=\mathcal{O}\left(\Delta z\right).$$

A combination of (4.57) and (4.58) yields:

$$\begin{aligned}\frac{\partial \mathbf{F}}{\partial \mathbf{U}}\left(\frac{\mathbf{U}_{i+1}^{j}+\mathbf{U}_{i}^{j}}{2}\right)\Delta\mathbf{U}_{i+\frac{1}{2}}^{j}&=-\frac{\Delta t}{2}\frac{\partial \mathbf{F}}{\partial \mathbf{U}}\left(\mathbf{U}_{i+\frac{1}{2}}^{j}\right)\frac{\partial \mathbf{F}}{\partial z}\left(z_{i+\frac{1}{2}},t_{j}\right)\\&\quad+\frac{\Delta t}{2}\frac{\partial \mathbf{F}}{\partial \mathbf{U}}\left(\mathbf{U}_{i+\frac{1}{2}}^{j}\right)\mathbf{S}\left(z_{i+\frac{1}{2}},t_{j}\right)+\mathcal{O}\left(\Delta t\Delta z^{2}\right).\end{aligned}$$

Finally, we obtain by using similar arguments as for (4.57):

$$\mathbf{H}_{\mathbf{F}}\left(\frac{\mathbf{U}_{i+1}^{j}+\mathbf{U}_{i}^{j}}{2}\right)\left(\Delta\mathbf{U}_{i+\frac{1}{2}}^{j}\right)^{2}=\mathbf{H}_{\mathbf{F}}\left(\mathbf{U}_{i+\frac{1}{2}}^{j}\right)\left(\Delta\mathbf{U}_{i+\frac{1}{2}}^{j}\right)^{2}+\mathcal{O}\left(\Delta t^{2}\Delta z^{2}\right).$$

As for the flux function $\mathbf{F}_{i+\frac{1}{2}}^{j+\frac{1}{2}}$ and the arithmetic mean $\frac{\mathbf{U}_{i+1}^{j}+\mathbf{U}_{i}^{j}}{2}$ one can compute Taylor expansions of $\mathbf{F}_{i-\frac{1}{2}}^{j+\frac{1}{2}}$ and $\frac{\mathbf{U}_{i}^{j}+\mathbf{U}_{i-1}^{j}}{2}$ about $\mathbf{U}_{i-\frac{1}{2}}^{j}$. All in all, one obtains for the spatial derivative the following representation:

$$\begin{aligned}\frac{\mathbf{F}_{i+\frac{1}{2}}^{j+\frac{1}{2}}-\mathbf{F}_{i-\frac{1}{2}}^{j+\frac{1}{2}}}{\Delta z}&=\frac{\mathbf{F}\left(\frac{\mathbf{U}_{i+1}^j+\mathbf{U}_i^j}{2}\right)-\mathbf{F}\left(\frac{\mathbf{U}_i^j+\mathbf{U}_{i-1}^j}{2}\right)}{\Delta z}\\&+\frac{\frac{\partial \mathbf{F}}{\partial \mathbf{U}}\left(\frac{\mathbf{U}_{i+1}^j+\mathbf{U}_i^j}{2}\right)\Delta\mathbf{U}_{i+\frac{1}{2}}^j-\frac{\partial \mathbf{F}}{\partial \mathbf{U}}\left(\frac{\mathbf{U}_i^j+\mathbf{U}_{i-1}^j}{2}\right)\Delta\mathbf{U}_{i-\frac{1}{2}}^j}{\Delta z}\\&+\frac{1}{2}\frac{\mathbf{H}_{\mathbf{F}}\left(\frac{\mathbf{U}_{i+1}^j+\mathbf{U}_i^j}{2}\right)\left(\Delta\mathbf{U}_{i+\frac{1}{2}}^j\right)^2-\mathbf{H}_{\mathbf{F}}\left(\frac{\mathbf{U}_i^j+\mathbf{U}_{i-1}^j}{2}\right)\left(\Delta\mathbf{U}_{i-\frac{1}{2}}^j\right)^2}{\Delta z}\\&+\mathcal{O}\left(\left(\Delta\mathbf{U}_{i+\frac{1}{2}}^j\right)^3/\Delta z\right)+\mathcal{O}\left(\left(\Delta\mathbf{U}_{i-\frac{1}{2}}^j\right)^3/\Delta z\right).\end{aligned}$$

From Eq. (4.56) and a similar formula for $z=z_{i-\frac{1}{2}}$, it follows:

$$\begin{aligned}\frac{\mathbf{F}\left(\frac{\mathbf{U}_{i+1}^j+\mathbf{U}_i^j}{2}\right)-\mathbf{F}\left(\frac{\mathbf{U}_i^j+\mathbf{U}_{i-1}^j}{2}\right)}{\Delta z}&=\frac{\partial \mathbf{F}}{\partial z}(z_i,t_j)\\&+\frac{\Delta z^2}{8}\frac{\partial}{\partial z}\left(\frac{\partial \mathbf{F}}{\partial \mathbf{U}}(\mathbf{U})\frac{\partial^2\mathbf{U}}{\partial z^2}\right)(z_i,t_j)+\mathcal{O}\left(\Delta z^2\right)\\&=\frac{\partial \mathbf{F}}{\partial z}(z_i,t_j)+\mathcal{O}\left(\Delta z^2\right).\end{aligned}$$

For the next summand one obtains with $\Delta t=\gamma\Delta z$ ($\gamma>0$ fixed):

$$\begin{aligned}&\frac{\frac{\partial \mathbf{F}}{\partial \mathbf{U}}\left(\frac{\mathbf{U}_{i+1}^j+\mathbf{U}_i^j}{2}\right)\Delta\mathbf{U}_{i+\frac{1}{2}}^j-\frac{\partial \mathbf{F}}{\partial \mathbf{U}}\left(\frac{\mathbf{U}_i^j+\mathbf{U}_{i-1}^j}{2}\right)\Delta\mathbf{U}_{i-\frac{1}{2}}^j}{\Delta z}=\\&=-\frac{\Delta t}{2}\frac{\partial}{\partial z}\left(\frac{\partial \mathbf{F}}{\partial \mathbf{U}}\frac{\partial \mathbf{F}}{\partial z}\right)(z_i,t_j)+\frac{\Delta t}{2}\frac{\partial}{\partial z}\left(\frac{\partial \mathbf{F}}{\partial \mathbf{U}}\mathbf{S}(\mathbf{U})\right)(z_i,t_j)+\mathcal{O}\left(\Delta z^2\right).\end{aligned}$$

Moreover, it holds that:

$$\mathcal{O}\left(\left(\Delta\mathbf{U}_{i+\frac{1}{2}}^j\right)^3/\Delta z\right)=\mathcal{O}\left(\left(\Delta\mathbf{U}_{i-\frac{1}{2}}^j\right)^3/\Delta z\right)=\mathcal{O}\left(\Delta z^2\right)$$

and

$$\begin{aligned}&\frac{\mathbf{H}_{\mathbf{F}}\left(\frac{\mathbf{U}_{i+1}^j+\mathbf{U}_i^j}{2}\right)\left(\Delta\mathbf{U}_{i+\frac{1}{2}}^j\right)^2-\mathbf{H}_{\mathbf{F}}\left(\frac{\mathbf{U}_i^j+\mathbf{U}_{i-1}^j}{2}\right)\left(\Delta\mathbf{U}_{i-\frac{1}{2}}^j\right)^2}{\Delta z}=\\&\qquad=\frac{\partial}{\partial z}\left(\mathbf{H}_{\mathbf{F}}(\Delta\mathbf{U})^2\right)(z_i,t_j)+\mathcal{O}\left(\Delta z^2\right)=\mathcal{O}\left(\Delta z^2\right).\end{aligned}$$

All in all, the spatial approximation of $\mathbf{F}$ is given by:

$$\begin{aligned}\frac{\mathbf{F}_{i+\frac{1}{2}}^{j+\frac{1}{2}} - \mathbf{F}_{i-\frac{1}{2}}^{j+\frac{1}{2}}}{\Delta z} &= \frac{\partial \mathbf{F}}{\partial z}\left(z_i, t_j\right) - \frac{\Delta t}{2}\frac{\partial}{\partial z}\left(\frac{\partial \mathbf{F}}{\partial \mathbf{U}}\frac{\partial \mathbf{F}}{\partial z}\right)\left(z_i, t_j\right)\\ &+ \frac{\Delta t}{2}\frac{\partial}{\partial z}\left(\frac{\partial \mathbf{F}}{\partial \mathbf{U}}\mathbf{S}\left(\mathbf{U}\right)\right)\left(z_i, t_j\right) + \mathcal{O}\left(\Delta z^2\right). \end{aligned} \tag{4.59}$$

To complete the Lax-Wendroff method, it remains to consider the term

$$\frac{1}{2}\left(\mathbf{S}_{i+\frac{1}{2}}^{j+\frac{1}{2}} + \mathbf{S}_{i-\frac{1}{2}}^{j+\frac{1}{2}}\right)$$

For the source term it holds:

$$\begin{aligned}\mathbf{S}_{i+\frac{1}{2}}^{j+\frac{1}{2}} &= \mathbf{S}\left(\frac{\mathbf{U}_{i+1}^j + \mathbf{U}_i^j}{2} - \frac{\Delta t}{2\Delta z}\left(\mathbf{F}_{i+1}^j - \mathbf{F}_i^j\right) + \frac{\Delta t}{4}\left(\mathbf{S}_{i+1}^j + \mathbf{S}_i^j\right)\right)\\ &= \mathbf{S}\left(\frac{\mathbf{U}_{i+1}^j + \mathbf{U}_i^j}{2} + \Delta\mathbf{U}_{i+\frac{1}{2}}^j\right)\\ &= \mathbf{S}\left(\frac{\mathbf{U}_{i+1}^j + \mathbf{U}_i^j}{2}\right) + \frac{\partial \mathbf{S}}{\partial \mathbf{U}}\left(\frac{\mathbf{U}_{i+1}^j + \mathbf{U}_i^j}{2}\right)\Delta\mathbf{U}_{i+\frac{1}{2}}^j + \mathcal{O}\left(\left(\Delta\mathbf{U}_{i+\frac{1}{2}}^j\right)^2\right)\end{aligned}$$

Using (4.55) we obtain:

$$\begin{aligned}\mathbf{S}\left(\frac{\mathbf{U}_{i+1}^j + \mathbf{U}_i^j}{2}\right) &= \mathbf{S}\left(\mathbf{U}_{i+\frac{1}{2}}^j\right) + \frac{\Delta z^2}{2}\frac{\partial \mathbf{S}}{\partial \mathbf{U}}\left(\mathbf{U}_{i+\frac{1}{2}}^j\right)\frac{\partial^2 \mathbf{U}_{i+\frac{1}{2}}^j}{\partial z^2} + \mathcal{O}\left(\Delta z^4\right)\\ &= \mathbf{S}\left(\mathbf{U}_{i+\frac{1}{2}}^j\right) + \mathcal{O}\left(\Delta z^2\right).\end{aligned}$$

For the second summand a similar expansion can be shown analogous to (4.57):

$$\begin{aligned}&\frac{\partial \mathbf{S}}{\partial \mathbf{U}}\left(\frac{\mathbf{U}_{i+1}^j + \mathbf{U}_i^j}{2}\right)\Delta\mathbf{U}_{i+\frac{1}{2}}^j =\\ &= -\frac{\Delta t}{2}\frac{\partial \mathbf{S}}{\partial \mathbf{U}}\left(\mathbf{U}_{i+\frac{1}{2}}^j\right)\left(\frac{\partial \mathbf{F}}{\partial z}\left(z_{i+\frac{1}{2}}, t_j\right) - \mathbf{S}\left(z_{i+\frac{1}{2}}, t_j\right) + \mathcal{O}\left(\Delta z^2\right)\right) + \mathcal{O}\left(\Delta z^2\right)\\ &= -\frac{\Delta t}{2}\frac{\partial \mathbf{S}}{\partial \mathbf{U}}\left(\mathbf{U}_{i+\frac{1}{2}}^j\right)\left(\frac{\partial \mathbf{F}}{\partial z}\left(z_{i+\frac{1}{2}}, t_j\right) - \mathbf{S}\left(z_{i+\frac{1}{2}}, t_j\right)\right) + \mathcal{O}\left(\Delta z^2\right).\end{aligned}$$

This results in the following representation for $\mathbf{S}_{i+\frac{1}{2}}^{j+\frac{1}{2}}$:

$$\mathbf{S}_{i+\frac{1}{2}}^{j+\frac{1}{2}} = \mathbf{S}\left(\mathbf{U}_{i+\frac{1}{2}}^{j}\right) - \frac{\Delta t}{2}\frac{\partial \mathbf{S}}{\partial \mathbf{U}}\frac{\partial \mathbf{F}}{\partial z}\left(z_{i+\frac{1}{2}}, t_j\right) + \frac{\Delta t}{2}\frac{\partial \mathbf{S}}{\partial \mathbf{U}}\mathbf{S}\left(z_{i+\frac{1}{2}}, t_j\right) + \mathcal{O}\left(\Delta z^2\right).$$

For $\mathbf{S}_{i-\frac{1}{2}}^{j+\frac{1}{2}}$ the expansion reads as follows:

$$\mathbf{S}_{i-\frac{1}{2}}^{j+\frac{1}{2}} = \mathbf{S}\left(\mathbf{U}_{i-\frac{1}{2}}^{j}\right) - \frac{\Delta t}{2}\frac{\partial \mathbf{S}}{\partial \mathbf{U}}\frac{\partial \mathbf{F}}{\partial z}\left(z_{i-\frac{1}{2}}, t_j\right) + \frac{\Delta t}{2}\frac{\partial \mathbf{S}}{\partial \mathbf{U}}\mathbf{S}\left(z_{i-\frac{1}{2}}, t_j\right) + \mathcal{O}\left(\Delta z^2\right).$$

Combining both expansions, we have:

$$\frac{1}{2}\left(\mathbf{S}_{i+\frac{1}{2}}^{j+\frac{1}{2}} + \mathbf{S}_{i-\frac{1}{2}}^{j+\frac{1}{2}}\right) = \mathbf{S}\left(\mathbf{U}_i^j\right) - \frac{\Delta t}{2}\frac{\partial \mathbf{S}}{\partial \mathbf{U}}\frac{\partial \mathbf{F}}{\partial z}\left(z_i, t_j\right) + \frac{\Delta t}{2}\frac{\partial \mathbf{S}}{\partial \mathbf{U}}\mathbf{S}\left(z_i, t_j\right) + \mathcal{O}\left(\Delta z^2\right). \tag{4.60}$$

Inserting (4.59) and (4.60) into (4.52), the grid transfer function for the Lax-Wendroff method is given by:

$$\begin{aligned} H\left(\mathbf{U}_{i-1}^j, \mathbf{U}_i^j, \mathbf{U}_{i+1}^j\right) = \mathbf{U}_i^j &- \Delta t\frac{\partial \mathbf{F}}{\partial z}\left(z_i, t_j\right) + \frac{\Delta t^2}{2}\frac{\partial}{\partial z}\left(\frac{\partial \mathbf{F}}{\partial \mathbf{U}}\frac{\partial \mathbf{F}}{\partial z}\right)\left(z_i, t_j\right) \\ &- \frac{\Delta t^2}{2}\frac{\partial}{\partial z}\left(\frac{\partial \mathbf{F}}{\partial \mathbf{U}}\mathbf{S}\left(\mathbf{U}\right)\right)\left(z_i, t_j\right) + \Delta t\mathbf{S}\left(\mathbf{U}_i^j\right) \\ &- \frac{\Delta t^2}{2}\frac{\partial \mathbf{S}}{\partial \mathbf{U}}\frac{\partial \mathbf{F}}{\partial z}\left(z_i, t_j\right) + \frac{\Delta t^2}{2}\frac{\partial \mathbf{S}}{\partial \mathbf{U}}\mathbf{S}\left(z_i, t_j\right) + \mathcal{O}\left(\Delta z^2 \Delta t\right) \end{aligned} \tag{4.61}$$

On the other hand it results from (4.53) and (4.54):

$$\begin{aligned} \mathbf{U}_i^{j+1} &= \mathbf{U}\left(z_i, t_j + \Delta t\right) = \mathbf{U}_i^j + \Delta t\frac{\partial \mathbf{U}}{\partial t}\left(z_i, t_j\right) + \frac{\Delta t^2}{2}\frac{\partial^2 \mathbf{U}}{\partial t^2}\left(z_i, t_j\right) + \mathcal{O}\left(\Delta t^3\right) \\ &= \mathbf{U}_i^j + \Delta t\frac{\partial \mathbf{U}}{\partial t}\left(z_i, t_j\right) + \frac{\Delta t^2}{2}\frac{\partial \mathbf{S}}{\partial \mathbf{U}}\mathbf{S}\left(z_i, t_j\right) - \frac{\Delta t^2}{2}\frac{\partial \mathbf{S}}{\partial \mathbf{U}}\frac{\partial \mathbf{F}}{\partial z}\left(z_i, t_j\right) \\ &\quad - \frac{\Delta t^2}{2}\frac{\partial}{\partial z}\left(\frac{\partial \mathbf{F}}{\partial \mathbf{U}}\mathbf{S}\left(\mathbf{U}\right)\right)\left(z_i, t_j\right) + \frac{\Delta t^2}{2}\frac{\partial}{\partial z}\left(\frac{\partial \mathbf{F}}{\partial \mathbf{U}}\frac{\partial \mathbf{F}}{\partial z}\right)\left(z_i, t_j\right) + \mathcal{O}\left(\Delta t^3\right). \end{aligned} \tag{4.62}$$

By means of (4.61) and (4.62) one can show that the consistency error is of second order in space and time:

$$\begin{aligned} \frac{\mathbf{U}_i^{j+1} - H\left(\mathbf{U}_{i-1}^j, \mathbf{U}_i^j, \mathbf{U}_{i+1}^j\right)}{\Delta t} &= \underbrace{\frac{\partial \mathbf{U}}{\partial t}\left(z_i, t_j\right) + \frac{\partial \mathbf{F}}{\partial z}\left(z_i, t_j\right) - \mathbf{S}\left(z_i, t_j\right)}_{=0} \\ &\quad + \mathcal{O}\left(\Delta z^2\right) + \mathcal{O}\left(\Delta t^2\right) = \mathcal{O}\left(\Delta z^2\right) + \mathcal{O}\left(\Delta t^2\right). \end{aligned} \tag{4.63}$$

The equation

$$\frac{\partial \mathbf{U}}{\partial t}\left(z_i, t_j\right) + \frac{\partial \mathbf{F}}{\partial z}\left(z_i, t_j\right) - \mathbf{S}\left(z_i, t_j\right) = 0$$

holds, since we assumed that $\mathbf{U}$ is a smooth solution of the Cauchyproblem (4.45).

Stability of the Lax-Wendroff Method

In order to investigate the numerical stability of the Lax-Wendroff method, we use von Neumann's stability criterion. The following steps are similar to the stability analysis discussed in [1, Sect. 2].

According to Definition 4.7 we have to compute the amplification matrix G of the Lax-Wendroff method. For this purpose we consider (4.52) at a grid node $\left(z_l, t_j\right)$. Here, only the homogeneous case i.e., $\mathbf{S} \equiv \mathbf{0}$ is investigated. In order to linearize (4.52) the non-linear fluxes $\mathbf{F}_{l+\frac{1}{2}}^{j+\frac{1}{2}}$ and $\mathbf{F}_{l+\frac{1}{2}}^{j-\frac{1}{2}}$ are approximated by a linear Taylorpolynom wrt. the vector $\mathbf{U}_l^j$. For this purpose we use the Taylor expansion of vector-valued functions (see Sect. 7.11):

$$\begin{aligned}
\mathbf{F}_{l+\frac{1}{2}}^{j+\frac{1}{2}} &= \mathbf{F}\left(\mathbf{U}_{l+\frac{1}{2}}^{j+\frac{1}{2}}\right) = \mathbf{F}\left(\mathbf{U}_l^j + \left(\mathbf{U}_{l+\frac{1}{2}}^{j+\frac{1}{2}} - \mathbf{U}_l^j\right)\right) \\
&= \mathbf{F}\left(\mathbf{U}_l^j\right) + \frac{\partial \mathbf{F}}{\partial \mathbf{U}}\left(\mathbf{U}_l^j\right)\left(\mathbf{U}_{l+\frac{1}{2}}^{j+\frac{1}{2}} - \mathbf{U}_l^j\right) + \frac{1}{2}\mathbf{H}_{\mathbf{F}}\left(\tilde{\mathbf{U}}_{l+\frac{1}{2}}^j\right)\left(\mathbf{U}_{l+\frac{1}{2}}^{j+\frac{1}{2}} - \mathbf{U}_l^j\right)^2 .
\end{aligned}$$

The vector $\tilde{\mathbf{U}}_{l+\frac{1}{2}}^j$ is consisting of intermediate values of the components of $\mathbf{U}_l^j$ and $\mathbf{U}_l^j + \left(\mathbf{U}_{l+\frac{1}{2}}^{j+\frac{1}{2}} - \mathbf{U}_l^j\right)$. Analogous to that, we obtain for $\mathbf{F}_{l-\frac{1}{2}}^{j+\frac{1}{2}}$:

$$\begin{aligned}
\mathbf{F}_{l-\frac{1}{2}}^{j+\frac{1}{2}} &= \mathbf{F}\left(\mathbf{U}_{l-\frac{1}{2}}^{j+\frac{1}{2}}\right) = \mathbf{F}\left(\mathbf{U}_l^j + \left(\mathbf{U}_{l-\frac{1}{2}}^{j-\frac{1}{2}} - \mathbf{U}_l^j\right)\right) \\
&= \mathbf{F}\left(\mathbf{U}_l^j\right) + \frac{\partial \mathbf{F}}{\partial \mathbf{U}}\left(\mathbf{U}_l^j\right)\left(\mathbf{U}_{l-\frac{1}{2}}^{j+\frac{1}{2}} - \mathbf{U}_l^j\right) + \frac{1}{2}\mathbf{H}_{\mathbf{F}}\left(\tilde{\mathbf{U}}_{l-\frac{1}{2}}^j\right)\left(\mathbf{U}_{l-\frac{1}{2}}^{j+\frac{1}{2}} - \mathbf{U}_l^j\right)^2 .
\end{aligned}$$

Using (4.52) it follows:

$$\begin{aligned}
\mathbf{U}_l^{j+1} &= \mathbf{U}_l^j - \frac{\Delta t}{\Delta z}\left(\mathbf{F}_{l+\frac{1}{2}}^{j+\frac{1}{2}} - \mathbf{F}_{l-\frac{1}{2}}^{j+\frac{1}{2}}\right) = \\
&= \mathbf{U}_l^j - \frac{\Delta t}{\Delta z}\frac{\partial \mathbf{F}}{\partial \mathbf{U}}\left(\mathbf{U}_l^j\right)\left(\frac{\mathbf{U}_{l+1}^j - \mathbf{U}_{l-1}^j}{2} - \frac{\Delta t}{2\Delta z}\left(\mathbf{F}_{l+1}^j - 2\mathbf{F}_l^j + \mathbf{F}_{l-1}^j\right)\right) \\
&\quad + \frac{\Delta t}{2\Delta z}\mathbf{H}_{\mathbf{F}}\left(\tilde{\mathbf{U}}_{l+\frac{1}{2}}^j\right)\left(\mathbf{U}_{l+\frac{1}{2}}^{j+\frac{1}{2}} - \mathbf{U}_l^j\right)^2 - \frac{\Delta t}{2\Delta z}\mathbf{H}_{\mathbf{F}}\left(\tilde{\mathbf{U}}_{l-\frac{1}{2}}^j\right)\left(\mathbf{U}_{l-\frac{1}{2}}^{j+\frac{1}{2}} - \mathbf{U}_l^j\right)^2 .
\end{aligned} \tag{4.64}$$

Moreover it holds:

$$\begin{aligned}\mathbf{F}_{l+1}^{j} &= \mathbf{F}\left(\mathbf{U}_{l+1}^{j}\right) = \mathbf{F}\left(\mathbf{U}_{l}^{j} + \left(\mathbf{U}_{l+1}^{j} - \mathbf{U}_{l}^{j}\right)\right) \\ &= \mathbf{F}\left(\mathbf{U}_{l}^{j}\right) + \frac{\partial \mathbf{F}}{\partial \mathbf{U}}\left(\mathbf{U}_{l}^{j}\right)\left(\mathbf{U}_{l+1}^{j} - \mathbf{U}_{l}^{j}\right) + \frac{1}{2}\mathbf{H}_{\mathbf{F}}\left(\tilde{\mathbf{U}}_{l+1}^{j}\right)\left(\mathbf{U}_{l+1}^{j} - \mathbf{U}_{l}^{j}\right)^{2}\end{aligned}$$

and

$$\begin{aligned}\mathbf{F}_{l-1}^{j} &= \mathbf{F}\left(\mathbf{U}_{l-1}^{j}\right) = \mathbf{F}\left(\mathbf{U}_{l}^{j} + \left(\mathbf{U}_{l-1}^{j} - \mathbf{U}_{l}^{j}\right)\right) \\ &= \mathbf{F}\left(\mathbf{U}_{l}^{j}\right) + \frac{\partial \mathbf{F}}{\partial \mathbf{U}}\left(\mathbf{U}_{l}^{j}\right)\left(\mathbf{U}_{l-1}^{j} - \mathbf{U}_{l}^{j}\right) + \frac{1}{2}\mathbf{H}_{\mathbf{F}}\left(\tilde{\mathbf{U}}_{l-1}^{j}\right)\left(\mathbf{U}_{l-1}^{j} - \mathbf{U}_{l}^{j}\right)^{2}.\end{aligned}$$

Combined with (4.64) one obtains:

$$\begin{aligned}\mathbf{U}_{l}^{j+1} &= \mathbf{U}_{l}^{j} - \frac{\Delta t}{\Delta z}\frac{\partial \mathbf{F}}{\partial \mathbf{U}}\left(\mathbf{U}_{l}^{j}\right)\left(\frac{\mathbf{U}_{l+1}^{j} - \mathbf{U}_{l-1}^{j}}{2}\right) \\ &+ \frac{\Delta t^{2}}{2\Delta z^{2}}\left(\frac{\partial \mathbf{F}}{\partial \mathbf{U}}\left(\mathbf{U}_{l}^{j}\right)\right)^{2}\left(\mathbf{U}_{l+1}^{j} - 2\mathbf{U}_{l}^{j} + \mathbf{U}_{l-1}^{j}\right) \\ &+ \frac{\Delta t^{2}}{4\Delta z^{2}}\frac{\partial \mathbf{F}}{\partial \mathbf{U}}\left(\mathbf{U}_{l}^{j}\right)\mathbf{H}_{\mathbf{F}}\left(\tilde{\mathbf{U}}_{l+1}^{j}\right)\left(\mathbf{U}_{l+1}^{j} - \mathbf{U}_{l}^{j}\right)^{2} \\ &+ \frac{\Delta t^{2}}{4\Delta z^{2}}\frac{\partial \mathbf{F}}{\partial \mathbf{U}}\left(\mathbf{U}_{l}^{j}\right)\mathbf{H}_{\mathbf{F}}\left(\tilde{\mathbf{U}}_{l-1}^{j}\right)\left(\mathbf{U}_{l-1}^{j} - \mathbf{U}_{l}^{j}\right)^{2} \\ &+ \frac{\Delta t}{2\Delta z}\mathbf{H}_{\mathbf{F}}\left(\tilde{\mathbf{U}}_{l+\frac{1}{2}}^{j}\right)\left(\mathbf{U}_{l+\frac{1}{2}}^{j+\frac{1}{2}} - \mathbf{U}_{l}^{j}\right)^{2} - \frac{\Delta t}{2\Delta z}\mathbf{H}_{\mathbf{F}}\left(\tilde{\mathbf{U}}_{l-\frac{1}{2}}^{j}\right)\left(\mathbf{U}_{l-\frac{1}{2}}^{j+\frac{1}{2}} - \mathbf{U}_{l}^{j}\right)^{2}.\end{aligned} \tag{4.65}$$

By means of (4.65), the linearized part H_L of the grid transfer function H is given by:

$$\begin{aligned}H_L\left(\mathbf{U}_{l+1}^{j}, \mathbf{U}_{l}^{j}, \mathbf{U}_{l-1}^{j}\right) &= \mathbf{U}_{l}^{j} - \frac{\Delta t}{\Delta z}\frac{\partial \mathbf{F}}{\partial \mathbf{U}}\left(\mathbf{U}_{l}^{j}\right)\left(\frac{\mathbf{U}_{l+1}^{j} - \mathbf{U}_{l-1}^{j}}{2}\right) \\ &+ \frac{\Delta t^{2}}{2\Delta z^{2}}\left(\frac{\partial \mathbf{F}}{\partial \mathbf{U}}\left(\mathbf{U}_{l}^{j}\right)\right)^{2}\left(\mathbf{U}_{l+1}^{j} - 2\mathbf{U}_{l}^{j} + \mathbf{U}_{l-1}^{j}\right).\end{aligned} \tag{4.66}$$

Further Taylor expansions show that the non-linear terms in (4.65) are contained in $\mathcal{O}(\Delta z)$. Now we consider for t_j and t_{j+1} the respective Fourier series:

$$\mathbf{U}^{j+1}(z) = \sum_{k=-\infty}^{\infty} \xi_k^{j+1} e^{ikz} \text{ und } \mathbf{U}^{j}(z) = \sum_{k=-\infty}^{\infty} \xi_k^{j} e^{ikz}.$$

Inserting the Fourier series in (4.66), it results for the differents components of the series:

$$\xi_k^{j+1} e^{ikl\Delta z} = \xi_k^j e^{ikl\Delta z} - \frac{\Delta t}{\Delta z}\frac{\partial \mathbf{F}}{\partial \mathbf{U}}\left(\mathbf{U}_l^j\right)\left(\frac{\xi_k^j e^{ik(l+1)\Delta z} - \xi_k^j e^{ik(l-1)\Delta z}}{2}\right)$$
$$+ \frac{\Delta t^2}{2\Delta z^2}\left(\frac{\partial \mathbf{F}}{\partial \mathbf{U}}\left(\mathbf{U}_l^j\right)\right)^2\left(\xi_k^j e^{ik(l+1)\Delta z} - 2\xi_k^j e^{ikl\Delta z} + \xi_k^j e^{ik(l-1)\Delta z}\right).$$

This relationship can be written down in compact form as follows:

$$\xi_k^{j+1} e^{ikl\Delta z} = G \cdot \xi_k^j e^{ikl\Delta z},$$

where we denote by G the amplification matrix:

$$G = I - i\sin(k\Delta z)\frac{\Delta t}{\Delta z}\frac{\partial \mathbf{F}}{\partial \mathbf{U}}\left(\mathbf{U}_l^j\right) + (\cos(k\Delta z) - 1)\frac{\Delta t^2}{\Delta z^2}\left(\frac{\partial \mathbf{F}}{\partial \mathbf{U}}\left(\mathbf{U}_l^j\right)\right)^2.$$

To determine the norm of the amplification matrix, we compute the absolute values of the eigenvalues of G. Since $\mathbf{F}$ is the flux function of a hyperbolic partial differential equation system, the corresponding Jacobian matrix is diagonalizable. We denote the corresponding eigenvalues by $\mu_1, \ldots, \mu_n$. By this, the absolute values of the eigenvalues $\lambda_1, \ldots, \lambda_n$ of G are given by [11]:

$$\begin{aligned}
|\lambda_i|^2 &= \left(1 - (1 - \cos(k\Delta z))\frac{\Delta t^2}{\Delta z^2}\mu_i^2\right)^2 + \left(\sin(k\Delta z)\frac{\Delta t}{\Delta z}\mu_i\right)^2 \\
&= 1 - 2(1 - \cos(k\Delta z))\frac{\Delta t^2}{\Delta z^2}\mu_i^2 + (1 - \cos(k\Delta z))^2\frac{\Delta t^4}{\Delta z^4}\mu_i^4 \\
&+ \left(1 - \cos^2(k\Delta z)\right)\frac{\Delta t^2}{\Delta z^2}\mu_i^2 \\
&= 1 - \left(1 - \cos^2(k\Delta z)\right)^2\left(\frac{\Delta t^2}{\Delta z^2}\mu_i^2 - \frac{\Delta t^4}{\Delta z^4}\mu_i^4\right) \\
&= 1 - \left(1 - \cos^2(k\Delta z)\right)^2\frac{\Delta t^2}{\Delta z^2}\mu_i^2\left(1 - \frac{\Delta t^2}{\Delta z^2}\mu_i^2\right).
\end{aligned}$$

Thus it holds:

$$|\lambda_i|^2 < 1 \Leftrightarrow \frac{\Delta t^2}{\Delta z^2}\mu_i^2 < 1,\ i \in \{1, \ldots, n\}$$

or

$$|\lambda_i| < 1 \Leftrightarrow \frac{\Delta t}{\Delta z}|\mu_i| < 1,\ i \in \{1, \ldots, n\}. \tag{4.67}$$

The inequality in (4.67) means that the spatial and temporal grid size can not be chosen independently from each other. Such a time step restriction is characteristic for explicit Onestep-methods and is referred to as CFL condition (Courant-Friedrichs-Lewy condition) in literature. However, this restriction is only a necessary, but not a sufficient condition for the numerical stability of the Lax-Wendroff method. Due the formation of shocks fronts there might be regions, in which the solution is no longer smooth enough. In the vicinity of shock fronts the Lax-Wendroff method is prone to numerical instabilities (see Sect. 4.4.2, Experiment 1)

Convergence Behavior of the Lax-Wendroff Method

Up to now, we have shown that the Lax-Wendroff method is an explicit, consistent, and conservative Finite Difference method. It has also been shown that the consistency order in both space and time is 2. The numerical stability of the method could be investigated using the von Neumann stability analysis. As expected, we obtained a stability condition depending on the choice of the spatial grid width and time stepsize.

However, all these observations do not guarantee that the Lax-Wendroff method is convergent for non-linear hyperbolic differential equation systems. A mathematical convergence analysis of the non-linear Lax-Wendroff method would still require further mathematical tools. Therefore, we refer here to the corresponding literature [2, 12].

4.3.3 Analysis of the Implicit Finite Volume Method

Discretizing the transport Eq. (3.24):

$$\frac{\partial \Gamma}{\partial t} + \frac{\partial}{\partial z}\left(\omega \frac{Q\Gamma}{A}\right) = 0,\ z \in \mathbb{R},\ t > 0,$$
$$\Gamma(z, 0) = \Gamma_0(z),\ z \in \mathbb{R}, \tag{4.68}$$

we have used the implicit Finite Volume method from Sect. 4.1. In this section, we now analyze the mathematical properties of this method. In particular, we show the convergence of the method. For this purpose we assume that the initial condition Γ_0 has a compact support i.e., there is an interval I such that:

$$\Gamma_0(z) = \begin{cases} \Gamma_0(z),\ z \in I, \\ 0.0,\ z \notin I. \end{cases}$$

Furthermore, we assume that the solution Γ is contained in $[0, L] \subset \mathbb{R}$ for the whole time period $[0, T]$, such that we can assume homogeneous boundary conditions for $t \in [0, T]$:

$$\Gamma(0, t) = \Gamma(L, t) = 0,\ t \in [0, T].$$

The interval $[0, L]$ is discretized with $N + 1$ equidistant grid nodes $z_i = i \cdot \Delta z$. For the time integration, the time step size Δt is used. According to (4.3)

$$\Gamma_i^{n+1} = \Gamma_i^n - \frac{\Delta t}{\Delta z} \left(\omega \frac{Q_{i+1}^{n+1}}{A_{i+1}^{n+1}} \Gamma_{up}(z_{i+1}, t_{n+1}) - \omega \frac{Q_i^{n+1}}{A_i^{n+1}} \Gamma_{up}(z_i, t_{n+1}) \right)$$

yields a linear system of Eq. (4.4) to determine the unknowns Γ^{n+1} at $t = t_{n+1}$:

$$A_\Gamma^{n+1} \Gamma^{n+1} = \mathbf{b}_\Gamma^n.$$

Since we consider homogeneous boundary conditions, $\mathbf{b}_\Gamma^n = \Gamma^n$ holds. All in all, the method can be considered as an Implicit Finite Difference method. Now we study the discretization error

$$e(t_{n+1}) = e_{n+1} = \left\| \Gamma(t_{n+1}) - \Gamma^{n+1} \right\|_{\Delta,\infty},$$

where the vector $\Gamma(t_{n+1})$ is defined as follows:

$$\Gamma(t_{n+1}) = \left(\Gamma\left(\frac{\Delta z}{2}, t_{n+1}\right), \ldots, \Gamma\left(z_{N-1} + \frac{\Delta z}{2}, t_{n+1}\right) \right)^T \in \mathbb{R}^N.$$

Γ is the exact solution of (4.68). From this follows:

$$\begin{aligned} e_{n+1} &\le \left\| \Gamma(t_{n+1}) - \left(A_\Gamma^{n+1}\right)^{-1} \Gamma(t_n) \right\|_{\Delta,\infty} + \left\| \left(A_\Gamma^{n+1}\right)^{-1} \Gamma(t_n) - \Gamma^{n+1} \right\|_{\Delta,\infty} \\ &= \left\| \left(A_\Gamma^{n+1}\right)^{-1} \left(A_\Gamma^{n+1} \Gamma(t_{n+1}) - \Gamma(t_n)\right) \right\|_{\Delta,\infty} + \left\| \left(A_\Gamma^{n+1}\right)^{-1} \left(\Gamma(t_n) - \Gamma^n\right) \right\|_{\Delta,\infty} \\ &\le \left\| \left(A_\Gamma^{n+1}\right)^{-1} \left(A_\Gamma^{n+1} \Gamma(t_{n+1}) - \Gamma(t_n)\right) \right\|_{\Delta,\infty} + \left\| \left(A_\Gamma^{n+1}\right)^{-1} \right\|_\infty e_n \\ &\le \underbrace{\left\| \left(A_\Gamma^{n+1}\right)^{-1} \right\|_\infty}_{\text{stability}} \underbrace{\left\| A_\Gamma^{n+1} \Gamma(t_{n+1}) - \Gamma(t_n) \right\|_{\Delta,\infty}}_{\text{consistency error}} + \underbrace{\left\| \left(A_\Gamma^{n+1}\right)^{-1} \right\|_\infty}_{\text{stability}} e_n. \end{aligned} \tag{4.69}$$

Considering the upper bound of the error e_{n+1}, it can be seen that it consists of two parts. The second part can be estimated by the error e_n and the matrix norm of $\left(A_\Gamma^{n+1}\right)^{-1}$. If this matrix norm can be uniformly bounded, a recursion for the discretization error can be derived. Thus, the boundedness of the matrix norm can be considered as the numerical stability of the method. On the other hand, the first part is a product of the matrix norm of $\left(A_\Gamma^{n+1}\right)^{-1}$ and an error term that can be considered as a consistency error. All in all, this means that from consistency and stability of the method convergence follows.

In literature, a similar statement can be found in *Lax's equivalence theorem* [4], which implies that the convergence of a linear method is equivalent to the consistency and numerical stability of the method. In the following we assume that it holds:

$$v(z) = \omega \frac{Q}{A}(z) > 0, \ z \in [0, L] .$$

The proof of the remaining cases can be done in a similar way. For $\Gamma_{up}\left(z_{i+1}, t_{n+1}\right)$ and $\Gamma_{up}\left(z_i, t_{n+1}\right)$ it follows:

$$\Gamma_{up}\left(z_{i+1}, t_{n+1}\right) = \Gamma\left(z_i + \frac{\Delta z}{2}, t_{n+1}\right) \text{ and } \Gamma_{up}\left(z_i, t_{n+1}\right) = \Gamma\left(z_i - \frac{\Delta z}{2}, t_{n+1}\right) .$$

In order to estimate the consistency error, we consider the following Taylor expansions:

$$\Gamma\left(z_i + \frac{\Delta z}{2}, t_{n+1}\right) = \Gamma\left(z_i + \frac{\Delta z}{2}, t_n\right) + \frac{\partial \Gamma}{\partial t}\left(z_i + \frac{\Delta z}{2}, t_n\right) \Delta t + \mathcal{O}\left(\Delta t^2\right),$$

and

$$\Gamma\left(z_i - \frac{\Delta z}{2}, t_{n+1}\right) = \Gamma\left(z_i + \frac{\Delta z}{2}, t_{n+1}\right) - \frac{\partial \Gamma}{\partial z}\left(z_i + \frac{\Delta z}{2}, t_{n+1}\right) \Delta z + \mathcal{O}\left(\Delta z^2\right).$$

In the following, we use the abbreviation: $v_i^n = v\left(z_i, t_n\right)$. By means of the first expansion, it results:

$$\begin{aligned}
&\left|\left(1 + \frac{\Delta t}{\Delta z} v_{i+1}^{n+1}\right) \Gamma_i\left(t_{n+1}\right) - \frac{\Delta t}{\Delta z} v_i^{n+1} \Gamma_{i-1}\left(t_{n+1}\right) - \Gamma_i\left(t_n\right)\right| = \\
&\left|\frac{\partial \Gamma}{\partial t}\left(z_i + \frac{\Delta z}{2}, t_n\right) \Delta t + \frac{\Delta t}{\Delta z}\left(v_{i+1}^{n+1} \Gamma_i\left(t_{n+1}\right) - v_i^{n+1} \Gamma_{i-1}\left(t_{n+1}\right)\right) + \mathcal{O}\left(\Delta t^2\right)\right| .
\end{aligned} \tag{4.70}$$

Using the second expansion, it can be shown that:

$$\begin{aligned}
v_{i+1}^{n+1} \Gamma_i\left(t_{n+1}\right) - v_i^{n+1} \Gamma_{i-1}\left(t_{n+1}\right) &= v_{i+1}^{n+1} \Gamma_i\left(t_{n+1}\right) - v_i^{n+1} \Gamma_i\left(t_{n+1}\right) \\
&\quad + v_i^{n+1} \frac{\partial \Gamma_i}{\partial z}\left(t_{n+1}\right) \Delta z + \mathcal{O}\left(\Delta z^2\right) \\
&= \Delta z \frac{\partial v}{\partial z}\left(z_i + \frac{\Delta z}{2}, t_{n+1}\right) \Gamma_i\left(t_{n+1}\right) \\
&\quad + v_i^{n+1} \frac{\partial \Gamma_i}{\partial z}\left(t_{n+1}\right) \Delta z + \mathcal{O}\left(\Delta z^2\right)
\end{aligned} \tag{4.71}$$

Further Taylor expansions yield:

$$\frac{\partial \Gamma_i}{\partial z}\left(t_{n+1}\right) = \frac{\partial \Gamma_i}{\partial z}\left(t_n\right) + \mathcal{O}(\Delta t), \ v_i^{n+1} = v\left(z_i + \frac{\Delta z}{2}, t_n\right) + \mathcal{O}(\Delta t) + \mathcal{O}(\Delta z)$$

and

$$\frac{\partial v}{\partial z}\left(z_i+\frac{\Delta z}{2},t_{n+1}\right)=\frac{\partial v}{\partial z}\left(z_i+\frac{\Delta z}{2},t_n\right)+\mathcal{O}(\Delta t),\ \Gamma_i(t_n+\Delta t)=\Gamma_i(t_n)+\mathcal{O}(\Delta t).$$

Summarizing the previous results, we obtain:

$$\Delta z\frac{\partial v}{\partial z}\left(z_i+\frac{\Delta z}{2},t_{n+1}\right)\Gamma_i(t_{n+1})=\Delta z\frac{\partial v}{\partial z}\left(z_i+\frac{\Delta z}{2},t_n\right)\Gamma\left(z_i+\frac{\Delta z}{2},t_n\right)+\mathcal{O}(\Delta z\Delta t)$$

and

$$v_i^{n+1}\frac{\partial \Gamma_i}{\partial z}(t_{n+1})\Delta z=\Delta z v\left(z_i+\frac{\Delta z}{2},t_n\right)\frac{\partial \Gamma}{\partial z}\left(z_i+\frac{\Delta z}{2},t_n\right)+\mathcal{O}(\Delta z\Delta t).$$

Substituting these expansions into (4.71) and using (4.70), it results:

$$\begin{aligned}
&\frac{\partial \Gamma}{\partial t}\left(z_i+\frac{\Delta z}{2},t_n\right)\Delta t+\frac{\Delta t}{\Delta z}\left(v_{i+1}^{n+1}\Gamma_i(t_{n+1})-v_i^{n+1}\Gamma_{i-1}(t_{n+1})\right)+\mathcal{O}\left(\Delta t^2\right)=\\
&\quad\Delta t\left(\frac{\partial \Gamma}{\partial t}\left(z_i+\frac{\Delta z}{2},t_n\right)+\frac{\partial v}{\partial z}\left(z_i+\frac{\Delta z}{2},t_n\right)\Gamma\left(z_i+\frac{\Delta z}{2},t_n\right)\right.\\
&\qquad\left.+v\left(z_i+\frac{\Delta z}{2},t_n\right)\frac{\partial \Gamma}{\partial z}\left(z_i+\frac{\Delta z}{2},t_n\right)\right)+\mathcal{O}(\Delta z\Delta t)+\mathcal{O}\left(\Delta t^2\right)=\\
&\quad\underbrace{\Delta t\left(\frac{\partial \Gamma}{\partial t}\left(z_i+\frac{\Delta z}{2},t_n\right)+\frac{\partial}{\partial z}\left(v\left(z_i+\frac{\Delta z}{2},t_n\right)\Gamma\left(z_i+\frac{\Delta z}{2},t_n\right)\right)\right)}_{=0}+\mathcal{O}(\Delta z\Delta t)+\mathcal{O}\left(\Delta t^2\right)\\
&\quad=\mathcal{O}(\Delta z\Delta t)+\mathcal{O}\left(\Delta t^2\right).
\end{aligned}$$

In the last step we have used the fact that Γ solves the transport Eq. (4.68). Thus, it holds for the consistency error:

$$\left|\left(1+\frac{\Delta t}{\Delta z}v_{i+1}^{n+1}\right)\Gamma_i(t_{n+1})-\frac{\Delta t}{\Delta z}v_i^{n+1}\Gamma_{i-1}(t_{n+1})-\Gamma_i(t_n)\right|\leq C_n\left(\Delta z\Delta t+\Delta t^2\right). \tag{4.72}$$

Next we try to estimate the matrix norm of $\left(A_\Gamma^{n+1}\right)^{-1}$. C_n is a positive constant associated with the Landau notation. On closer examination, we realize that A_Γ^{n+1} has the following structure:

$$A_\Gamma^{n+1}=\begin{pmatrix}
1+a_1 & 0 & 0 & 0 & \dots & 0\\
-a_1 & 1+a_2 & 0 & 0 & \dots & 0\\
0 & -a_2 & 1+a_3 & 0 & \dots & 0\\
0 & \ddots & -a_3 & 1+a_4 & \dots & 0\\
\vdots & & \ddots & \ddots & \ddots & 0\\
0 & \dots & \dots & 0 & -a_{N-1} & 1+a_N
\end{pmatrix}.$$

It holds that: $a_i = \frac{\Delta t}{\Delta z} v(z_i, t_{n+1})$. Using [13], it results for the matrix norm of the inverse matrix:

$$\left\| \left(A_\Gamma^{n+1}\right)^{-1} \right\|_\infty \leq \frac{1}{\min\left(\min_{2\leq i\leq N}\left(1+a_i-a_{i-1}\right), 1+a_1\right)}$$
$$= \frac{1}{\min\left(\min_{2\leq i\leq N}\left(1+\frac{\Delta t}{\Delta z}\left(v\left(z_i, t_{n+1}\right)-v\left(z_{i-1}, t_{n+1}\right)\right)\right), 1+\frac{\Delta t}{\Delta z} v\left(z_1, t_{n+1}\right)\right)}$$
$$\leq \frac{1}{\min\left(\min_{2\leq i\leq N}\left(1+\Delta t \frac{\partial v}{\partial z}\left(\xi, t_{n+1}\right)\right), 1+\frac{\Delta t}{\Delta z} v\left(z_1, t_{n+1}\right)\right)}, \quad \xi \in \left(z_{i-1}, z_i\right), \tag{4.73}$$

where in the last step of (4.73) the meanvalue theorem of calculus is used. In order to ensure well-posedness, we assume that:

$$\left|\Delta t \frac{\partial v}{\partial z}(z, t)\right| < 1, \; z \in [0, L] \text{ and } t \geq 0. \tag{4.74}$$

Furthermore, let $C > 0$ be a constant, such that it holds:

$$-C < \frac{\partial v}{\partial z}(z, t) \text{ and } -C < \frac{v(z, t)}{\Delta z}, \; z \in [0, L] \text{ and } t \geq 0.$$

Thus it follows that:

$$\left\| \left(A_\Gamma^{n+1}\right)^{-1} \right\|_\infty \leq \frac{1}{1 - C\Delta t}. \tag{4.75}$$

Combining (4.69), (4.72) and (4.75) we obtain:

$$\begin{aligned} e_{n+1} &\leq \left\| \left(A_\Gamma^{n+1}\right)^{-1} \right\|_\infty \left\| A_\Gamma^{n+1} \Gamma\left(t_{n+1}\right) - \Gamma\left(t_n\right) \right\|_{\Delta,\infty} + \left\| \left(A_\Gamma^{n+1}\right)^{-1} \right\|_\infty e_n \\ &\leq \frac{C_n \Delta t}{1 - C\Delta t}(\Delta t + \Delta z) + \frac{1}{1 - C\Delta t} e_n. \end{aligned}$$

Assuming that $e_0 = 0$ holds, it follows by means of induction:

$$\begin{aligned} e_{n+1} &\leq \Delta t \sum_{k=1}^{n+1} \frac{C_{n-k+1}}{(1 - C\Delta t)^k}(\Delta t + \Delta z) \\ &\leq \Delta t \max_{1\leq k\leq n+1} C_{n-k+1} \sum_{k=1}^{n+1} \left(\frac{1}{1 - C\Delta t}\right)^k (\Delta t + \Delta z) \\ &\leq T \max_{1\leq k\leq n+1} C_{n-k+1} \left(\frac{1}{1 - C\Delta t}\right)^M (\Delta t + \Delta z), \; \Delta t = \frac{T}{M}, \; n+1 \leq M. \end{aligned}$$

Now it holds:

$$\left(\frac{1}{1-C\Delta t}\right)^M = \left(\frac{1}{1+\frac{-TC}{M}}\right)^M \underbrace{\to}_{M\to\infty} e^{TC}.$$

Thus there is $\varepsilon > 0$ and $M_1 \in \mathbb{N}$ such that:

$$\left|\left(\frac{1}{1-C\Delta t}\right)^M - e^{TC}\right| < \varepsilon,\ M \geq M_1.$$

From this it follows for $\varepsilon = 0.5 \cdot e^{TC}$ and a sufficiently small Δt or a sufficiently large M:

$$e_{n+1} \leq T \cdot \frac{3}{2} \cdot e^{TC} \max_{1\leq k\leq n+1} C_{n-k+1} \left(\Delta t + \Delta z\right).$$

This means that there is a constant

$$D \geq T \cdot \frac{3}{2} \cdot e^{TC} \max_{1\leq k\leq n+1} C_{n-k+1} > 0,$$

such that for a sufficiently small Δt it holds:

$$e_{n+1} \leq D\left(\Delta t + \Delta z\right).$$

Thus this method is convergent of order 1 in space and time.

4.4 Numerical Experiments

In this section, we will support the theoretical results obtained in the previous sections and chapters with numerical experiments. In total, three different experiments will be considered: In our first experiment, the results on the existence of classical solutions from Sect. 3.4 are verified in a numerical way. The convergence and numerical stability of the Lax-Wendroff method as well as the implicit Finite Volume method are considered in the second experiment (see Subsect. 4.4.3). Finally, we show that using a 1D-0D coupled model, realistic blood pressures and flow rates can be simulated in a network consisting of the 43 main arteries of the human body. In addition, a transport process through this network is simulated.

Table 4.1 List of fluid parameters

Parameter	Symbol	Value	Unit
Density	ρ	$1.085 \cdot 10^{-3}$	kg/cm^3
Viscosity	μ	$4.500 \cdot 10^{-5}/\rho$	cm^2/s
Poisson number	ν	0.500	---

4.4.1 *Choice of the Parameters*

The parameters that determine the property of blood can be found in Table 4.1. Furthermore, we use data of vessels from the network already considered in Chap. 1, Fig. 1.7. The structure of the network used for the simulations that now follow can be found in Fig. 4.8. The lengths and diameters of the corresponding vessels, as well as the E-modules and wall thicknesses, can be found in Table 7.2 (Sect. 7.14). Since this network consists of 43 vessels, as mentioned earlier, we refer to it as *Network 43*.

4.4.2 *Experiment 1: Formation of Shocks*

In our first experiment, we consider a fictitious vessel with a total length of 5.0 m. As radius, wall thickness and E-modulus, we choose the arithmetic mean of the data for the vessels Abdominal Aorta I-V from Table 7.2. Therefore, we obtain a vessel representing the center of *Network* 43. The corresponding data are summarized in Table 4.2. The reason why we want to consider a vessel with a completely unrealistic length is to test at which point in time and at which position in the vessel a shock occurs. This test is of great importance for the choice of the numerical solution method. If shocks are occuring, only low order methods or methods with stabilization could be applied. A method like the Lax-Wendroff method would not be applicable, since its derivation is based on higher order Taylor expansions (see Sect. 4.3). The initial conditions for our solution variables are given by:

$$A(z,0) = A_0 = R^2\pi, \;\; p(z,0) = 0 \text{ und } Q(z,0) = 0, \;\; z \in (0,L).$$

As a boundary condition at the inlet $z = 0$, we choose the inflow profile (3.69). The missing parameters $Q_{\max}$ and T can again be found in Table 4.2. The second boundary condition at $z = L$ is given by the values of the initial condition. The simulation is stopped before the pulse wave reaches the boundary $z = L$. As in [14, Sect. 4.3], we try to roughly estimate the time t_s at which the shock occurs. Neglecting the parameter δ in (3.53) and setting

$$v_{AQ} = 0 \text{ as well as } A = A_0,$$

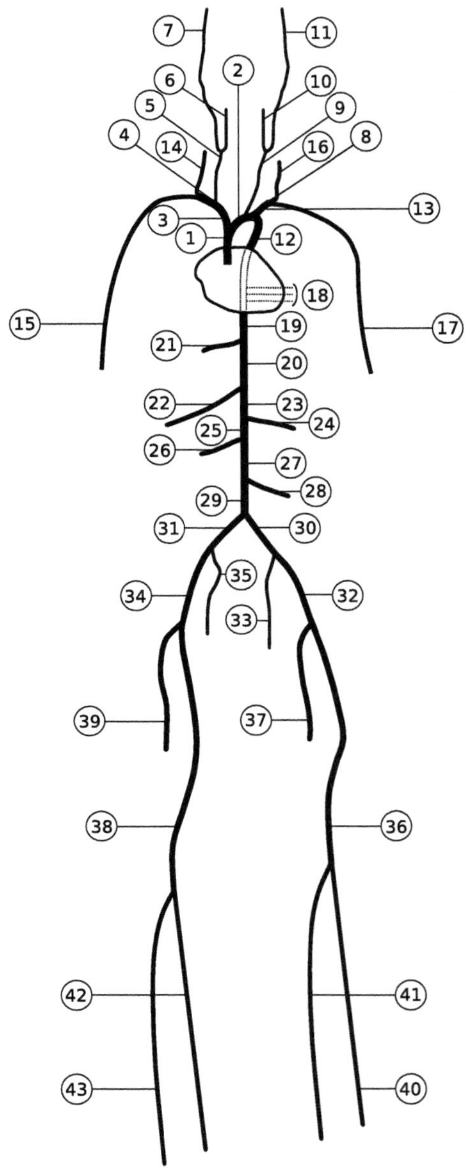

Fig. 4.8 Network used for the simulations in Sect. 4.4

Table 4.2 Parameter for experiment 1

Parameter	Symbol	Value	Unit
Radius	R	0.7688	cm
Length	L	5.0	m
Wall thickness	h	0.078	cm
E-modul	E	$0.4 \cdot 10^6$	Pa
Maximal flow rate	$Q_{\max}$	150.0	cm^3/s
Duration of the heart beat	T	1.0	s

it results

$$t_s \approx \frac{\sqrt{\frac{G_0}{2\rho}}}{\frac{5}{4}v'_{in}(t)} \text{ with } v_{in}(t) = \frac{Q(0,t)}{A_0}.$$

To estimate the smallest value for t_s, we try to find the largest value for $v'_{in}(t)$. Deriving (3.69) with respect to time, it turns out that the derivative attains its maximum at $t_h = 0$. This means:

$$v'_{in}(t) \leq \left| \frac{Q'(0,t)}{A_0} \right| = \frac{Q_{\max} \cdot \pi}{A_0 \cdot 0.3 \cdot T}$$

Using the parameter set in Tables 4.2 and (3.54) we obtain:

$$t_s \geq 0.48004 \text{ s and } z_s \geq 243.673 \text{ cm}.$$

In fact, it can be seen from the solutions at the time points

$$t \in \{0.2 \text{ s}, \ 0.4 \text{ s}, \ 0.6 \text{ s}, \ 0.8 \text{ s}\},$$

that between $t = 0.4$ s and $t = 0.6$ s a shock is formed (see Figs. 4.9 and 4.10). If one considers the solutions at the time $t = t_s$ as in Fig. 4.11, it can be seen in an enlarged part of the figure that there is already a small shock at the time t_s and at z_s. As mentioned at the beginning of the subsection, numerical oscillations arise near the shock, since the required smoothness of the solution is not given. For the numerical solution, we use the grid width $\Delta z = 500.0 \cdot 10^{-4}$ cm and the time stepsize $\Delta t = t_s \cdot 10^{-4}$ s. The only exception are the solutions in the lower part of Fig. 4.11. Here, we have used $\Delta z = 500.0 \cdot 10^{-4}/16.0$ cm and $\Delta t = t_s \cdot 10^{-4}/16.0$ s to represent the shock more accurately.

All in all, the numerical results confirm the theoretical considerations from Sect. 3.3. Furthermore, it is pleasing to note that shocks form after a very long flow path. Adding the lengths of the vessels Abdominal Aorta I-V from Table 7.2, we obtain 18.9 cm. This means that in the abdominal vessel no shock is formed. Moreover, flow paths exceeding a distance of 2.40 m do not occur in the networks that are considered

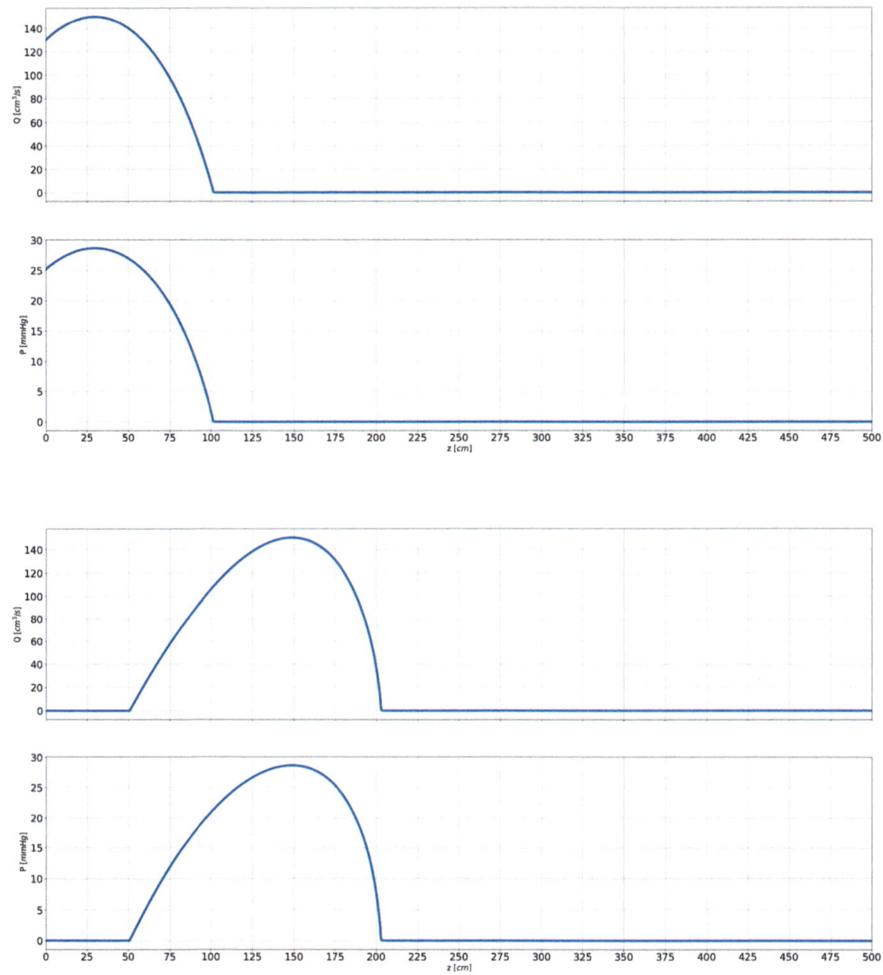

Fig. 4.9 Numerical solutions of the flow problem for a vessel of length 5.0 m. The graphs show the solutions for Q and p at $t = 0.2$ s und $t = 0.4$ s. It can be seen that the form of the sine shaped inflow profile is slowly deformed

in this book. From this, one can conclude: Choosing a sufficiently smooth boundary condition, one can assume that in terms of application relevant scenarios the solutions of (3.21) are smooth, despite of the fact that the flux function is not linear. Since the solution is expected to be sufficiently smooth a method of higher order like the Lax-Wendroff method can be applied without suffering a loss of convergence order.

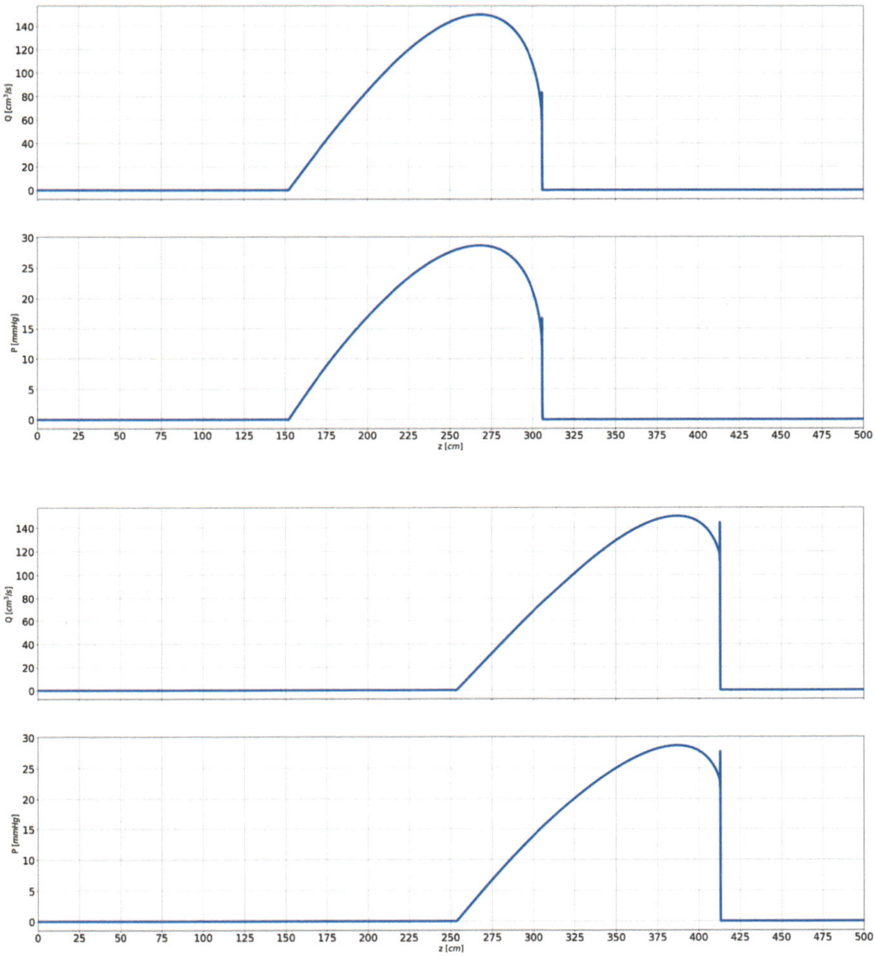

Fig. 4.10 Numerical solutions of the flow problem for a vessel of length 5.0 m. The graphs show the solutions for Q and p at $t = 0.6$ s and $t = 0.8$ s. Contrary to the solutions for $t = 0.2$ s and $t = 0.4$ s (see Fig. 4.9) a shock with numerical oscillations can be detected

4.4.3 Experiment 2: Stability and Convergence Tests

The second experiment is concerned with the numerical methods that are considered for solving the model equations. These are the Lax-Wendroff method and the implicit Finite Volume method. In Sect. 4.3 some results on the convergence behavior and stability of these methods have been derived. The goal of this subsection is to support the theoretical results by numerical simulations.

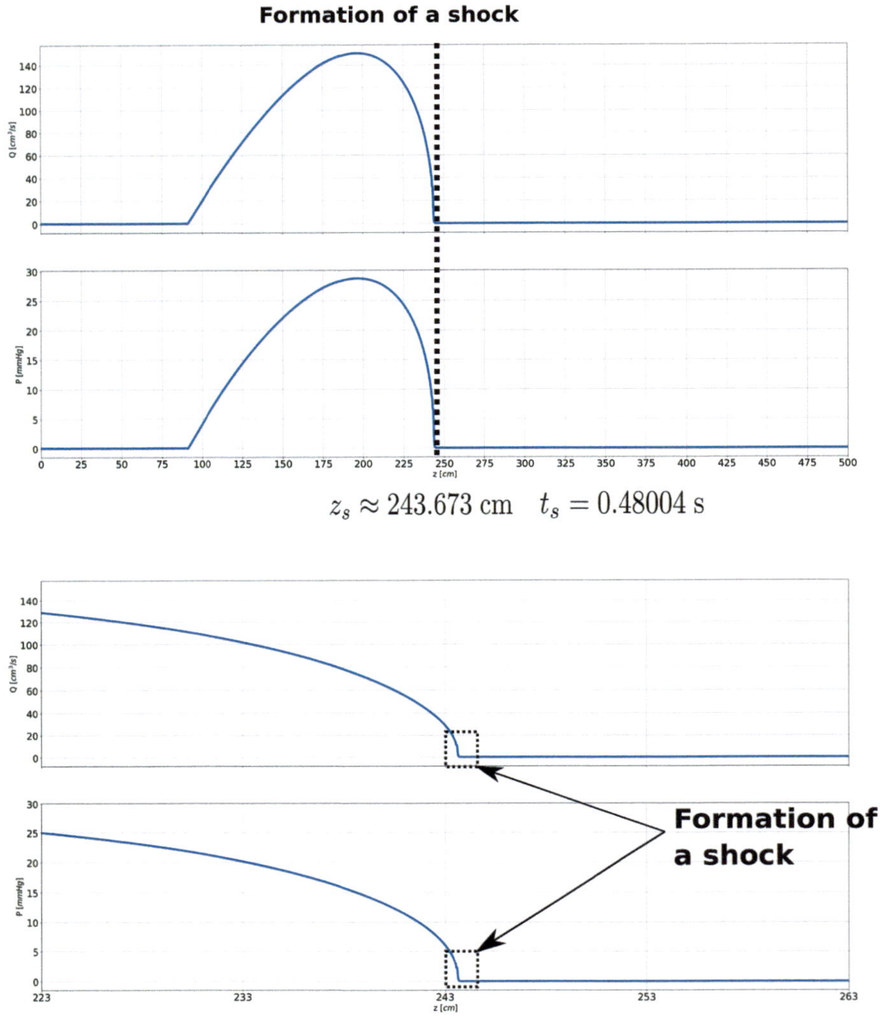

Fig. 4.11 Q and p at the time point t_s. In the vicinity of z_s the formation of a shock can be observed

Stability and Convergence of the Lax-Wendroff Method

To investigate the convergence and stability properties of the Lax-Wendroff method, we consider the benchmark problem previously discussed in [15, Sect. 6.1]. For this purpose let us consider a single vessel of length $L = 20.0$ cm and cross-sectional area $A_0 = 1.0\ \text{cm}^2$. The elasticity parameter G_0 has the following value: $G_0 = 737.341 \cdot 10^2$ Pa (see also Table 4.3). This yields a characteristic velocity of $c_0 = 592.549\ {}^{cm}/_{s}$. A solution of the blood flow model (3.23) is provided in [15] and is given by:

Table 4.3 Parameters for experiment 2 (Convergence Lax-Wendroff)

Parameter	Symbol	Value	Unit
Section area	A_0	1.0	cm^2
Length	L	20.0	m
Elasticity parameter	G_0	$737.341 \cdot 10^2$	Pa

$$A(z,t) = \left(1 + te^{-10t} \sin\left(\frac{\pi z}{L}\right)\right)^2 \text{ and } Q(z,t) = 0 \tag{4.76}$$

where $(z, t) \in [0.0 \text{ cm}, 20.0 \text{ cm}] \times [0 \text{ s}, 1.0 \text{ s}]$. A graph of the solution variable A can be found on Fig. 4.12. Obviously, (4.76) can only be a solution of (3.23), if parts of the model equations are adapted. It can be shown that the source term

$$\mathbf{S}(\mathbf{U}) = \begin{pmatrix} -2 \sin\left(\frac{\pi z}{L}\right) e^{-10t} (10t - 1) \left(t \sin\left(\frac{\pi z}{L}\right) e^{-10t} + 1\right) \\ \frac{2\pi c_0^2 t \cos\left(\frac{\pi z}{L}\right) e^{-10t} \left(t \sin\left(\frac{\pi z}{L}\right) e^{-10t} + 1\right)^2}{\sqrt{A_0} L} \end{pmatrix} \tag{4.77}$$

enforces the solution (4.76). The initial and boundary conditions are also adapted. In order to investigate the stability of the method, it is recommended to consider the CFL number:

$$K_{CFL} = \frac{c_0 \Delta t}{\Delta z} \tag{4.78}$$

According to the stability condition (4.67) the Lax-Wendroff is stable, if it holds:

$$\frac{|\lambda_i| \Delta t}{\Delta z} < 1$$

λ_i are the eigenvalues of the Jacobian matrix derived from the flux function (3.38):

$$\lambda_1 = v - c_0 \text{ and } \lambda_2 = v + c_0.$$

Since $Q \equiv 0$ holds for the solution (4.76), it follows that $v \equiv 0$. Thus in this case the Lax-Wendroff method is stable, if it holds:

$$K_{CFL} < 1$$

In Table 4.4 the discrete L^1-norms (see Sect. 4.3.1) e_{LW} of the absolute approximation error at time $t = 0.1$ are listed. The grid widths Δz for different time stepsizes Δt can be computed by means of the following equation:

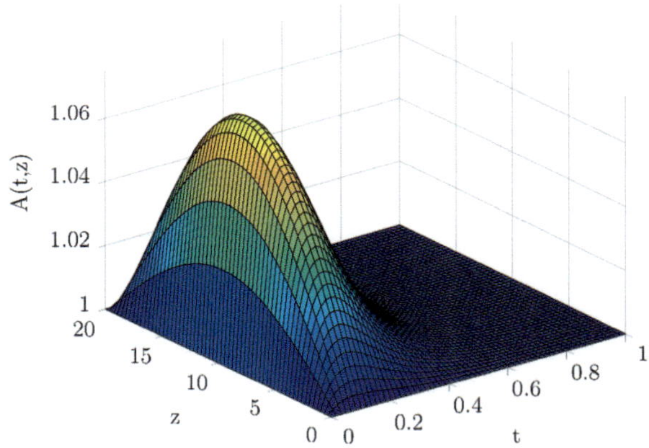

Fig. 4.12 Representation of the solution variable A in the space time domain $[0.0\text{ cm}, 20.0\text{ cm}] \times [0\text{ s}, 1.0\text{ s}]$

$$\Delta z = \frac{c_0 \Delta t}{K_{CFL}}.$$

Due to the fact that the solution (4.76) is sufficiently smooth, we expect the method to exhibit the full consistency and convergence order in case of this problem. As we have shown in Sect. 4.3.2, the Lax-Wendroff method has consistency order 2 in both time and space. This is reflected in the convergence rates in Tables 4.4 and 4.5. Besides the expected convergence rate, it is also demonstrated that the Lax-Wendroff method is stable only if $K_{CFL} < 1$ holds. This confirms the theoretical considerations from Sect. 4.3.2.

Stability and Convergence of the Implicit Finite Volume Method

In a next step, we consider the transport problem (3.24):

Table 4.4 Convergence and stability of the Lax-Wendroff method for smaller CFL numbers. The abbreviation *e.c.o* stands for *estimated convergence order*

	$K_{CFL} = 0.296275$		$K_{CFL} = 0.444412$	
Δt [s]	e_{LW}	*e.c.o*	e_{LW}	*e.c.o*
$2.000 \cdot 10^{-4}$	$9.000 \cdot 10^{-3}$	----	$4.052 \cdot 10^{-3}$	----
$1.000 \cdot 10^{-4}$	$1.894 \cdot 10^{-3}$	2.25	$8.556 \cdot 10^{-4}$	2.24
$5.000 \cdot 10^{-5}$	$4.275 \cdot 10^{-4}$	2.15	$1.933 \cdot 10^{-4}$	2.15
$2.500 \cdot 10^{-5}$	$1.014 \cdot 10^{-4}$	2.08	$4.595 \cdot 10^{-5}$	2.07
$1.250 \cdot 10^{-5}$	$2.470 \cdot 10^{-5}$	2.04	$1.119 \cdot 10^{-5}$	2.04
$6.250 \cdot 10^{-6}$	$6.089 \cdot 10^{-6}$	2.02	$2.760 \cdot 10^{-6}$	2.02
$3.125 \cdot 10^{-6}$	$1.513 \cdot 10^{-6}$	2.01	$6.854 \cdot 10^{-7}$	2.01

Table 4.5 Convergence and stability of the Lax-Wendroff method for larger CFL numbers. nan abbreviates the expression *not a number* and indicates the numerical instability and divergence of the method. The abbreviation $e.c.o$ stands for *estimated convergence order*

	$K_{CFL} = 0.888824$		$K_{CFL} = 1.1851$	
Δt [s]	e_{LW}	$e.c.o$	e_{LW}	$e.c.o$
$2.000 \cdot 10^{-4}$	$1.083 \cdot 10^{-3}$	----	nan	----
$1.000 \cdot 10^{-4}$	$2.319 \cdot 10^{-4}$	2.22	nan	----
$5.000 \cdot 10^{-5}$	$5.290 \cdot 10^{-5}$	2.13	nan	----
$2.500 \cdot 10^{-5}$	$1.263 \cdot 10^{-5}$	2.06	nan	----
$1.250 \cdot 10^{-5}$	$3.084 \cdot 10^{-6}$	2.03	nan	----
$6.250 \cdot 10^{-6}$	$7.618 \cdot 10^{-7}$	2.02	nan	----
$3.125 \cdot 10^{-6}$	$1.893 \cdot 10^{-7}$	2.01	nan	----

$$\frac{\partial \Gamma}{\partial t} + \frac{\partial}{\partial z}\left(\omega \frac{Q\Gamma}{A}\right) = 0, \;\; z \in (0, 1)\,, t \in (0, t_{end})\,. \tag{4.79}$$

The velocity field v is given by:

$$v = \omega \frac{Q}{A} \equiv 1.$$

Furthermore, two different initial conditions $\Gamma_0(z) = \Gamma(z, 0)$ are considered. In the first case we set Γ_0 to zero:

$$\Gamma_0(z) = 0, \;\; z \in (0, 1)\,, \tag{4.80}$$

while in the second case a continuous differentiable function is chosen:

$$\Gamma_0(z) = \begin{cases} 0, \;\; 0 \leq z \leq 0.05, \\ 0.5 \cdot \cos(4\pi(z + 0.2)) + 0.5, \;\; \text{if } 0.05 < z < 0.55, \\ 0, \;\; 0.55 \leq z \leq 1.0. \end{cases} \tag{4.81}$$

In addition to that, we require suitable boundary conditions for the boundaries $z = 0$ and $z = 1$. Due to $v > 0$ a boundary condition is needed at the inlet $z = 0$. The initial condition (4.80) is combined with the following boundary condition:

$$\Gamma(0, t) = 1.0, \;\; t \in (0, t_{end}) \tag{4.82}$$

The second initial condition (4.81) is combined with a homogeneous boundary condition:

$$\Gamma(0, t) = 0.0, \;\; t \in (0, t_{end})\,. \tag{4.83}$$

According to the considerations from Sect. 7.8

$$\Gamma(z,t) = \begin{cases} 1.0, \; z - vt \leq 0, \\ \\ \Gamma_0(z - vt), \; z - vt \in (0,1). \end{cases} \tag{4.84}$$

solves problem (4.79), (4.80), (4.82) in a weak sense. We denote this problem by *Problem I*. The consideration of weak solutions is necessary, since on space time lines $g(z) = \frac{1}{v}z$ a shock or discontinuity propagates. Therefore, the solution (4.84) is only differentiable in a local sense, but not in a global sense. Contrary to that

$$\Gamma(z,t) = \begin{cases} 0.0, \; z - vt \leq 0, \\ \\ \Gamma_0(z - vt), \; z - vt \in (0,1). \end{cases} \tag{4.85}$$

solves (4.79), (4.81), (4.83) (*Problem II*) and is globally continuous differentiable. The boundary conditions at the boundary $z = 1$ are adapted according to the solutions (4.84) and (4.85). *Problem I* and *Problem II* are now solved using the implicit Finite Volume method (see Sect. (4.1)) combinded with an Upwinding at the interfaces between two elements. The grid width in space and the time stepsizes are denoted by Δz and Δt, respectively. Analogous to (4.78), we define a CFL number for these experiments:

$$K_{CFL} = v\frac{\Delta t}{\Delta z}. \tag{4.86}$$

For a given CFL number K_{CFL} and a time stepsize Δt the grid width Δz is chosen as follows:

$$\Delta z = v\frac{\Delta t}{K_{CFL}}.$$

Additionally, we solve the two problems by means of the Lax-Wendroff method. Considering the derivation of the method in Sect. 4.1, we realize that this method can easily be applied to (4.79). For a grid node $z_i = i \cdot \Delta z$ and a time point $t_j = j \cdot \Delta t$ one obtains:

$$\Gamma_i^{j+1} = \Gamma_i^j - \frac{1}{2}\cdot\left(\frac{v\Delta t}{\Delta z}\right)\left(\Gamma_{i+1}^j - \Gamma_{i-1}^j\right) + \frac{1}{2}\cdot\left(\frac{v\Delta t}{\Delta z}\right)^2\left(\Gamma_{i+1}^j - \Gamma_i^j + \Gamma_{i-1}^j\right)$$

The approximation errors are denoted by e_{UP} for the Finite Volume method with Upwinding and e_{LW} for the Lax-Wendroff method. Both errors are given the discrete L^1-norm (4.51) of the absolute errors. Thereby, the errors for the Finite Volume method are determined wrt. the dual grid consisting of the grid nodes: $z_{j+\frac{1}{2}} = \left(j + \frac{1}{2}\right)\Delta z$. Although we have shown the convergence of the Finite Volume method with respect to the L^∞-norm, we expect the same behavior for the

Table 4.6 Convergence and stability of the implicit Finite Volume method for small CFL numbers K_{CFL} (*Problem I*). The abbreviation $e.c.o$ stands for *estimated convergence order*

	$K_{CFL} = 0.32$		$K_{CFL} = 0.64$	
Δt [s]	e_{UP}	$e.c.o$	e_{UP}	$e.c.o$
$8.00 \cdot 10^{-4}$	$4.097 \cdot 10^{-2}$	----	$3.228 \cdot 10^{-2}$	----
$4.00 \cdot 10^{-4}$	$2.898 \cdot 10^{-2}$	0.50	$2.228 \cdot 10^{-2}$	0.50
$2.00 \cdot 10^{-4}$	$2.049 \cdot 10^{-2}$	0.50	$1.611 \cdot 10^{-2}$	0.50
$1.00 \cdot 10^{-4}$	$1.449 \cdot 10^{-2}$	0.50	$1.119 \cdot 10^{-2}$	0.53
$5.00 \cdot 10^{-5}$	$9.668 \cdot 10^{-3}$	0.58	$7.640 \cdot 10^{-3}$	0.55
$2.50 \cdot 10^{-5}$	$6.074 \cdot 10^{-3}$	0.67	$4.442 \cdot 10^{-3}$	0.78
$1.25 \cdot 10^{-5}$	$3.998 \cdot 10^{-3}$	0.60	$2.593 \cdot 10^{-3}$	0.78

Table 4.7 Convergence and stability of the implicit Finite Volume method for CFL numbers K_{CFL} (*Problem I*) that are larger than one. The abbreviation $e.c.o$ stands for *estimated convergence order*

	$K_{CFL} = 1.28$		$K_{CFL} = 2.56$	
Δt [s]	e_{UP}	$e.c.o$	e_{UP}	$e.c.o$
$8.00 \cdot 10^{-4}$	$2.691 \cdot 10^{-2}$	----	$2.350 \cdot 10^{-2}$	----
$4.00 \cdot 10^{-4}$	$1.903 \cdot 10^{-2}$	0.50	$1.575 \cdot 10^{-2}$	0.58
$2.00 \cdot 10^{-4}$	$1.343 \cdot 10^{-2}$	0.50	$9.352 \cdot 10^{-3}$	0.75
$1.00 \cdot 10^{-4}$	$9.366 \cdot 10^{-3}$	0.52	$4.787 \cdot 10^{-3}$	0.97
$5.00 \cdot 10^{-5}$	$6.235 \cdot 10^{-3}$	0.59	$2.373 \cdot 10^{-3}$	1.01
$2.50 \cdot 10^{-5}$	$3.648 \cdot 10^{-3}$	0.77	$1.201 \cdot 10^{-3}$	0.98
$1.25 \cdot 10^{-5}$	$8.570 \cdot 10^{-4}$	2.09	$7.882 \cdot 10^{-4}$	0.61

discrete L^1-norm. This is due to the fact that both norms are equivalent. The main motivation for using the L^1-norm is that it accounts for the solution behavior in the entire simulation domain.

The errors e_{UP} for *Problem I* can be seen in Tables 4.6 and 4.7. In addition, the respective convergence rates are listed. Since the solution (4.84) is not even continuous, we observe a reduced convergence rate of about 0.5. According to the analysis in Sect. 4.3.3, the full consistency error can only be observed, if the solution is not smooth enough. For this reason, one does not observe the full order of convergence or consistency in this case. Furthermore, the tables show that the method is also convergent for CFL numbers larger than one. This observation confirms the theoretical considerations in Sect. 4.3.2. Since $\frac{\partial v}{\partial z} \equiv 0$, the stability condition (4.74) does not induce any restriction regarding the choice of Δt. Explicit methods like the Lax-Wendroff method are subject to the restriction $K_{CFL} < 1$ (consider e.g., the convergence test in terms of the blood flow model). However, it turns out that this condition is only *necessary* and not *sufficient* for the stability of the Lax-Wendroff method. In Fig. 4.13 it can be seen that the Lax-Wendroff method exhibits numerical oscillations near the jump even though $K_{CFL} < 1$ is satisfied. This behavior is typ-

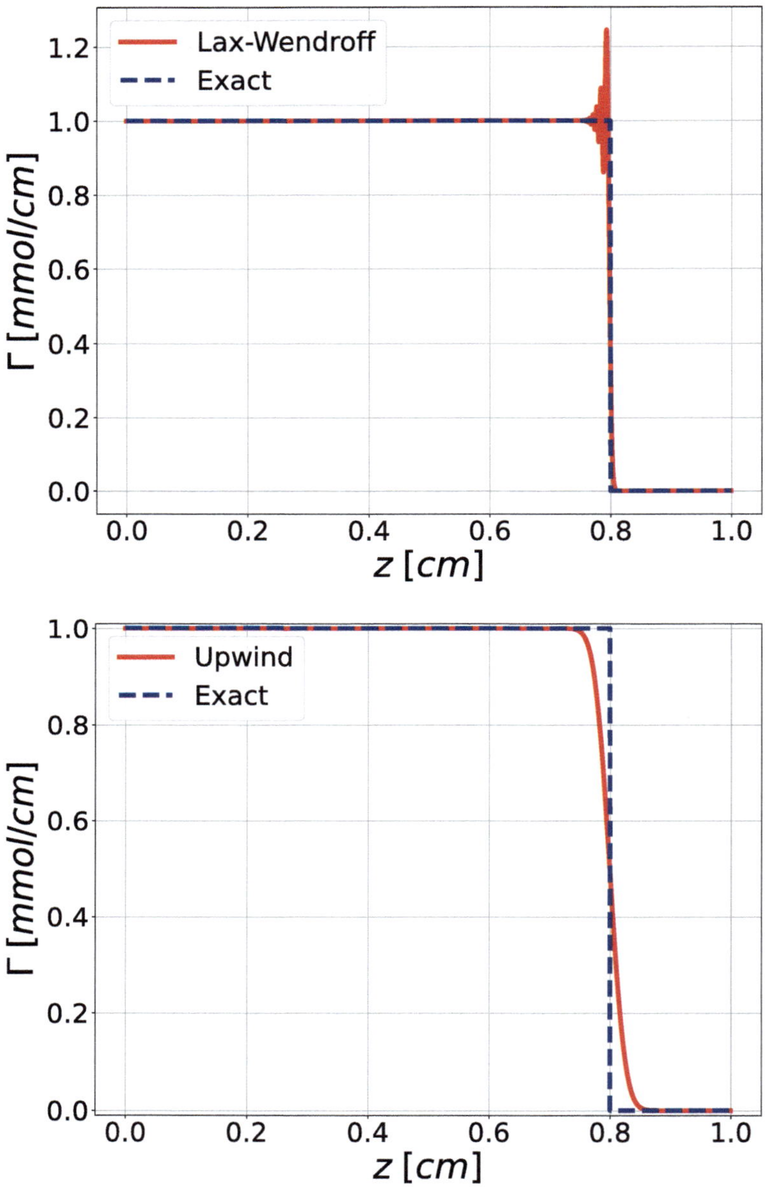

Fig. 4.13 Comparison of the implicit Finite Volume method with Upwind and the Lax-Wendroff method for *Problem I* and $K_{CFL} < 1$. The curves show the solutions for $t = 0.8$ s

ical for higher order [16] methods, which demonstrates that numerical instabilities can also be induced by a boundary condition even though the step size constraint $K_{CFL} < 1$ is satisfied. Since the Lax-Wendroff method is unstable in case of *Problem I*, we omitted the computation of e_{LW}.

Tables 4.8 and 4.9 contain the errors and convergence rates of the Finite Volume method for *Problem II*. Since the corresponding solution (4.85) is continuous differentiable the full order of convergence can be observed. As already mentioned, the solution must be continuous differentiable according to the analysis in Sect. 4.3.2. If one includes the derivatives, which are hidden by the Landau symbols, the solution should be even smoother. Nevertheless, the full order of convergence can be observed numerically. As we have already observed in terms of *Problem I*, the method is also convergent for $K_{CFL} \geq 1$. Contrary to that, the Lax-Wendroff method diverges for $K_{CFL} = 1.2$ (see Tables 4.10 and 4.11). Otherwise, the method converges only with order 1.0 and not with order 2.0. In order to obtain the optimal convergence behavior for the Lax-Wendroff method, the solution has to be at least twice continuous differentiable (see Sect. 4.3.2). Solution (4.85), however, is only one times continuous differentiable. The optimal convergence order of the Lax-Wendroff method can be demonstrated by means of the blood flow model with source term (4.77).

Table 4.8 Convergence and stability of the implicit Finite Volume method for small CFL numbers K_{CFL} (*Problem II*). The abbreviation *e.c.o* stands for *estimated convergence order*

	$K_{CFL} = 0.2$		$K_{CFL} = 0.4$	
Δt [s]	e_{UP}	*e.c.o*	e_{UP}	*e.c.o*
$4.00 \cdot 10^{-4}$	$1.160 \cdot 10^{-2}$	----	$6.875 \cdot 10^{-3}$	----
$2.00 \cdot 10^{-4}$	$3.971 \cdot 10^{-3}$	1.55	$1.501 \cdot 10^{-3}$	2.20
$1.00 \cdot 10^{-4}$	$1.998 \cdot 10^{-3}$	0.99	$7.522 \cdot 10^{-4}$	1.00
$5.00 \cdot 10^{-5}$	$1.002 \cdot 10^{-3}$	0.99	$3.765 \cdot 10^{-4}$	1.00
$2.50 \cdot 10^{-5}$	$5.019 \cdot 10^{-4}$	1.00	$1.883 \cdot 10^{-4}$	1.00
$1.25 \cdot 10^{-5}$	$2.511 \cdot 10^{-4}$	1.00	$9.421 \cdot 10^{-5}$	1.00
$6.25 \cdot 10^{-6}$	$1.256 \cdot 10^{-4}$	1.00	$4.711 \cdot 10^{-5}$	1.00

Table 4.9 Convergence and stability of the implicit Finite Volume method for $K_{CFL} = 0.8$ and $K_{CFL} = 1.2$ (*Problem II*). The abbreviation *e.c.o* stands for *estimated convergence order*

	$K_{CFL} = 0.8$		$K_{CFL} = 1.2$	
Δt [s]	e_{UP}	*e.c.o*	e_{UP}	*e.c.o*
$4.00 \cdot 10^{-4}$	$2.996 \cdot 10^{-3}$	----	$3.640 \cdot 10^{-3}$	----
$2.00 \cdot 10^{-4}$	$2.515 \cdot 10^{-4}$	3.57	$1.832 \cdot 10^{-3}$	0.99
$1.00 \cdot 10^{-4}$	$1.257 \cdot 10^{-4}$	1.00	$9.188 \cdot 10^{-4}$	1.00
$5.00 \cdot 10^{-5}$	$6.284 \cdot 10^{-5}$	1.00	$4.593 \cdot 10^{-4}$	1.00
$2.50 \cdot 10^{-5}$	$3.142 \cdot 10^{-5}$	1.00	$2.235 \cdot 10^{-4}$	1.04
$1.25 \cdot 10^{-5}$	$1.570 \cdot 10^{-5}$	1.00	$4.189 \cdot 10^{-5}$	2.42
$6.25 \cdot 10^{-6}$	$7.854 \cdot 10^{-6}$	1.00	$1.320 \cdot 10^{-5}$	1.67

Table 4.10 Convergence and stability of the Lax-Wendroff method for small CFL numbers K_{CFL} (*Problem II*). The abbreviation $e.c.o$ stands for *estimated convergence order*

	$K_{CFL} = 0.2$		$K_{CFL} = 0.4$	
Δt [s]	e_{LW}	$e.c.o$	e_{LW}	$e.c.o$
$4.00 \cdot 10^{-4}$	$2.117 \cdot 10^{-3}$	----	$1.025 \cdot 10^{-3}$	----
$2.00 \cdot 10^{-4}$	$1.030 \cdot 10^{-3}$	1.04	$5.067 \cdot 10^{-4}$	1.02
$1.00 \cdot 10^{-4}$	$5.077 \cdot 10^{-4}$	1.02	$2.517 \cdot 10^{-4}$	1.01
$5.00 \cdot 10^{-5}$	$2.519 \cdot 10^{-4}$	1.01	$1.254 \cdot 10^{-4}$	1.01
$2.50 \cdot 10^{-5}$	$1.255 \cdot 10^{-4}$	1.01	$6.261 \cdot 10^{-5}$	1.00
$1.25 \cdot 10^{-5}$	$6.263 \cdot 10^{-5}$	1.00	$3.128 \cdot 10^{-5}$	1.00
$6.25 \cdot 10^{-6}$	$3.128 \cdot 10^{-5}$	1.00	$1.563 \cdot 10^{-5}$	1.00

Figure 4.14 shows the exact solution of *Problem II* as well as the solutions of the Finite Volume and Lax-Wendroff method. Despite of the fact that both methods converge with order 1.0, the Lax-Wendroff method exhibits a much better approximation property compared to the Finite Volume method. While hardly any difference can be seen between the exact solution and the Lax-Wendroff method, a clear difference can be seen in the case of the Finite Volume method. Figure 4.13 shows a similar phenomenon: The Lax-Wendroff method resolves the jump in an accurate way, whereas the Finite volume method approximates the jump in an inexact way.

This is due to the fact that the convergence analysis yields only meaningful statements for $\Delta t,\ \Delta z \to 0$ and a sufficiently smooth solution. The numerical simulation of a problem, however, is based on fixed discretization parameters Δt and Δz. Therefore additional error measures are needed to characterize the numerical error for a fixed Δt and Δz. These error measures, are called *dissipation* and *dispersion*. To define them mathematically, we consider the Fourier representation of a 2π-periodic initial condition Γ_0 of (4.79):

$$\Gamma(z, 0) = \Gamma_0(z) = \sum_k \alpha_k e^{ikz}. \tag{4.87}$$

From this it follows for the exact solution of (4.79) on an unbounded domain:

$$\Gamma\left(z_j, t^n\right) = \Gamma_0\left(z_j - vn \cdot \Delta t\right) = \sum_k \alpha_k e^{ik(z_j - vn\Delta t)} = \sum_k \alpha_k e^{ikj\cdot\Delta z} g_k^n$$

where $g_k = e^{-ivk\Delta t}$. Introducing the phase angle $\varphi_k = k\Delta z$, the following holds:

$$g_k = e^{-iv\lambda\varphi_k},\ \lambda = \frac{\Delta t}{\Delta z}.$$

Applying an explicit method such as the Lax-Wendroff method to (4.79) combined with (4.87), it can be represented as follows:

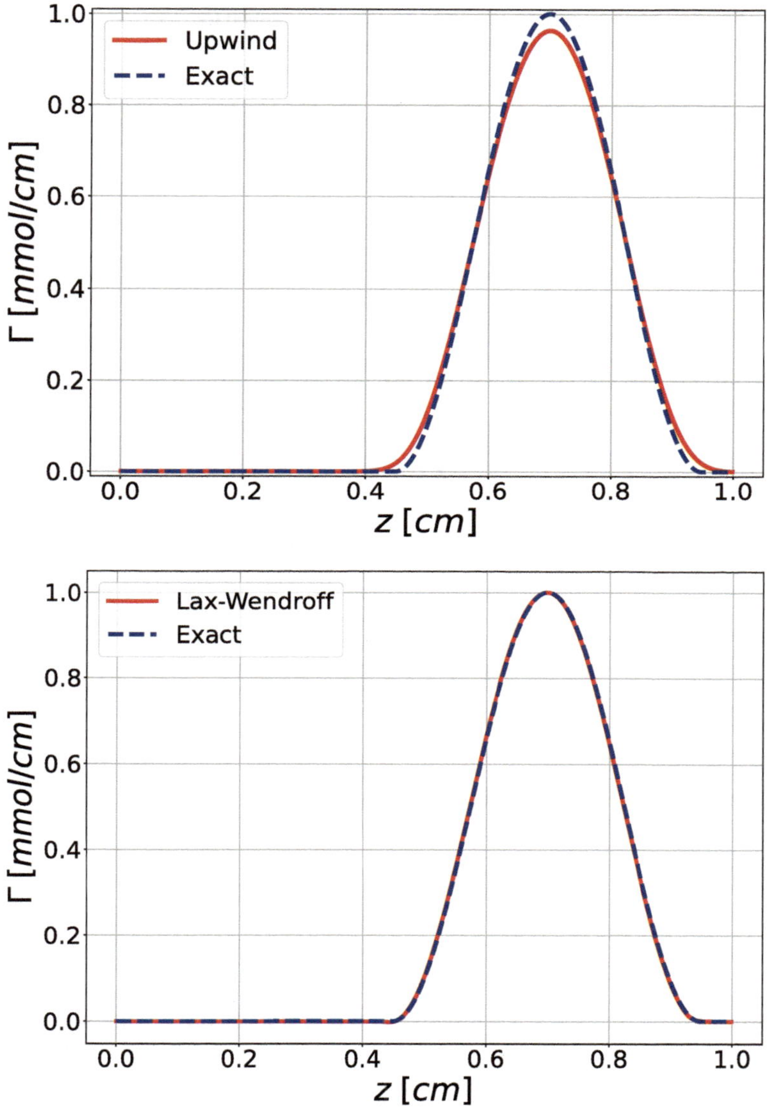

Fig. 4.14 Comparison of the implicit finite upwinding method and Lax-Wendroff method for a smooth solution. The curves show the solutions for $t = 0.4$ s

Table 4.11 Convergence and stability of the Lax-Wendroff method for $K_{CFL} = 0.8$ and $K_{CFL} = 1.2$ (*Problem II*). The abbreviation *e.c.o* stands for *estimated convergence order*

	$K_{CFL} = 0.8$		$K_{CFL} = 1.2$	
Δt [s]	e_{LW}	$e.c.o$	e_{LW}	$e.c.o$
$4.00 \cdot 10^{-4}$	$5.027 \cdot 10^{-4}$	----	nan	----
$2.00 \cdot 10^{-4}$	$2.507 \cdot 10^{-4}$	1.00	nan	----
$1.00 \cdot 10^{-4}$	$1.252 \cdot 10^{-4}$	1.00	nan	----
$5.00 \cdot 10^{-5}$	$6.255 \cdot 10^{-5}$	1.00	nan	----
$2.50 \cdot 10^{-5}$	$3.126 \cdot 10^{-5}$	1.00	nan	----
$1.25 \cdot 10^{-5}$	$1.563 \cdot 10^{-5}$	1.00	nan	----
$6.25 \cdot 10^{-6}$	$7.813 \cdot 10^{-6}$	1.00	nan	----

$$\Gamma_j^n = \sum_k \alpha_k e^{ikj\Delta z} \cdot \gamma_k^n.$$

The factor γ_k depends on Δt and Δz and is denoted as *amplification factor*:

$$\gamma_k = |\gamma_k| \, e^{-i\omega_k \Delta t}.$$

This gives rise to the concepts dissipation and dispersion.

Definition 4.8 (Dissipation). The ratio

$$\epsilon_a(k) = \frac{|\gamma_k|}{|g_k|} \text{ is denoted as dissipation error of the } k\text{-th frequency.}$$

Definition 4.9 (Dispersion). The ratio

$$\epsilon_d(k) = \frac{|\omega_k|}{|v \cdot k|} \text{ is denoted as dispersion error of the } k\text{-th frequency.}$$

The *dissipation error* can be considered as the factor between numerical and exact amplitude, while the *dispersion error* denotes the difference between numerical and exact propagation velocity. For our methods it holds:

(1) The Lax-Wendroff method has only small dissipation and dispersion errors.
(2) The Finite Volume method exhibits small numerical dispersion, but high dissipation errors.
(3) If the solution has a jump or high gradients, low order methods show a dissipative behavior. Higher order methods approximate the jumps quite well, but exhibit numerical oscillations.

The points listed above suggest that one needs further concepts besides consistency and stability to investigate a numerical method. Three important concepts read as follows:

- **Monotonicity:** If for two initial conditions Γ_0 and $\tilde{\Gamma}_0$ it holds: $\Gamma_0 \geq \tilde{\Gamma}_0$, then the same holds for the solutions of (4.79):

$$\Gamma(t) \geq \tilde{\Gamma}(t), \ \forall t > 0.$$

  A method that accounts for this property is called **monotonous**. For these methods, the following implication holds:

$$\Gamma_i^n \geq \tilde{\Gamma}_i^n \Rightarrow \Gamma_i^{n+1} \geq \tilde{\Gamma}_i^{n+1}, \ \forall n > 1.$$

- **TVD:** It can be shown that the total variation TV of the solution Γ

$$TV(\Gamma) = \int_{-\infty}^{\infty} \left| \frac{\partial \Gamma}{\partial z} \right| dz$$

  is monotonous decreasing:

$$TV(\Gamma(\cdot, t_n)) \geq TV(\Gamma(\cdot, t_{n+1})), \ \forall n > 1.$$

  The discrete counterpart of TV is defined as follows:

$$TV\left(\Gamma^n\right) = \sum_i \left|\Gamma_{i+1}^n - \Gamma_i^n\right|.$$

  A numerical method is denoted as TVD (total variation diminishing), if it holds [17, 18]:

$$TV\left(\Gamma^n\right) \geq TV\left(\Gamma^{n+1}\right), \ \forall n > 1.$$

- **Local mass conservation:** If Γ is a concentration variable, mass in an interval $[z_0, z_1]$ can be computed as follows:

$$m(t) = \int_{z_0}^{z_1} \Gamma(z, t)\, dz.$$

  A method approximating this variable up to a small discretization error, is referred to as **locally mass conservative**.

The first two properties prevent numerical oscillations in the vicinity of a jump, while the third property prevents large dissipation and dispersion errors.

Mass Conservation at a Bifurcation

After studying the convergence behavior of the Lax-Wendroff method and the implicit finite volume method, we now combine the two methods to simulate blood flow and transport processes through a single bifurcation. For this purpose, we use a small part of the network that we have already presented in Sect. 4.4.1.

Since long flow and transport paths should be studied, we consider three arteries in the lower right leg of the body, which are referred to in medical terminology as the femoral and tibial arteries (deep femoral artery, anterior tibial artery, and posterior tibial artery). In Fig. 4.8, these vessels are numbered by 40, 42, and 43. The corresponding model parameters can be found in Table 7.2. For simplicity, we choose a new numbering in this experiment: The main vessel deep femoral artery is numbered by 1, while the remaining vessels are numbered by 2 and 3. Figure 4.15 shows the structure of the bifurcation. The solution variables for Vessel i are denoted by A_i, Q_i, and Γ_i. At the beginning of the simulation period $t = 0$, all solution variables are set to zero. As boundary conditions for the blood flow model, we choose free outflow boundary conditions at the outlets of the tibial arteries. At the inlet of the femoral artery, the boundary condition (3.69) with $T = 1.0$ s and $Q_{\max} = 14.0\ {cm^3}/{s}$ is used. In case of the transport problem, we impose a volumetric concentration $c_1 = \Gamma_1/A_1$ at the inlet of the femoral artery, which has the following form:

$$c_1(0,t) = \begin{cases} 1.0\ \frac{mmol}{cm^3},\ t < 1.0\ \text{s}, \\ \\ 0.0\ \frac{mmol}{cm^3},\ t \geq 1.0\ \text{s}. \end{cases} \tag{4.88}$$

At the outlets of the tibial arteries, the averaged concentration $\Gamma_i,\ i \in \{2, 3\}$ is determined by upwinding. If numerical reflections produce negative values for Q_i, we use $\Gamma_i = 0$ as a surrogate value. At the bifurcation, we use the results from Sect. 3.6.3. The incorporation of the boundary conditions into the numerical methods is described in Sect. 4.2.

In order to evaluate the results of this experiment, we plot the volumetric concentrations $c_i = \Gamma_i/A_i$ at the midpoints $z_{i,m}$ of the vessels (see Fig. 4.16). As expected, the larger part of the solute flows into the second vessel, since its cross-sectional area is slightly larger than the cross-sectional area of Vessel 3. Due to the alternation of pulsating flow and periodes of rest, the concentration profile increases and decreases stepwise, respectively. In a next step, we investigate whether the mass transported through the inlet of Vessel 1 is also transported through the tibial vessels. For this purpose, we compute the number of particles $N_{femoral}$ that are transported into the system using (3.69) and (4.88):

$$\begin{aligned} N_{femoral} &= \int_0^{t_{end}} \frac{Q_1(0,t) \cdot \Gamma_1(0,t)}{A_1(0,t)}\, dt = \int_0^1 Q_1(0,t) \cdot c_1(0,t)\, dt \\ &= 2 \cdot 14.0 \cdot \frac{0.3}{\pi}\ \text{mmol} = \frac{8.4}{\pi}\ \text{mmol}, \end{aligned}$$

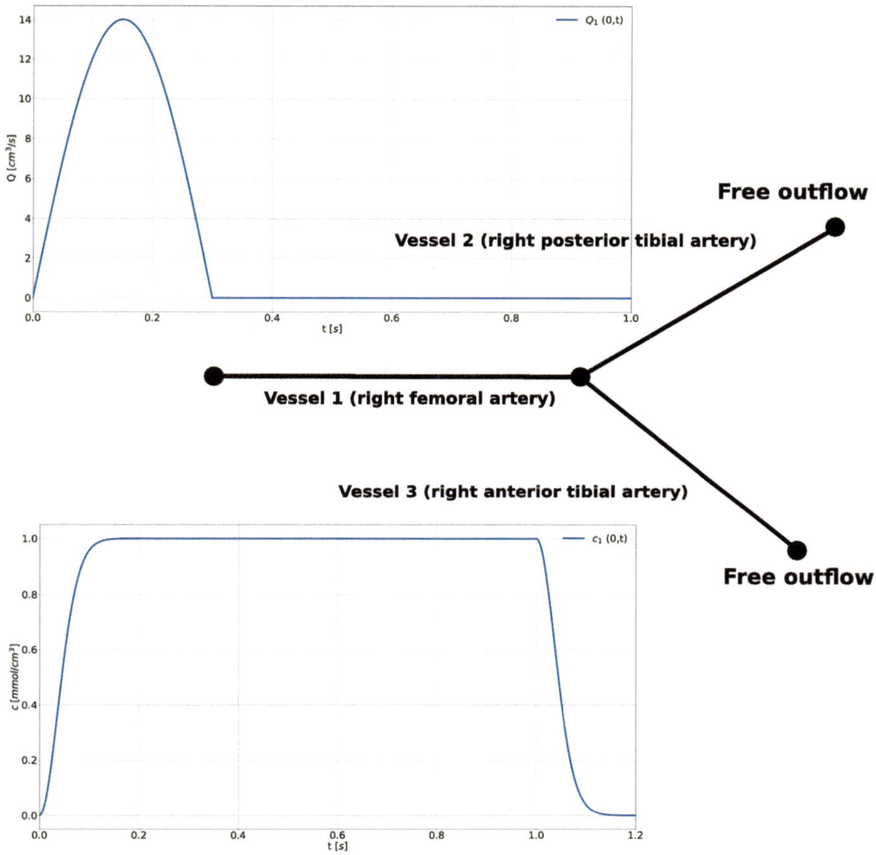

Fig. 4.15 Structure of the bifurcation. At the inlet of the main artery (right femoral artery) we prescribe for the flow rate Q a sine shaped profile. Furthermore, we set for $t \leq 1.0$ the volumetric concentration c to 1.0. At the outlets of the two branches we simulate a free outflow boundary. In the branching point the respective models are coupled such that the fluid mass and the mass of a solute are conserved

where $t_{end} = 35.0$ s denotes the end of the simulation period. Analogous to that we compute in the midpoints of the tibial arteries:

$$N_{atibial} = \int_0^{t_{end}} \frac{Q_2\left(z_{m,2}, t\right) \cdot \Gamma_2\left(z_{m,2}, t\right)}{A_2\left(z_{m,2}, t\right)}\, dt$$

and

$$N_{ptibial} = \int_0^{t_{end}} \frac{Q_3\left(z_{m,3}, t\right) \cdot \Gamma_3\left(z_{m,3}, t\right)}{A_3\left(z_{m,3}, t\right)}\, dt.$$

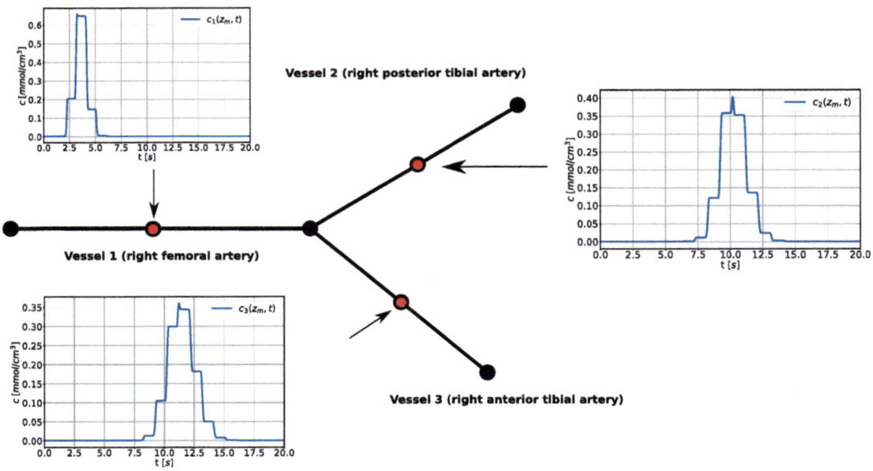

Fig. 4.16 This figure shows the volumetic concentrations c_i, $i \in \{1, 2, 3\}$, that are reported in the midpoints z_m of the respective vessels. As grid width and time stepsize we have used the following values: $\Delta z = 0.5$ cm and $\Delta t = 2.0 \cdot 10^{-4}$ s

Table 4.12 Relative errors e_N for different time stepsizes Δt and grid widths Δz

Δt [s]	Δz [cm]	e_N
$4.00 \cdot 10^{-4}$	2.0000	$7.75171 \cdot 10^{-4}$
$2.00 \cdot 10^{-4}$	1.0000	$1.90569 \cdot 10^{-4}$
$1.00 \cdot 10^{-4}$	0.5000	$2.98272 \cdot 10^{-5}$
$5.00 \cdot 10^{-5}$	0.2500	$2.16557 \cdot 10^{-7}$
$2.50 \cdot 10^{-5}$	0.1250	$1.53228 \cdot 10^{-8}$

By this, a relative particle or mass conservation error can be defined:

$$e_N = \frac{\left|N_{atibial} + N_{ptibial} - N_{femoral}\right|}{N_{femoral}}.$$

In Table 4.12 the error e_N is listed for different grid widths Δz and time stepsizes Δt. One can observe that e_N is decreasing, if Δz and Δt are decreasing. Thus the numerical model is consistent wrt. to e_N.

Concluding Remarks on Numerical Solution Methods

In this subsection, we have conducted three different convergence tests on the Lax-Wendroff and implicit upwind procedures. The following insights can be drawn from these experiments:

- **Lax-Wendroff method**
  - Advantages:
    (1) Due to the fact that the Lax-Wendroff method is a two-stage method and the spatial derivatives are approximated by central differences, a second-order method can be constructed in a relatively simple way.
    (2) The boundary conditions can be easily incorporated into the numerical scheme. An explicit formula for the characteristic variables is not required.
  - Disadvantages:
    (1) The consistency order/convergence order is fixed to 2. An order control with respect to the time and space discretization is not possible. A higher regularity of the solution can not be exploited by the Lax-Wendroff method.
    (2) In case of discontinuous solutions, numerical oscillations appear in the vicinity of jumps, which can no longer be removed once they occur.
    (3) The Lax-Wendroff method is subject to a stepsize constraint (CFL condition (4.67)). In particular for fine grids and high flow velocities, small time stepsizes have to be chosen.
- **Implicit Finite Volume method**
  - Advantages:
    (1) The implicit Finite Volume method is not subject to any CFL-like step size restriction. This means that it is possible to choose the stepsize indepently from the spatial grid width. Thus the stepsize does not have to be adapted to the stepsize of a solver for the blood flow model.
    (2) The construction of the implicit Finite Volume method is based on a discretization of a mass balance equation. Consequently, the method is *globally* mass conservative.
  - Disadvantages:
    (1) Due to the low order of approximation, high dissipation and dispersion errors occur. Thus, the method is *locally* not mass conservative.
    (2) Since the method is implicit, a system of equations must be solved for each timestep. The assembly of the system matrices is a huge challenge. In particular, if a transport process in a network has to be simulated.

Considering the disadvantages of the two methods, it becomes clear that more advanced solution techniques for hyperbolic differential equations have to developed. For this purpose, several approaches have been investigated in the past decades. Here, we will briefly discuss three different concepts:

- **Discontinuous Galerkin method**

  This method class is characterized by the fact that not the differential equation is discretized, but the weak formulation of the differential equation (see Sect. 7.8). The simulation domain is split up into elements (in 1D intervals). On each interval, the solution is approximated by a polynomial of degree p, where the different polynomials do not have to coincide at the interval boundaries, i.e., the solution can be discontinuous. This has the advantage that the solution does not have to be locally smooth and even discontinuities are allowed. The degree of the polynomials is not fixed, which makes order control possible [19, 20]. However, it turns out that if the solution is not smooth inside an element, numerical oscillations can occur. This can be prevented by damping the numerical oscillations.

One possible strategy to accomplish this is to use adaptive slope limiter procedures [16, 18, 21, 22]. This correction method makes it possible to approximate the solution in smooth regions with high accuracy or high polynomial degrees, while in rough regions one can use a low order approximation and damped derivatives [23].

Performing a spatial discretization, we obtain a system of ordinary differential equations. In order to solve the ordinary differential equations there is a variety of time integrators. One suitable class of time integrators, consists of higher-order Runge-Kutta methods with TVD properties [16, 23]. Since they do not deteriorate a high approximation order in space. The disadvantage of such a discretization method is its complexity. Efficient implementation, especially of high-order DG methods on a network, poses a variety of challenges. In order to couple the different elements and incorporate boundary conditions, one needs the explicit form of the characteristic variables. Furthermore, numerical analysis of these methods requires the use of techniques from functional analysis.

- **Method of characteristics**

A further approach to solve the blood flow model (3.21) numerically is to transform it into its characteristic form (3.42) and to use formulas for the characteristic variables (3.43). The spatial part of the computational domain is discretized by a uniform grid. In each time step $\left[t_j, t_{j+1}\right]$ the forward and backward moving characteristic curve are traced back in the space time domain from each grid node on at $t = t_{j+1}$ until $t = t_j$ is reached. Thereby the linearized characteristic curve is used instead of the actual characteristic curve. As a result, the point in which the linearized characteristic curve is intersecting the spatial domain at $t = t_j$ can be determined without great effort. Using linear interpolation the numerical solution values can be computed at these points. These values are used to compute the numerical solution at the grid nodes for $t = t_{j+1}$.

A big advantage of this method is that it is easy to implement and that there is no CFL-type timestep restriction. This means that using fine grids in space, larger timesteps can be exerted [15]. The drawback of ths method is that the convergence order is 1 in both in space and time. This means that as in case of the implicit Finite Volume method large dissipation errors can be observed. A problem is caused by characteristic curves leaving the simulation domain. Since they leave the domain, the solution can not be interpolated in space. In order to determine surrogate values, one can compute the time points at which the characteristic curves are leaving the respective boundary. At these time points the corresponding boundary condition is interpolated. By this, all the data for computing the solution at $t = t_{j+1}$ are available [24, 25]. Finally, one needs to note that the method is only applicable for flat velocity profiles, since an algebraic expression for the characteristic variables is required.

- **Data based learning**

Data-based learning has been successfully used in recent years not only to solve classification problems, but also to approximate non-linear mappings or solve partial differential equations. Well-known methods which have been successfully used in this context are e.g., *Reduced basis method* [26], *Kernel methods* [25, 27] or *Physics informed neural network (PINN)* [28]. By means of well-trained surrogate models, solutions of partial differential equations can be provided within a very short time. The PINN method has already been successfully applied to the blood flow problem. In [3] one can find a lot of information on how to train a PINN for the blood flow model (3.21) and how to normalize input data. The PINN trained in [3], describes a mapping of the space and time variable to the solution variables of the blood flow model. This mapping can be extended by further input or output parameters. A disadvantage of this method is that the architecture of the neural network is unknown i.e., it is unclear a priori how many neurons and hidden layers the neural network should contain. Moreover, the initialization of the weights and other hyperparameters is unclear. However, an appropriate choice of these parameters is necessary for a successful training process.

4.4.4 Experiment 3: Simulation of Realistic Flow and Transport Processes in a Network

At the end of this section we want to demonstrate that we can use the 1D-0D blood flow models and numerical solution methods to simulate realistic pressure and velocity curves. For this purpose, we now consider the entire network from Sect. 4.4.1. The corresponding network data can be found in Sect. 7.14. As initial conditions, we choose $Q \equiv 0$, $A \equiv A_0$, and $\Gamma \equiv 0$ for all vessels in the network. At the inlet of the aorta (Vessel 1), we use the boundary condition (3.69) with $T = 1.0$ and $Q_{max} = 485.0\ ^{cm^3}/_{s}$. To couple the models at the bifurcations, we use the concepts introduced in Sect. 3.6.3. Simulating physiologically relevant blood flows, the Windkessel model from Sect. 3.6.2 is considered. The values for the corresponding resistances and capacities are listed in Table 7.2.

In a first step, we first test to which extend the computational grid has to be refined in space and time in order to produce a representative solution. After that the influence of the Coriolis parameter α in (3.21) is the subject of our investigations. Finally, we compare our simulation data with values from literature and study the significance of the Windkessel effect.

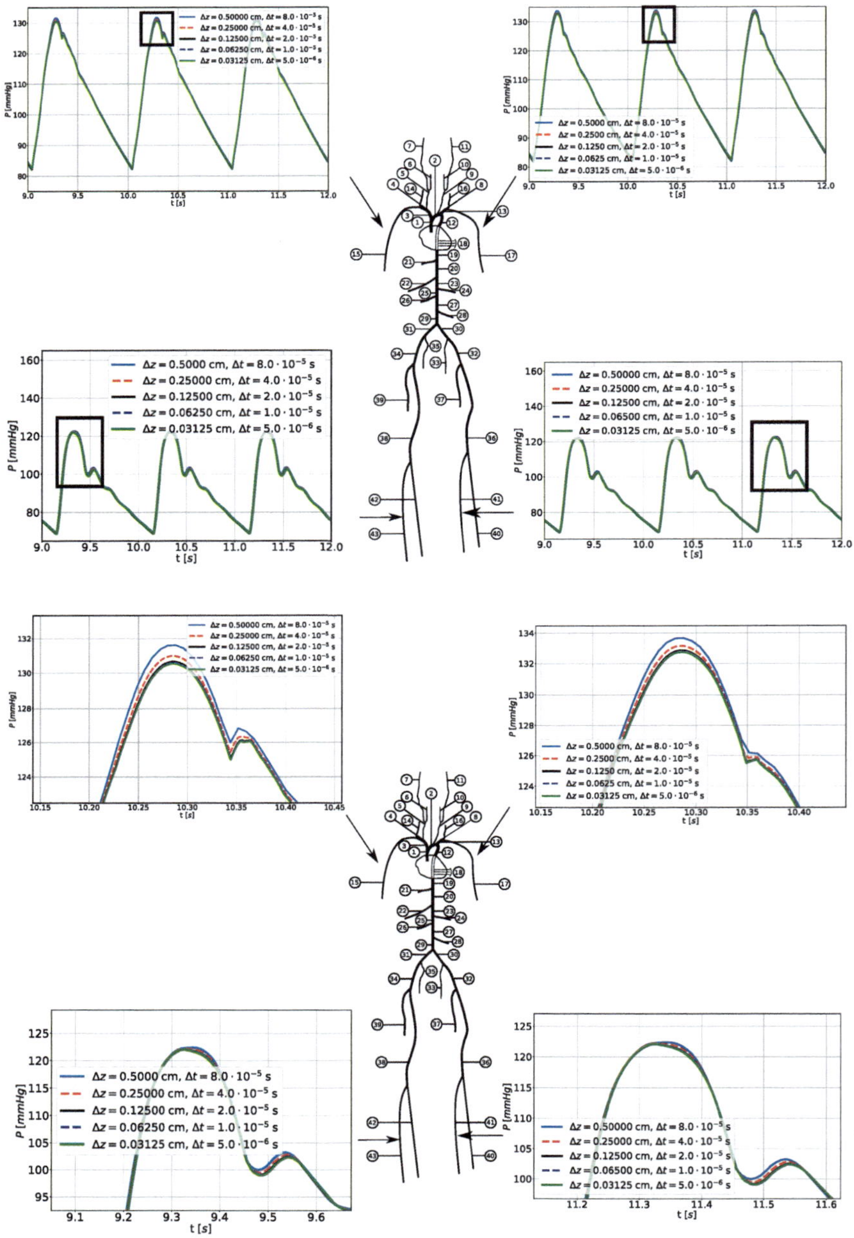

Fig. 4.17 The figure shows pressure curves for different discretization parameters reported in the middle of Vessel 15, 17, 41 and 43. The black rectangles mark regions, that are magnified in the lower part of the figure

Sensitivity wrt. Grid Width and Time Stepsize

Studying the influence of the discretization parameters Δz (spatial grid width) and Δt (time stepsize) on the simulation results, we plot pressure and flow rates curves in the middle of the two main arteries in the arms and two tibial arteries. For this experiment, we choose Δz and Δt as follows:

$$\Delta z = \frac{0.5 \text{ cm}}{2^i} \text{ and } \Delta t = \frac{8.0 \cdot 10^{-5} \text{ s}}{2^i}, \quad i \in \{0, 1, 2, 3, 4\}.$$

It can be seen in Fig. 4.17 that the pressure curves hardly change when the computational grid is refined. Considering the systolic (maximal) pressure, differences can be observed at the transition from $i = 0$ to $i = 1$ and $i = 1$ to $i = 2$. After that, the different curves can hardly be distinguished. This means that the results for $i = 2$ are already representative. For this reason, we use $\Delta z = 0.125$ cm and $\Delta t = 2.0 \cdot 10^{-5}$ s for the simulations in the remainder of this subsection.

Influence of the Coriolis Coefficient

Considering the blood flow model (3.21), it can be noted that a parameter α appears in the flux function. Its value depends on the shape of the velocity profile, which plays an important role in the derivation of the dimensionally reduced models (see also Sect. 3.2). In literature [29, Chap. 10] this parameter is also referred to as Coriolis coefficient (see also Eq. (3.12)). For a flat velocity profile, $\alpha = 1$ is obtained. However, in terms of realistic blood flows, one observes a profile that is flat in the center of the vessel and decreases very fast near the vessel wall (see Fig. 3.4). An approximation of such a profile is obtained by setting the parameter γ to 9 in (3.13). The corresponding Coriolis coefficient is 1.1.

So far, we have considered only the flat velocity profile with $\alpha = 1.0$ for our considerations. Thus, in Subsect. 3.4, the homogeneous blood flow model (3.22) has been used to prove the existence of classical solutions and to estimate where and when shocks are occuring. For this purpose, we required an analytical expression for the characteristic variables. However, it turns out that an analytical expression can only be derived for $\alpha = 1$. On the other hand, choosing $\gamma = 9$ yields a velocity profile that corresponds to the real flow conditions in a vessel. This means that one should actually choose $\alpha = 1.1$ to design a realistic model. However, in this case the characteristic variables can not be calculated in an analytical way, which makes a transformation between characteristic and primary variables difficult. For this reason, many approaches requiring an algebraic relationship between characteristic and primary variables such as (3.43), are based on a mixed model (3.23) [15, 30–33]. Thereby, $\alpha = 1.0$ is used in the flux function and $\alpha = 1.1$ in the viscous part of the source term. By means of (3.15) and (3.16), it can be seen that the choice of α significantly influences the viscosity parameter K_R. While K_R is not defined for $\alpha = 1.0$, one obtains for $\alpha = 1.1$: $K_R = 22 \cdot \pi \cdot \mu$. However, to be able to use a

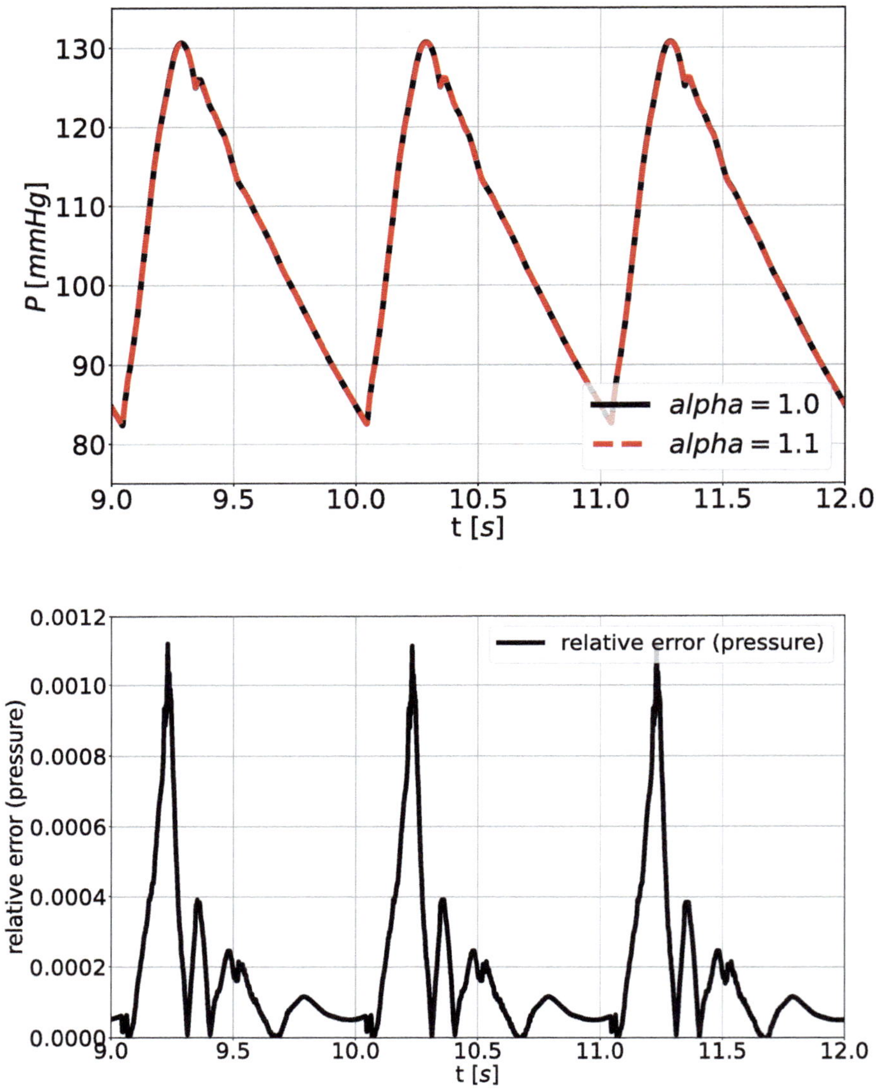

Fig. 4.18 The graphs at the top show pressure curves reported in the middle of Vessel 15 and the time period [9.0 s, 12.0 s]. Thereby simulation results for the Coriolis parameter $\alpha = 1.0$ and $\alpha = 1.1$ are compared. The lower graph shows the relative difference between these curves

realistic viscous term and at the same time the relationship (3.43), the mixed model (3.23) is quite often used.

One advantage of the Lax-Wendroff method is that no algebraic relationship between the characteristic and primary variables as in (3.43) is needed. For this reason, it can be used to investigate the difference between (3.21) and (3.23). Following

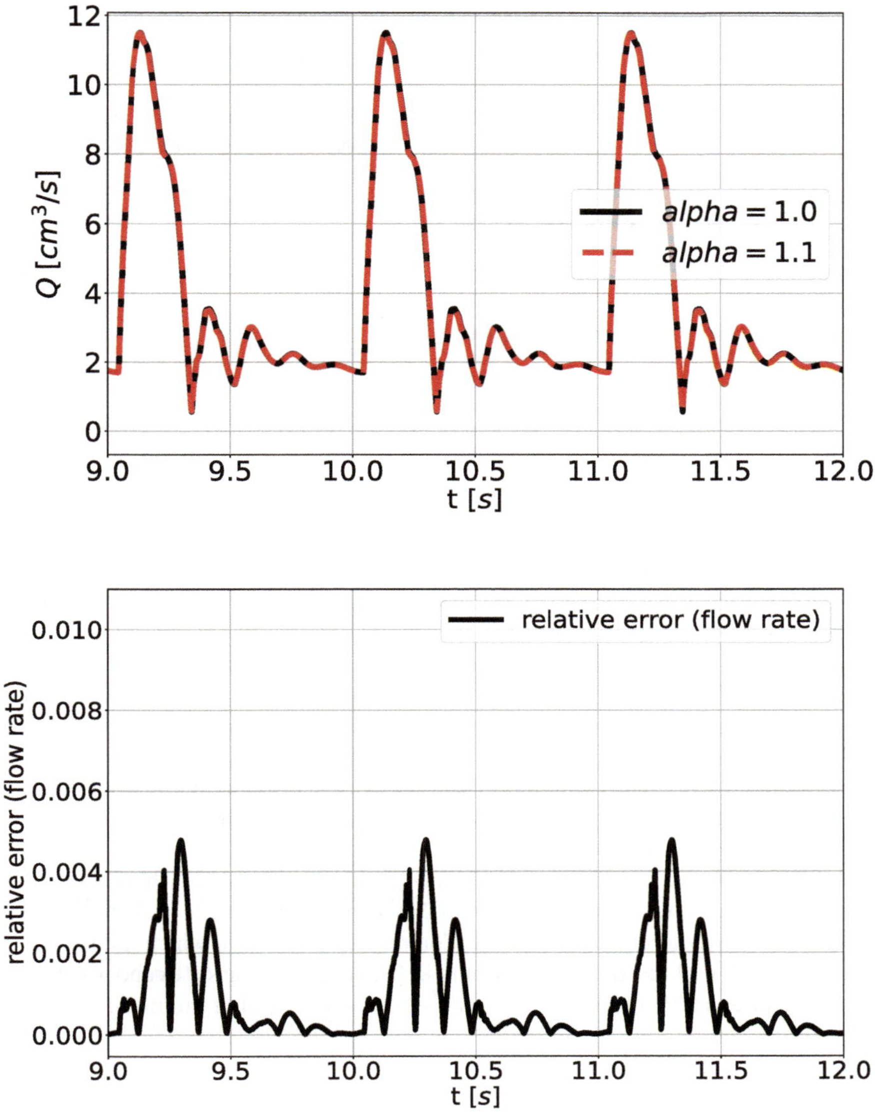

Fig. 4.19 The figure shows flow rates, reported in the middle of Vessel 15 and the time period [9.0 s, 12.0 s]. Thereby simulation results for the Coriolis parameter $\alpha = 1.0$ and $\alpha = 1.1$ are compared. The lower graph shows the relative difference between these curves

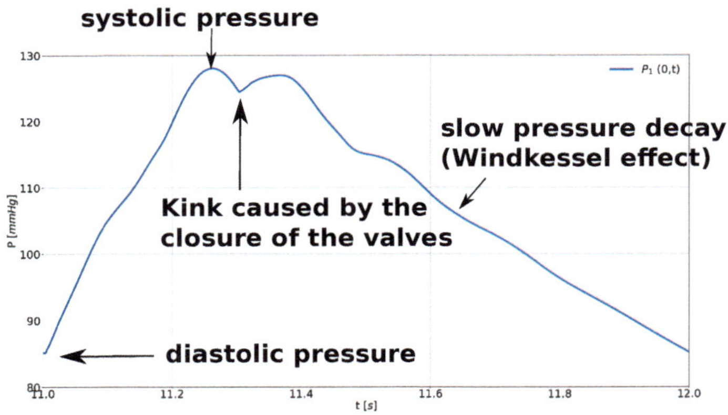

Fig. 4.20 Pressure curve reported at the inlet of the aorta (Vessel 1, see Fig. 4.21) and the time period [11.0 s, 12.0 s]

the insights from the last experiment, we use $\Delta z = 0.125$ cm and $\Delta t = 2.0 \cdot 10^{-5}$ s. In Figs. 4.18 and 4.19 one can find a comparsion of the pressure and flow rate curves that are reported during three heart beats in the middle of Vessel 15 and 41. One notes that the curves in Vessel 15 show little differences between $\alpha = 1.0$ and $\alpha = 1.1$. To make the difference clearer, we compute the relative error of the model for $\alpha = 1.0$ compared to $\alpha = 1.1$. The relative error is given by the absolute value of the differences normalized by the systolic pressure for $\alpha = 1.1$ in case of the pressure and $12.0\ {}^{cm^3}\!/_s$ in case of the flow rate. The relative differences are smaller than $1.2 \cdot 10^{-3}$ and $6.0 \cdot 10^{-3}$ respectively. Similar observations can be made in the other vessels. This means that both the blood flow model (3.21) and (3.23) can be used to simulate realistic pressure and flow rate curves.

In [32] a similar comparison can be found. The main difference is that for $\alpha = 1.0$ the viscous term is set to zero, while in our experiment both for $\alpha = 1.0$ and $\alpha = 1.1$ the same viscosity term has been used. Contrary to the results, which are shown in Figs. 4.18 and 4.19 the simulation results in [32] exhibit clear differences between $\alpha = 1.0$ and $\alpha = 1.1$.

Comparison with Medical Data

After investigating the influence of discretization and modeling errors on the simulation results in the last two experiments, we now want to compare the simulation results with data from the medical literature. Thereby, the medical data compiled in Chap. 1 are reconsidered.

Considering the pressure curve in Fig. 4.20, which is reported during one heart beat, one observes that the form of the pressure curve is similar to those in Figs. 1.10 and 1.11 from Chap. 1. It is revealed that both the simulated and schematic curves exhibit a kink caused by closing the heart valves. After that the pressure

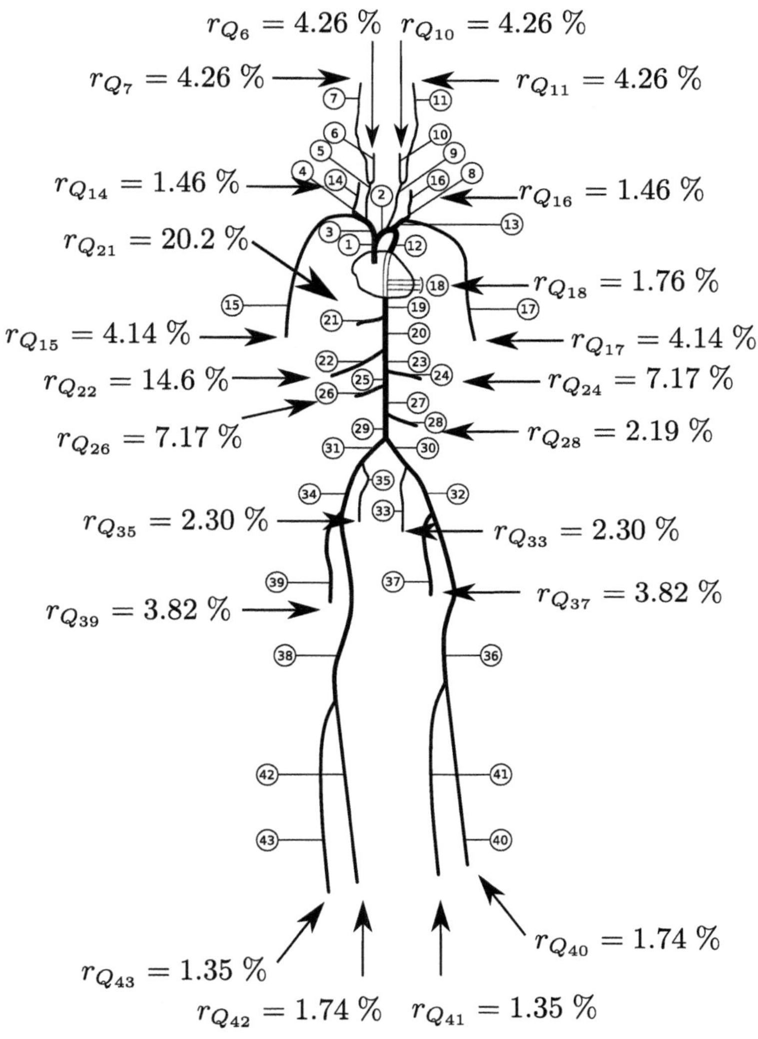

Fig. 4.21 The percentages r_{Q_i} at the outlets of Network 43 indicate the distribution of blood volume flowing through the inlet of the aorta into the network

does not drop down to zero. As in medical literature a slow decay can be observed. The systolic pressure varies between 120 mmHg and 130 mmHg. This is in accordance with physiological values (see Fig. 1.10). The diastolic pressure is around 85 mmHg. Comparing this value with data from Sect. 1.1.2 one can see that this value is a little bit larger compared to physiological values. According to Sect. 1.1.2, the diastolic pressure is in the range between 60 mmHg and 80 mmHg. In Table 4.13, one can find further simulation data reported at $z_{i,m} = 0.5 \cdot l_i$ and for the time period [11.0 s, 12.0 s].

$$l_i,\ i \in \{1, \dots, 43\}$$

are the lengths of the different vessels. These are the systolic and diastolic pressures at $z_{i,m}$ as well as the averaged pressure $\overline{P}$:

$$\overline{P}\left(z_{i,m}\right) = \frac{1}{12.0\ \text{s} - 11.0\ \text{s}} \int_{11}^{12} p_i\left(z_{i,m}, t\right)\, dt.$$

In the same way the averaged velocity $\overline{v}$ is computed:

$$\overline{v}\left(z_{i,m}\right) = \frac{1}{12.0\ \text{s} - 11.0\ \text{s}} \int_{11}^{12} \frac{|Q_i|}{A_i}\left(z_{i,m}, t\right)\, dt$$

Using $\overline{v}$ the Reynolds number Re is computed according to (1.2), while the Womersley number Wo is governed by Eq. (1.1).

Considering the data in Table 4.13, it can be observed that the systolic pressure is increasing towards the peripheral vessels. Contrary to that the diastolic pressure is slightly decreasing. This can be observed e.g., in the brachial arteries 15 and 17 as well as the arteries 37 and 39 contained in the legs. The reason for this phenomenon is due to the resistance of the small vessels and the blood volume pushed towards the peripheral vessels. Both mechanisms result in an increased systolic pressure on the vessel wall. The pressure curve, which can be found in Fig. 1.11 exhibits a similar behavior. For most of the vessels the averaged pressure $\overline{P}$ is given by 107 mmHg. Towards the tibial arteries the averaged pressure is decreasing. This observation is in accordance with the curve for the averaged pressure that can be seen in Fig. 1.11. Since the diastolic and systolic pressure, which are produced by means of the numerical model are increased, the averaged pressure is increased compared to the values in Fig. 1.11. All in all, one can note that the simulated pressures are reproducing the qualitative behavior of pressure curves from literature. However, the simulated pressure values are increased compared to the values from Chap. 1.

According to Sect. 1.1.2, the velocity in the large arteries varies between 15 $^{cm}/_{s}$ and 20 $^{cm}/_{s}$. However, only the averaged velocities in the Vessels $2, 9, 18, 32 - 43$ are close to this range. All the other velocities obtained from the numerical model are significantly smaller. Consequently the resulting Reynolds numbers are too small. Considering Table 1.1 the Reynolds number in Vessel 1 should be around 4000.

Table 4.13 Simulation data, which are reported at the midpoints of the vessels contained in *Network 43*

ID	P_{sys} [mmHg]	P_{dia} [mmHg]	$\overline{P}$ [mmHg]	$\overline{v}$ [cm/s]	Re −−	Wo −−
1	127.356	85.207	107.736	8.71878	618.045	14.4363
2	128.445	85.207	107.594	13.3894	723.148	10.9991
3	128.630	85.206	107.793	7.35541	219.911	6.08878
4	129.194	85.213	107.788	6.56371	133.887	4.15412
5	129.098	85.211	107.650	12.3354	220.091	3.63363
6	131.072	85.135	107.024	7.01558	128.557	3.73183
7	131.072	85.135	107.024	7.01542	128.554	3.73183
8	129.402	85.207	107.709	9.90090	572.932	11.7847
9	129.221	85.204	107.642	13.7709	245.703	3.63363
10	133.725	84.809	106.692	6.99888	128.251	3.73183
11	133.725	84.809	106.690	6.93135	127.013	3.73183
12	130.472	85.209	107.700	10.5173	568.029	10.9991
13	130.564	85.210	107.814	5.21296	118.149	4.61569
14	129.488	85.174	107.705	11.3134	100.928	1.81681
15	129.425	85.193	107.664	9.97790	149.158	3.04439
16	130.841	85.176	107.743	11.3158	100.950	1.81681
17	130.767	85.194	107.694	9.97305	149.086	3.04439
18	131.646	85.183	107.739	14.0290	118.389	1.71861
19	131.634	85.213	107.658	12.4541	610.774	9.98756
20	133.834	85.235	107.670	10.8689	482.193	9.03496
21	133.856	85.216	107.727	11.0356	312.911	5.77452
22	134.681	85.228	107.686	12.0161	289.142	4.90049
23	134.712	85.247	107.681	9.27248	372.914	8.19039
24	134.808	85.225	107.683	12.3656	208.705	3.43721
25	134.853	85.251	107.682	8.48451	324.859	7.79757
26	134.930	85.227	107.677	12.3641	208.678	3.43721
27	134.952	85.254	107.638	9.79465	314.093	6.53071
28	135.999	85.236	107.570	13.5098	126.386	1.90520
29	136.058	85.264	107.599	8.70157	264.773	6.19681
30	136.144	85.263	107.596	8.46202	191.787	4.61569
31	136.144	85.263	107.596	8.46202	191.787	4.61569
32	136.207	85.195	107.348	16.9814	262.041	3.14260
33	136.309	85.179	107.431	15.8990	153.337	1.96412
34	136.207	85.195	107.348	16.9814	262.042	3.14260
35	136.309	85.179	107.431	15.8990	153.337	1.96412
36	135.852	84.419	105.888	14.5869	165.302	2.30784
37	136.013	84.405	105.929	14.4178	180.767	2.55336

Table 4.13 (continued)

ID	p_{sys} [mmHg]	p_{dia} [mmHg]	$\overline{P}$ [mmHg]	$\overline{v}$ $\left[\frac{cm}{s}\right]$	Re ––	Wo ––
38	135.852	84.419	105.888	14.5869	165.302	2.30784
39	136.013	84.405	105.929	14.4178	180.767	2.55336
40	129.557	79.638	98.2698	13.4701	123.416	1.86592
41	129.563	79.641	98.2751	13.0399	106.898	1.66950
42	129.557	79.638	98.2698	13.4701	123.416	1.86592
43	129.563	79.641	98.2751	13.0399	106.898	1.66950

The right column of Table 1.1 contains the values for the Womersley number. Our simulations indicate that the Womersley number in the large vessels is ranging between 7.7 and 10.5. Womersley numbers, which are larger that 7.7 or slightly smaller can only be found in Vessel i, where

$$i \in \{1, 2, 3, 8, 12, 19, 20, 23, 25, 28, 29\}$$

These are mainly the vessels representing the abdominal aorta. All the other vessels exhibit smaller Womersley numbers. In the following, we list some possible reasons for the differences between the simulated and medical data:

- The 1D models are based on averaging 3D quantities. Thereby, velocity peaks may disappear. This can result in lower velocities.
- Choosing the model parameters, it has been assumed that each vessel has a constant radius. In a vessel with decreasing radius, the flow velocity can increase. A different choice of other model parameters, such as the E-modulus, can also improve the results.
- All the simulations in this chapter are based on the assumption that the viscosity is constant. However, if one uses the curves from Fig. 1.14, Sect. 1.1.3, smaller viscosities may result in the smaller vessels of *Network 43*. This should yield larger Womersley numbers, in particular in the vessels with small Womersley numbers.
- Viscoelastic effects, which occur during the deformation of the vessel wall, are neglected.
- The resistances and capacities of the Windkessel models have been determined by means of the approach described in [31]. An alternative approach would be to use calibration algorithms as in [34].
- Modeling the influence of the heart a simple boundary condition has been used. Instead a 0D model such as (3.71) and (3.72) could be considered. Moreover, the duration of the heartbeat could be changed from $T = 1.0$ s to $T = 0.8$ s, which increases the frequency ω in (1.1) and thus the values of the Womersley numbers are increasing.
- Gravity has been neglected.

Concluding this subsection, we consider the data shown in Fig. 4.21. To the outlets of the network different percentages are assigned. They indicate the portion of blood volume leaving each outlet during a heartbeat. In order to calculate the percentages r_{Q_i}, we first determine the volume of blood flowing into the network through the inlet of Vessel 1:

$$V_1(0) = \int_{11}^{12} Q_1(0, t) \ dt.$$

For Vessels i of length l_i and an outlet i.e.

$$i \in \{6, 7, 10, 11, 14, 15, 16, 17, 18, 21, 22, 24, 26, 28, 33, 35, 37, 39, 40, 41, 42, 43\}$$

it holds:

$$V_i(l_i) = \int_{11}^{12} Q_i(l_i, t) \ dt.$$

Thus, we obtain: $r_{Q_i} = V_i(l_i) / V_1(0)$. As it can be seen in Fig. 4.21, approximately 4 % of the blood volume flows into the left and right arms (Vessels 15 and 17), while approximately 20 % are flowing into the head (Vessels 6, 7, 10, 11, 14 and 16). The remaining 72% of blood volume is distributed among the remaining outlets. These results are comparable to data from [31, Sect. 2.2] and [35]. According to these references, 5 % of the blood volume flows into one arm, 15 % − 20 % into the head, and 70 % − 75 % into the rest of the body. Contrary to the velocities, Reynolds and Womersley numbers, the distribution of blood volume is reproduced in accordance with medical literature.

Impact of the Windkessel Effect

At the end of this section, we want to investigate the influence of the Windkessel effect on the transport of a substance through a network. Compared to the previous simulations that take into account the Windkessel effect, we also perform simulations with free outflow boundary conditions. In both cases, the flow rate Q and the volumetric concentration $c = \Gamma / A$ are reported at selected points in the network. As a boundary condition for Γ and c we set at the inlet of the aorta:

$$c(0, t) = \frac{\Gamma(0, t)}{A(0, t)} = 1.0 \ {}^{mmol}/{}_{cm^3}, \ t \geq 1.0 \text{ s}.$$

For both cases the volumetric flow rates Q and concentrations c are considered in the midpoints of Vessel 27, 29, 30 and 31. The corresponding results are depicted in Figs. 4.22 to 4.25. It can be observed that the simulation results produced by the Windkessel model differ significantly from the simulation results generated with a free outlfow boundary. Neglecting the Windkessel effect (see Fig. 4.22) reflections and negative flow rates occur. Similar phenomena can be observed e.g., in [33]. The

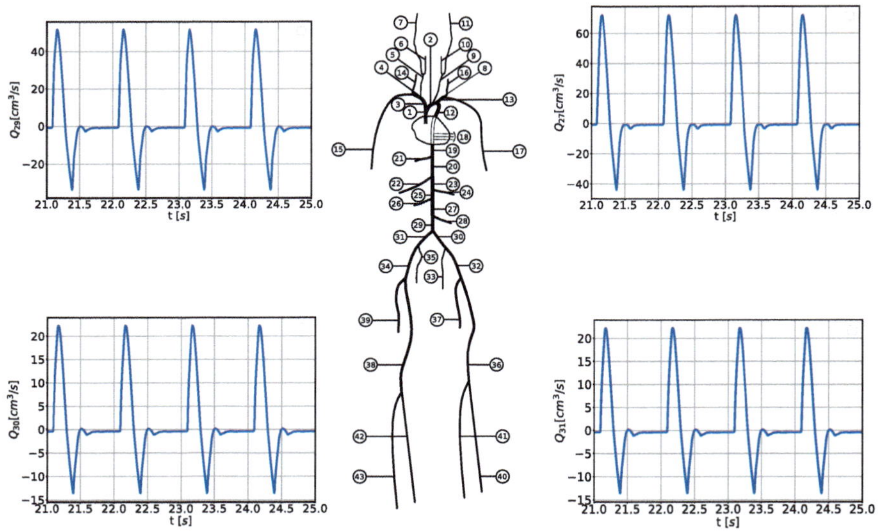

Fig. 4.22 Flow rates Q $\left[{}^{cm^3}/_{s}\right]$ reported at the midpoints of Vessel i, $i \in \{27,\ 29,\ 30,\ 31\}$. At the outlets of the network the Windkessel effect is *not* considered

reflections have the consequence that the concentration front is thrown back again and again, i.e., the transport of substances to the organs is slowed down, considerably.

Moreover, the maxima of the flow rates are smaller compared to the maxima resulting from the Windkessel effect (siehe Fig. 4.23). As in the free outflow case, the flow rates exhibit some reflections even if the Windkessel effect is applied. However, they are significantly smaller. A further difference is that the flow rates are almost zero after the reflections have occured. Apparently, the Windkessel effect produces flow rates Q that are positive apart from the reflections. This means that the concentration front is moving forward both during the diastole and systole. The reason for the different behavior is that due to the Windkessel effect the vessel walls are more enlarged compared as in case of the free outflow boundary condition. If no resistances are considered at the outlets, no pressure can build up in the vessels as it can be seen, e.g., in Fig. 4.17. Consequently the viscous resistance is higher without Windkessel effect. Due to the huge reflections the concentration curves exhibit oscillations before the maximal value $c_i = 1.0\ {}^{mmol}/_{cm^3},\ \ i \in \{27, 29, 30, 31\}$ is attained. The increased flow rates, caused by the Windkessel effect ensure that the concentration $c_i = 1.0\ {}^{mmol}/_{cm^3}$ is attained $9 - 10$ heartbeats earlier as without Windkessel effect. All in all, one can conclude that the Windkessel effect improves the supply of the organs with substances such as oxygen, significantly.

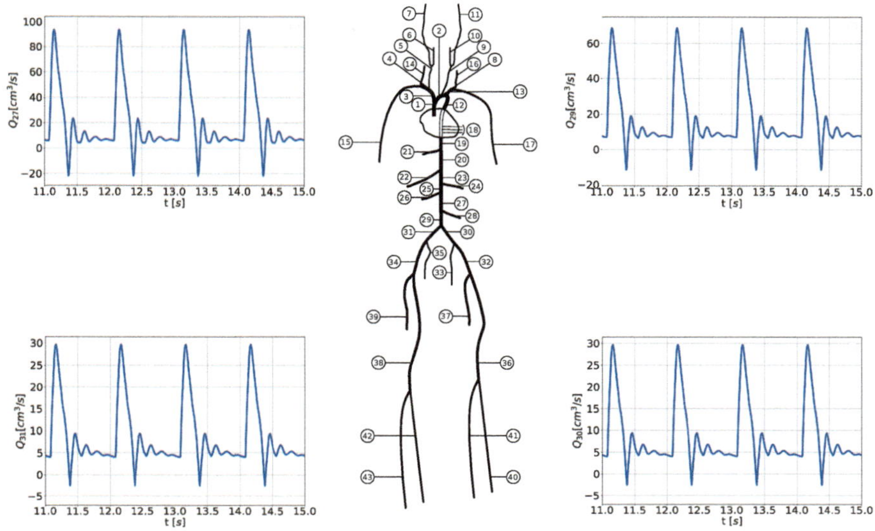

Fig. 4.23 Flow rates Q $\left[{cm^3}/{s}\right]$ reported at the midpoints of Vessel i, $i \in \{27,\ 29,\ 30,\ 31\}$. At the outlets of the network the respective 1D models are coupled with 0D Windkessel models

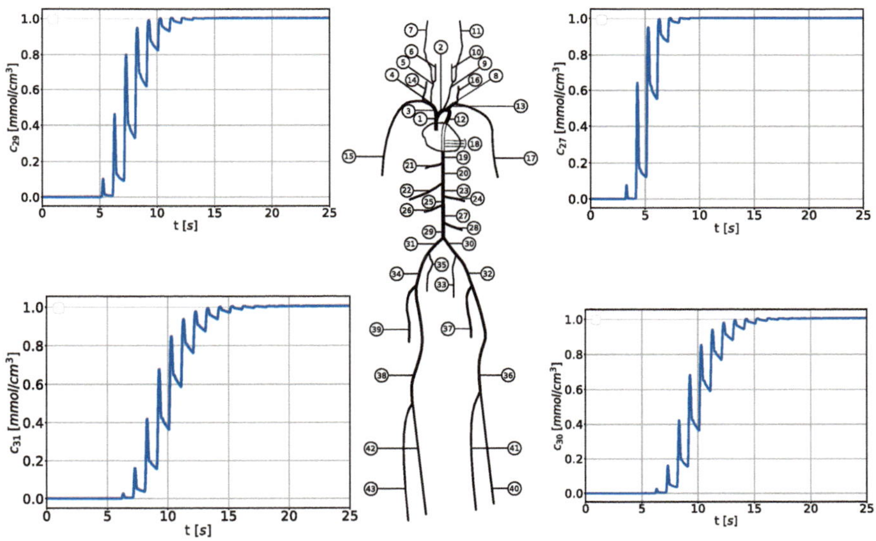

Fig. 4.24 Volumetric concentrations c $[{mmol}/{cm^3}]$ reported at the midpoints of Vessel i, $i \in \{27,\ 29,\ 30,\ 31\}$. At the outlets of the network the Windkessel effect is *not* considered

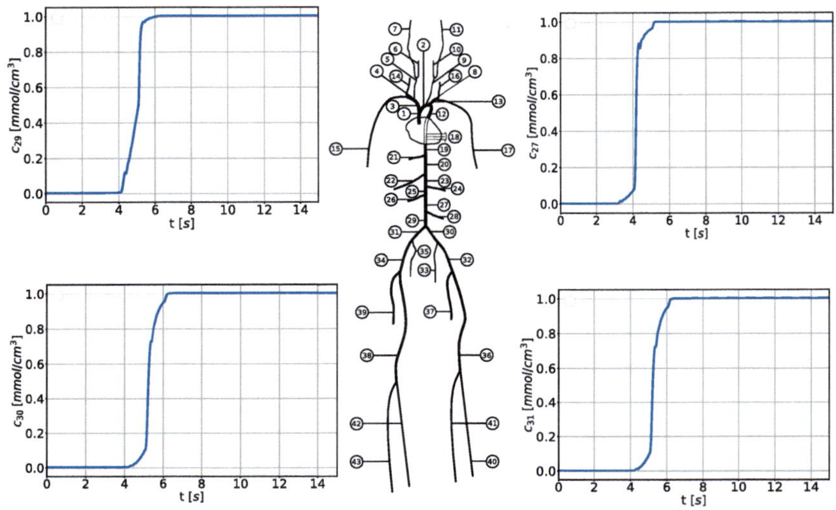

Fig. 4.25 Volumetric concentrations c [$^{mmol}/_{cm^3}$] reported at the midpoints of Vessel i, $i \in \{27, 29, 30, 31\}$. At the outlets of the network the respective 1D models are coupled with 0D Windkessel models

References

1. Lax P (1959) Systems of conservation laws. LOS ALAMOS NATIONAL LAB NM
2. Collins JB, Estep D, Tavener S (2014) A posteriori error estimation for the Lax–Wendroff finite difference scheme. J Comput Appl Math 263:299–311. Elsevier
3. Kissas G, Yang Y, Hwuang E, Witschey W, Detre J, Perdikaris P (2020) Machine learning in cardiovascular flows modeling: Predicting arterial blood pressure from non-invasive 4D flow MRI data using physics-informed neural networks. Comput Methods Appl Mech Eng 358:112623. Elsevier
4. LeVeque R (1992) Numerical methods for conservation laws. 132. Springer
5. Kolachalama V, Bressloff N, Nair P, Shearman C (2007) Predictive haemodynamics in a one-dimensional human carotid artery bifurcation. Part I: Application to stent design. IEEE Trans Biomed Eng 54(5):802–812. IEEE
6. Kolachalama V, Bressloff N, Nair P, Shearman C (2008) Predictive haemodynamics in a one-dimensional human carotid artery bifurcation. Part II: Application to graft design. IEEE Trans Biomed Eng 55(3):1176–1184. IEEE
7. He Y, Liu H, Himeno R (2004) A one-dimensional thermo-fluid model of blood circulation in the human upper limb. Int J Heat Mass Transf 47:12–13, 2735–2745. Elsevier
8. Deuflhard P, Hohmann A (1991) Numerische Mathematik I. De Gruyter, Berlin
9. Liang F, Takagi S, Himeno R, Liu H (2009) Multi-scale modeling of the human cardiovascular system with applications to aortic valvular and arterial stenoses. Med Biol Eng Comput 47(7):743–755. Springer
10. Watt A (1990) Fundamentals of three-dimensional computer graphics. Addison-Wesley Longman Publishing Co., Inc
11. Werner D (2018) VI Spektraltheorie kompakter Operatoren, Funktionalanalysis. Springer, Heidelberg, pp 279–343. isbn=978-3-662-55407-4, https://doi.org/10.1007/978-3-662-55407-4_6
12. Chen G-Q, Liu J-G (1997) Convergence of difference schemes with high resolution for conservation laws. Math Comput 66(219):1027–1053

13. Varah J (1975) A lower bound for the smallest singular value of a matrix. Linear Algebra Appl 11(1):3–5. Elsevier
14. Čanić S, Kim E (2003) Mathematical analysis of the quasilinear effects in a hyperbolic model blood flow through compliant axi-symmetric vessels. Math Methods Appl Sci 26(14):1161–1186. Wiley Online Library
15. Acosta S, Puelz C, Rivière B, Penny D, Rusin C (2015) Numerical method of characteristics for one-dimensional blood flow. J Comput Phys 294:96–109. Elsevier
16. Kuzmin D (2013) Slope limiting for discontinuous Galerkin approximations with a possibly non-orthogonal Taylor basis. Int J Numer Methods Fluids 71(9):1178–1190. Wiley Online Library
17. Cockburn B, Shu C (1989) TVB Runge-Kutta local projection discontinuous Galerkin finite element method for conservation laws. II. General framework. Math Comput 52(186):411–435
18. Krivodonova L (2007) Limiters for high-order discontinuous Galerkin methods. J Comput Phys 226(1):879–896. Elsevier
19. Masri R, Puelz C, Riviere B (2021) A reduced model for solute transport in compliant blood vessels with arbitrary axial velocity profile. Int J Heat Mass Transf 176:121379
20. Luo H, Baum JD, Löhner R (2008) A discontinuous Galerkin method based on a Taylor basis for the compressible flows on arbitrary grids, English. J Comput Phys 227:8875–8893
21. Krivodonova L, Xin J, Remacle J, Chevaugeon N, Flaherty J (2004) Shock detection and limiting with discontinuous Galerkin methods for hyperbolic conservation laws. Appl Numer Math 48(3):323–338. Elsevier
22. Michalak K, Ollivier-Gooch C (2008) Limiters for unstructured higher-order accurate solutions of the Euler equations. In: Proceedings of the AIAA Forty-Sixth Aerospace Meeting
23. Köppl T, Wohlmuth B, Helmig R (2013) Reduced one-dimensional modelling and numerical simulation for mass transport in fluids. Int J Numer Methods Fluids 72(2):135–156. Wiley Online Library
24. Drzisga D, Köppl T, Pohl U, Helmig R, Wohlmuth B (2016) Numerical modeling of compensation mechanisms for peripheral arterial stenoses. Comput Biol Med 7:190–201. Elsevier
25. Koeppl T, Santin G, Haasdonk B, Helmig R (2018) Numerical modelling of a peripheral arterial stenosis using dimensionally reduced models and kernel methods. Int J Numer Methods Biomed Eng 34(8):e3095. Wiley Online Library
26. Quarteroni A, Manzoni A, Negri F (2015) Reduced basis methods for partial differential equations: an introduction. 92, Springer
27. Santin G, Haasdonk B (2019) Kernel methods for surrogate modeling. Model Order Reduc 1:311–354
28. Cai S, Mao Z, Wang Z, Yin M, Karniadakis G (2022) Physics-informed neural networks (PINNs) for fluid mechanics: A review. Acta Mech Sinica 1–12. Springer
29. Formaggia L, Quarteroni A, Veneziani A (Eds) (2010) Cardiovascular mathematics: Modeling and simulation of the circulatory system, vol 1. Springer Science & Business Media
30. Acosta S, Puelz C, Rivière B, Penny D, Brady K, Rusin C (2017) Cardiovascular mechanics in the early stages of pulmonary hypertension: A computational study. Biomech Model Mechanobiol 16(6):2093–2112. Springer
31. Alastruey JPKH, Parker KH, Peiró J, Byrd SM, Sherwin SJ (2007) Modelling the circle of Willis to assess the effects of anatomical variations and occlusions on cerebral flows. J Biomech 40(8):1794–1805
32. Puelz C, Čanić S, Rivière B, Rusin C (2017) Comparison of reduced models for blood flow using Runge–Kutta discontinuous Galerkin methods. Appl Numer Math 115:114–141. Elsevier
33. Sherwin S, Formaggia L, Peiro J, Franke V (2003) Computational modelling of 1D blood flow with variable mechanical properties and its application to the simulation of wave propagation in the human arterial system. Int J Numer Methods Fluids 43(6–7):673–700. Wiley Online Library
34. Ismail M, Wall W, Gee M (2013) Adjoint-based inverse analysis of windkessel parameters for patient-specific vascular models. J Comput Phys 244:113–130. Elsevier
35. Burton AC (1965) Physiology and biophysics of the circulation. Acad Med 40:8. LWW

Chapter 5
Compensation of Unilateral Carotid Stenoses

Abstract In the previous chapters we have discussed a mathematical model for simulating blood flow and transport processes in large arteries. Since the model contains non-linear terms and consists of coupled partial differential equations (see Chap. 3) an analytical solution it is hard to compute it in general. Thus it is necessary to compute the solutions of the model equations in a numerical way. As solution methods, we have considered the Lax-Wendroff method for the blood flow model and the implicit Finite Volume method for the transport processes. In Chap. 4 the convergence and stability properties are examined both analytically and numerically. By means of an arterial network consisting of 43 arteries, it could be demonstrated that using the numerical model realistic blood pressure curves and blood volume distributions can be produced. Now we want to use the knowledge from the previous chapters to investigate a concrete issue.

In Sect. 1.2 of this book, various cardiovascular diseases have already been described. One of these diseases is called arteriosclerosis and refers to the narrowing of arteries by deposits. In medical literature, these deposits are also referred to as stenoses. A growing stenosis reduces blood flow to an increasing extent, so that the supply of organs with oxygen and nutrients can be endangered. A stenosis contained in one of the carotid arteries, is also known as carotid stenosis. If only one of the two carotid arteries is affected, it is called unilateral carotid stenosis. A carotid stenosis is particularly dangerous, since it affects the supply of brain tissue. Fortunately, there are several compensation mechanisms to restore the reduced blood flow. In the following, we will consider two important compensation mechanisms. One compensation mechanism is given by a ring shaped vasculature that is located beneath the brain and is connected with the two carotid arteries (see also Fig. 1.18). The ring shaped structure and further flow paths can partially compensate the reduced blood flow out of a carotid artery. Another way to compensate the reduced blood flow is to locally reduce the flow resistance of the arterioles, which can facilitate blood flow to an undersupplied region of the brain. A reduction of flow resistance by the arterioles is connected with the partial pressure of oxygen PO_2 in brain tissue. If PO_2 in brain tissue decreases, the section areas of arterioles are enlarged, lowering the flow resistance. The term describing this mechanism is called metabolic autoregulation. Now

T. Köppl and R. Helmig, *Dimension Reduced Modeling of Blood Flow in Large Arteries*, Mathematical Engineering, https://doi.org/10.1007/978-3-031-33087-2_5

we study the issue to which narrowing degree the reduced blood flow can be compensated. Furthermore, it is interesting to determine whether the compensation potential of the Circle of Willis and the arterioles must be fully exploited to compensate the missing blood volume. Using the simplified 1D-0D coupled blood flow model and other model equations, we attempt to find an answer. Similar investigations have been performed in [1–3]

The remainder of this chapter is divided into two further sections: In the first section, we describe the data set that is used for our simulations. We also present the model equations and list all parameter values. In particular, we extend the 1D-0D coupled model by model equations for the autoregulation. The simulation results are discussed in the second section of the chapter.

5.1 Mathematical Model and Choice of Parameters

Simulating blood flow and oxygen transport from the heart to the brain, we consider an arterial network, which has already been used in [2, 4]. This vessel system consists of 34 arteries and comprises the largest vessels, branching out of the heart and connecting it with the arms, head and the rest of the body. Moreover, the Circle of Willis is part of this network. The outline of the whole network can be seen in Fig. 5.1, while the model parameters are listed in Table 7.1 (see Sect. 7.13). In order to simulate the influence of a carotid stenosis, we consider a stenosis in the middle of the right carotid (Vessel 6). This means that Vessel 6 from the data set that can be found in [4] is split into two new Vessels 6 and 34. The lengths l_6 and l_{34} of the two vessels correspond approximatley to the half length of the old Vessel 6. Inbetween Vessel 6 and 34 there is a stenosis with a length of $l_s = 1.0\,\text{cm}$. Thus, it holds:

$$l_{34} = l_6 = \frac{l_{6,old} - l_s}{2}.$$

All the other parameters remain unchanged. Blood pressure p and volumetric flow rate Q in this network, are determined by means of the non-linear 1D-0D coupled model from Chap. 3 and 4. The influence of the stenosis between Vessel 6 and 34 is simulated by the 0D model described in Sects. 3.6.4 and 4.2.4. Thus it remains to model the autoregulation. For this reason, we enhance the 0D Windkessel model at the outlets of the Circle of Willis $i \in \{23, 24, 29, 30, 32, 33\} = I_{cer}$ by further equations. In the following, we denote these outlets as cerebral outlets and the resistances of the corresponding 0D models by cerebral resistances. Considering the Windkessel models (see e.g. Equation (3.79)), resistance R_2 is varied. Contrary to that R_1 is given by the characteristic impedance (3.81) of the 1D model coupled with the Windkessel model. Thus R_1 is fixed. This means that autoregulation could be achieved by adjusting the resistance R_2. In [1] autoregulation is governed by the concentration of carbon dioxide (CO_2) in brain tissue. Thereby, the cerebral resis-

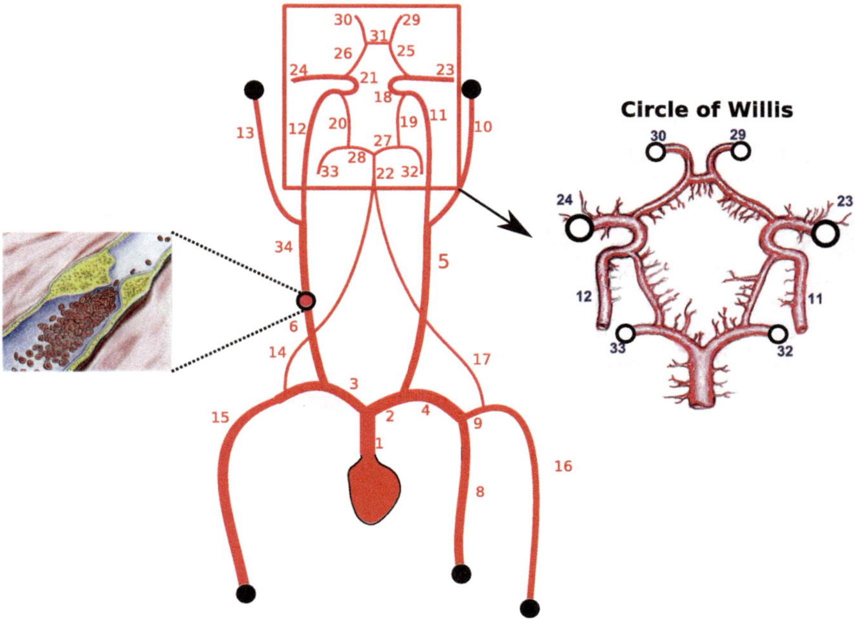

Fig. 5.1 This figure shows the outline of an arterial network comprising the brachial arteries as well as the Circle of Willis. At the inlet of Vessel 1 we prescribe a boundary condition for the volumetric flow rate Q, emulating the pulsation of the heart. Between Vessel 6 and 34 i.e. in the right carotid a stenosis is considered. At the black marked outlets we use a Windkessel model with fixed resistances to account for the omitted vessels. At the six outlets of the Circle of Willis we enhance the standard Windkessel model such that metabolic autoregulation is incorporated

tancies $R_{2,i},\ i \in I_{cer}$ are increased, if the concentration of carbon dioxide exceeds a certain threshold. Otherwise the resistances $R_{2,i}$ are decreased.

As it has already been mentioned, we use in this chapter the partial pressure of oxygen in brain tissue to determine the resistances $R_{2,i}$. For this purpose, we denote by $P_i O_2$ $\left[\text{mmHg}\right]$ the mean partial pressure in the part of the brain supplied by the cerebral outlet $i \in I_{cer}$. If $P_i O_2$ deviates from the required partial pressure $\overline{PO_2}$ $\left[\text{mmHg}\right]$, $R_{2,i}$ is adapted. As a first step, we compute a dimensionless coefficient $A_{R,i}$ using the following differential equation:

$$\frac{dA_{R,i}}{dt} = G_R\left(P_i O_2 - \overline{PO_2}\right), \ A_{R,i}(0) = 0, \tag{5.1}$$

where G_R $[1/\text{mmHg}\cdot\text{s}]$ influences the rate of change. The partial pressure is governed by Henry's law relating the partial pressure to the concentration of a substance. Using for the concentration of oxygen in brain tissue the symbol $(C_t O_2)_i$ $[\text{ml } O_2/\text{ml Gewebe}]$, Henry's law reads as follows:

$$(C_t O_2)_i = \alpha_G \cdot P_i O_2.$$

α_G is denoted as Henry's constant. Computing $(C_t O_2)_i$ we use an ODE, which can be found, e.g., in [6, Chap. 2, Sect. 2.5]:

$$V_i \frac{d\,(C_t O_2)_i}{dt}(t) = Q_i\,(l_i, t)\left(C_{O_2} - \sigma\,(C_t O_2)_i\,(t)\right) - v\,((C_t O_2)_i\,(t))\,. \qquad (5.2)$$

$V_i\ \left[\mathrm{cm}^3\right]$ is the tissue volume supplied by Artery i and C_{O_2} $[\mathrm{ml\ O_2/ml\ blood}]$ represents the oxygen concentration in blood. σ is a parameter that controls the outflow into the venous vasculature and v represents the rate of consumption of oxygen in brain tissue. The right-hand side of the differential equation consists of three terms. The first term represents the amount of oxygen transported through the outlet i, while the second term describes the transfer of oxygen into the venous system. Finally, the third term is governed by the consumption rate v of brain tissue:

$$v\,((C_t O_2)_i\,(t)) = 2 \cdot K_0 \cdot V_i \cdot \frac{(C_t O_2)_i\,(t)}{\overline{C_t O_2} + (C_t O_2)_i\,(t)}.$$

This relationship is also known as Michaelis-Menten law. K_0 $[\mathrm{ml\ O_2/ml\ tissue\cdot s}]$ is the mean consumption rate of brain tissue and $\overline{C_t O_2}$ $[\mathrm{ml\ O_2/ml\ tissue}]$ is the mean concentration of oxygen in brain tissue. The parameter σ is resulting from the following equation:

$$\overline{Q}\,(l_i)\left(C_{O_2} - \sigma \cdot \overline{C_t O_2}\right) - K_0 \cdot V_i = 0,\ i \in I_{cer}.$$

$\overline{Q}\,(l_i)$ is computed as follows:

$$\overline{Q}\,(l_i) = \frac{1}{T}\int_t^{t+T} Q\,(l_i, t)\ dt,$$

where T is the duration of a heartbeat. Using $A_{R,i}$ we compute the resistances $R_{2,i}$ as proposed in [1] by means of the following formula:

$$R_{2,i} = \frac{R_L + R_U e^{A_{R_i} - \hat{C}}}{1.0 + e^{A_{R_i} - \hat{C}}},\ \hat{C} = -\ln\left[\frac{R_{set,i} - R_L}{R_U - R_{set,i}}\right].$$

$R_{set,i}$ is the resistance corresponding to $A_{R,i} = 0$. This means that $R_{set,i}$ is equivalent to $R_{2,i}$ given by:

$$R_i = R_{1,i} + R_{2,i} = R_{1,i} + R_{set,i}.$$

The total resistances R_i can be found in Table 7.1, Sect. 7.13, while $R_{1,i}$ is given by (3.81). R_U and R_L denote the maximum and minimum for $R_{2,i}$. According to [1] they are chosen as follows:

Table 5.1 List of model parameters [1, 8, 9, Chap. 36]

Parameter	Symbol	Value	Unit
Density	ρ	1.085	g/cm^3
Viscosity	μ	$4.500 \cdot 10^{-5}/\rho$	cm^2/s
Poisson number	ν	0.500	–
Oxygen concentration	C_{O_2}	0.2	ml O_2/ml Blut
Mean PO_2	$\overline{PO_2}$	25.0	mmHg
Consumption rate	K_0	$6.067 \cdot 10^{-4}$	ml O_2/ml Gewebe·s
Henry coefficient	α_G	$2.894 \cdot 10^{-5}$	ml O_2/ml Gewebe·mmHg
Brain volume	V_{br}	1.071	dm^3
Reaction coefficient	G_R	2.0	1/mmHg
Time stepsize	Δt	$2.0 \cdot 10^{-5}$	s
Length (stenosis)	l_s	1.0	cm
Parameter (stenosis)	K_u	1.20	–
Parameter (stenosis)	K_t	1.52	–

$$R_U = 1.30 \cdot R_{set,i} \text{ und } R_L = 0.74 \cdot R_{set,i}.$$

Please note that the sum of the volumes V_i is equal to the total brain volume V_{br}. To compute V_i, we use the following formula:

$$\frac{V_i}{V_{br}} = \frac{\overline{Q}\,(l_i)}{\sum_{i \in I_{cer}} \overline{Q}\,(l_i)}.$$

Solving the ODEs (5.1) and (5.2) we use a semi-implicit Euler method. This method is essentially equal to the implicit Euler method [7]. The only difference is that non-linear terms as the consumption rate v are handled in an explicit manner. Table 5.1 contains the fixed values for the model parameters and the time stepsize. In Table 7.1, Sect. 7.13 for each vessel the number of grid nodes N_i is listed. By this and the lengths l_i the grid widths $\Delta z_i = l_i / (N_i - 1)$ can be computed. At the inlet of the aorta (Vessel 1) we use the boundary condition (3.69) with $T = 1.0$ s and $Q_{max} = 485.0$ cm^3/s. As initial condition, we choose for the solution variables A_i and Q_i, $i \in \{1, \ldots, 33\}$ the following values: $A_i = A_{0,i}$ und $Q_i = 0.0$. After 10 s the pressure and flow rate curves become periodic and the autoregulation is switched on. From 15 s on, the influence of the stenosis is considered.

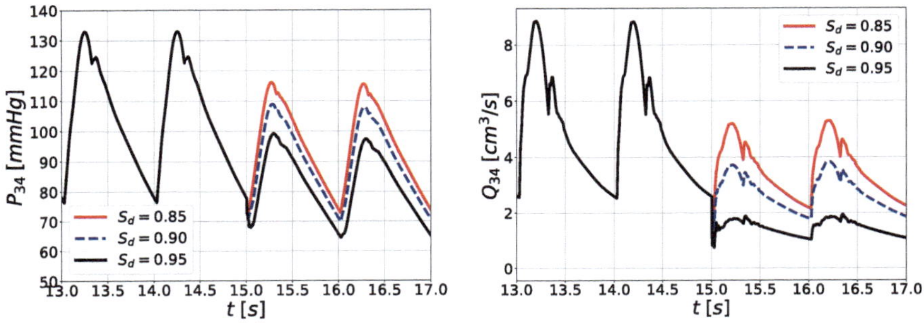

Fig. 5.2 The graphs show the pressure and flow rate curves, reported in the middle of Vessel 34. From $t = 15$ s on, the influence of the stenosis with degree S_d can be recognized

5.2 Discussion of the Simulation Results

In order to study the influence of a stenosis on oxygen supply of the brain, we vary the stenosis degree S_d (3.93). Thereby, we restrict ourselves to the following values:

$$S_d \in \{0.85, 0.90, 0.95\} .$$

Based on the choice of S_d the parameter K_v (3.92) is computed, since it depends on the diameter and the section area. Figure 5.2 shows the pressure and flow rate curves in the middle of the distal part of the carotid (Vessel 34). In the time interval [13.0 s, 15.0 s] a typical blood pressure curve can be seen. Blood pressure varies between 80.0 mmHg and 130.0 mmHg. Due to the Windkessel effect, we obtain a flow rate between 2.0 $^{\text{cm}^3}/_{\text{s}}$ and 9.0 $^{\text{cm}^3}/_{\text{s}}$ such that a continuous blood supply of the brain is ensured. The subsequent heartbeats show the influence of the stenosis between Vessel 6 and 34. As it could be expected a stenosis with increasing narrowing degree results in a decreasing blood pressure and flow rate. The systolic pressure is reduced by more than 30.0 mmHg for $S_d = 0.95$. Considering the flow rate, it can be observed that it is still positive, but smaller than 2.0 $^{\text{cm}^3}/_{\text{s}}$. The decreasing flow rate through the right carotid endangers the blood supply of the brain's right half, since there is only a little contribution by Vessel 12 (see Fig. 5.3). The missing blood volume can be partially compensated by redirecting the blood flow. We observe that the flow rates through Vessel 11 and 22 are increased as the stenosis degree is increased. The flow rates through the linking Vessels 20 and 31 are suddently increased. This means that the linking Vessels 20 and 31 play a crucial role in compensating the reduced blood flow through Vessel 12. Contrary to that the contribution of Vessel 19 can be neglected. It can be observed that its flow rates are low compared to those of Vessel 20 and 31. Moreover, they are hardly affected by the stenosis degree.

Redirecting the blood volume is not only enabled by a complete Circle of Willis, but also by reducing the flow resistances at the cerebral outlets. In Figs. 5.4 to 5.6

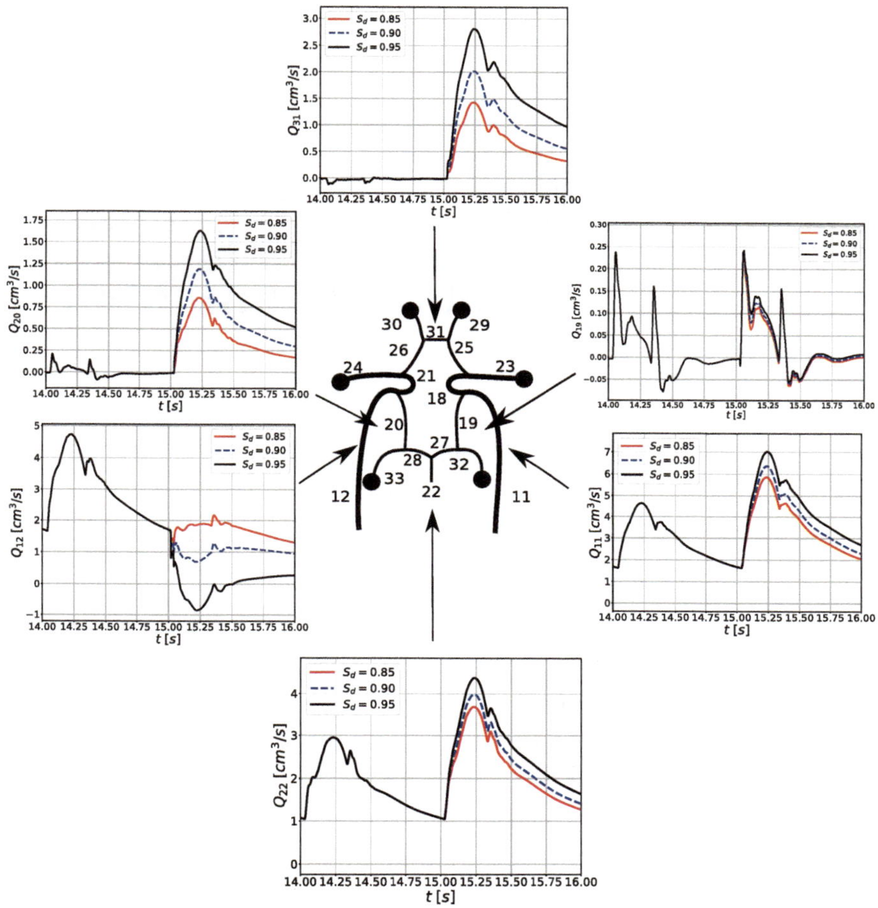

Fig. 5.3 Flow rate reported in the middle of Vessel 11, 12, 19, 20, 22 and 31 within two heartbeats. From $t = 15.0$ s on, flow in the Circle of Willis is remarkably affected by the stenosis

the normalized flow resistance is reported for the whole simulation time. After 10 s the autoregulation is activated, causing an increase of the flow resistance. In the time interval [13 s, 15 s] the normalized resistance is oscillating around 1, which can be expected under healthy conditions i.e. for $S_d = 0$.

After a narrowing of the right carotid the flow resistances are decreased. In particular at the outlets supplying the right half the brain the resistance has to be reduced remarkably. For $S_d = 0.95$ the resistance at Vessel 24 and 30 has to be minimized. For Vessel 32 only a small reduction is required, since Vessel 32 is located near Vessel 22 suppling the Circle of Willis with redirected blood volume.

Now the issue arises, whether the metabolic autoregulation and a complete Circle of Willis can fully restore the oxygen supply of brain tissue. In order to investigate

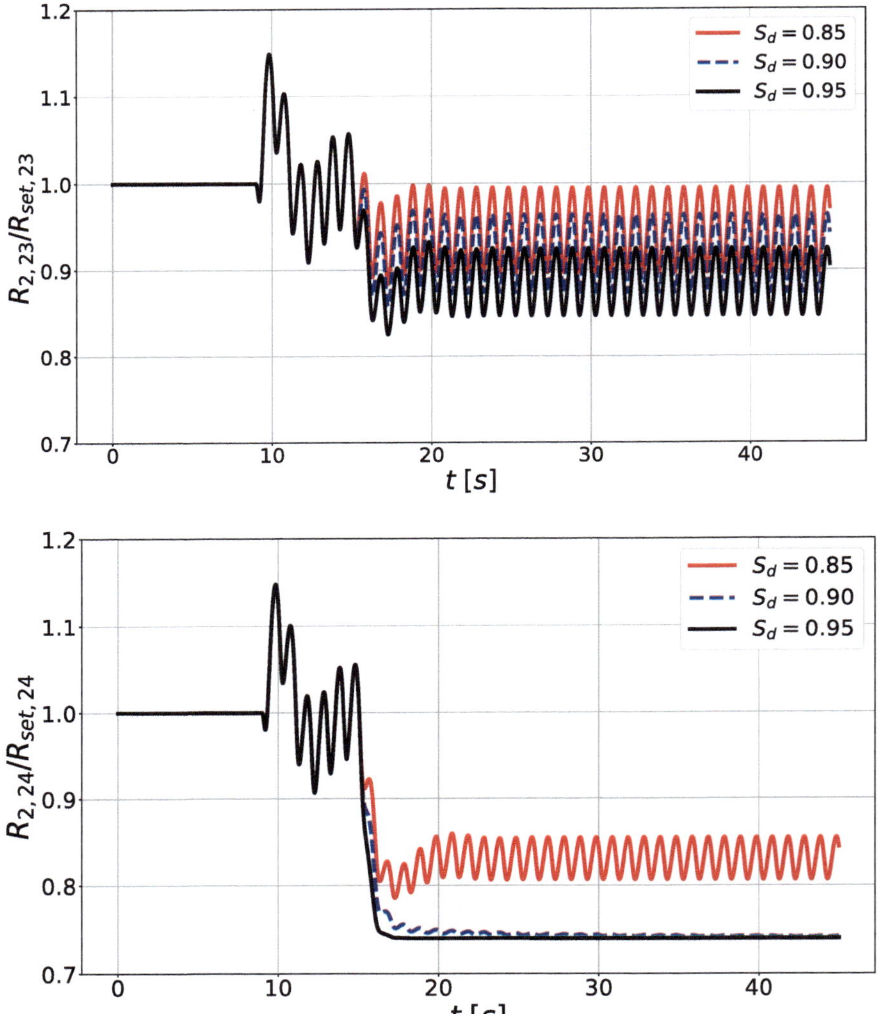

Fig. 5.4 Normalized resistances at the outlets of Vessel 23 and 24

this issue, we compute the relative deviation of the mean partial pressures $\overline{P_i O_2}$ wrt. $\overline{P O_2}$ (see Table 5.1). The mean partial pressures $\overline{P_i O_2}$ are determined as follows:

$$\overline{P_i O_2} = \frac{1}{T} \int_t^{t+T} P_i O_2 (t) \ dt$$

and the reduction rate of the partial pressure can be computed by:

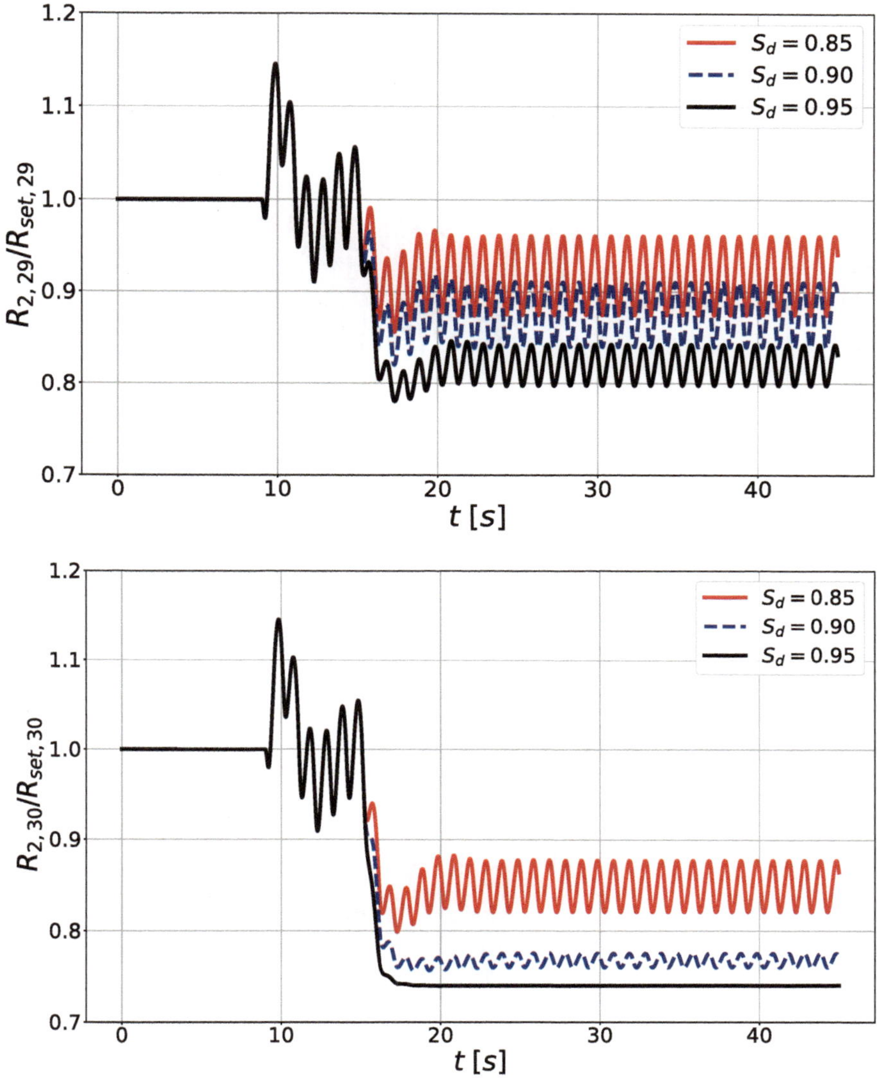

Fig. 5.5 Normalized resistances at the outlets of Vessel 29 and 30

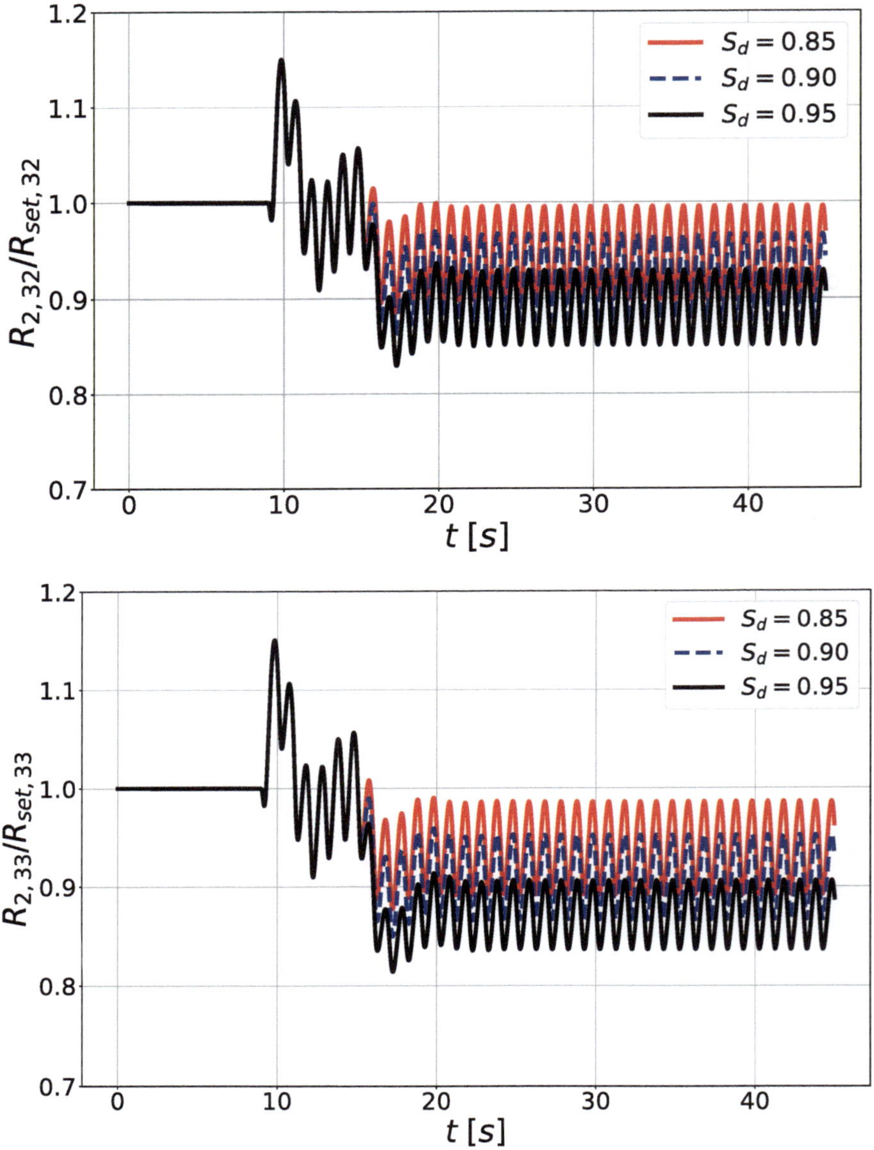

Fig. 5.6 Normalized resistances at the outlets of Vessel 32 and 33

Table 5.2 Reduction rates $r_{P_i O_2}$ of the mean PO_2. The Circle of Willis is complete and the metabolic autoregulation is active

	$S_d = 0.85$		$S_d = 0.90$		$S_d = 0.95$	
ID	[15 s, 16 s] (%)	[44 s, 45 s] (%)	[15 s, 16 s] (%)	[44 s, 45 s] (%)	[15 s, 16 s] (%)	[44 s, 45 s] (%)
23	0.87	0.24	1.18	0.24	1.65	0.24
24	2.54	0.25	3.77	0.34	5.68	4.19
29	1.23	0.24	1.73	0.25	2.48	0.25
30	2.19	0.25	3.20	0.25	4.78	2.67
32	0.82	0.24	1.11	0.24	1.54	0.24
33	0.93	0.24	1.28	0.24	1.80	0.24

$$r_{P_i O_2} = \left(1 - \frac{\overline{P_i O_2}}{\overline{PO_2}}\right) \cdot 100\ \%, \quad i \in \{23, 24, 29, 30, 32, 33\}\,.$$

In Table 5.2 the reduction rates for the time intervals [15 s, 16 s] and [44 s, 45 s] can be found. It can be observed that the reduction rates for the time interval [44 s, 45 s] is smaller than 0.40 %. This means that the partial pressure has essentially attained its old level. An exception is the outlet of Vessel 24 for $S_d = 0.95$. In this case $r_{P_{24} O_2}$ is around 4.19 %. All in all, one can conclude that a combination of a complete Circle of Willis and an active metabolic autoregulation can compensate the influence of a stenosis apart from the case $S_d = 0.95$. In order to support this statement, we deactivate in a further simulation the metabolic autoregulation. The reduction rates wrt. [44 s, 45 s] are shown in Fig. 5.7. It is revealed that without autoregulation there are significantly higher reduction rates. Even for $S_d = 0.85$ the right hand side of the Circle of Willis exhibits reduction rates between 3.0 % and 5.0 %. From this it can be concluded that the metabolic autoregulation plays a major role in compensating the reduced blood flow. As we have already noted in Sect. 1.2.5 the Circle of Willis can be incomplete i.e. some linking vessels such as Vessel 19 or 20 can be missing. In Fig. 1.19 we can see six different configurations of the Circle of Willis. Considering Fig. 5.3, one can realize that Vessel 31 plays an important role. For this reason, we investigate the case of a missing Vessel 31 as an example for an incomplete Circle of Wills. The corresponding reduction rates can be seen in Fig. 5.8. Only for $S_d = 0.85$ a complete compensation can be observed. In all the other cases there are high reduction rates for Vessel 24 and 30. If $S_d = 0.95$ holds, there are reduction rates of about 13.0 %. Summarizing the main results of this section, we obtain:

(1) Both the metabolic autoregulation and a complete Circle of Willis are necessary to compensate the impact of a carotid stenosis effectively.
(2) If there is a stenosis with $S_d = 0.85$ or $S_d = 0.90$ the reduction rates for a suppressed autoregulation and an incomplete Circle of Willis are less than 7 %. In case of an extreme stenosis i.e. $S_d = 0.95$ the reduction rates are much higher.

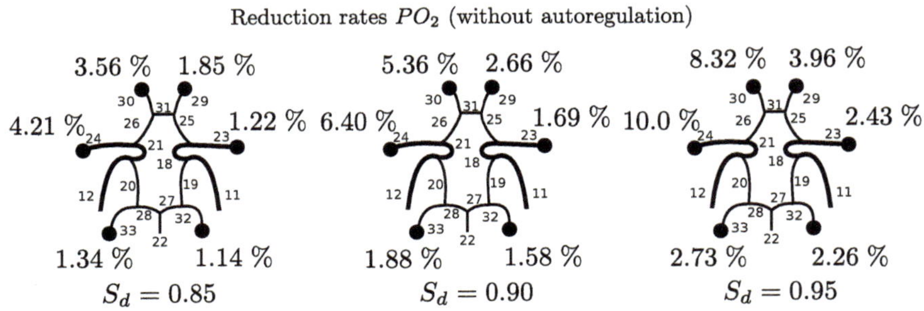

Fig. 5.7 Reduction rates $r_{P_i O_2}$ of the mean PO_2. The Circle of Willis is complete. However, the metabolic autoregulation is *not active*

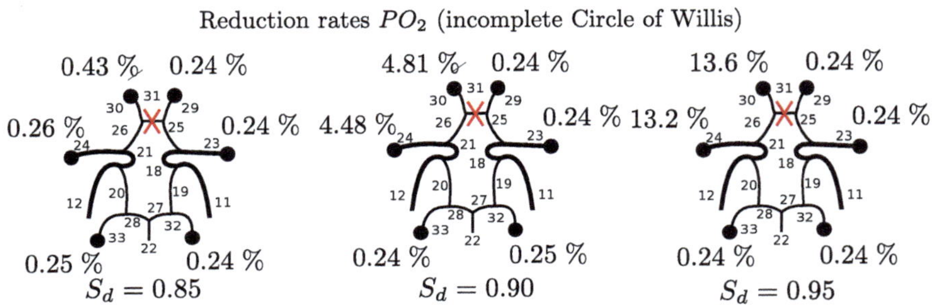

Fig. 5.8 Reduction rates $r_{P_i O_2}$ of the mean PO_2. The Circle of Willis is incomplete. However, the metabolic autoregulation is *active*

(3) For $S_d = 0.95$ even a combination of a complete Circle of Willis and metabolic autoregulation is not sufficient to obtain a complete compensation.

(4) The mean partial pressure of oxygen comprises a huge range of local PO_2 values in brain tissue [10]. The partial pressure varies between 5.0 mmHg and the arterial partial pressure of oxygen. This means that restoring the mean partial pressure of oxygen a local undersupply of brain tissue can not be excluded.

References

1. Alastruey J, Moore S, Parker K, David T, Peiró J, Sherwin S (2008) Reduced modelling of blood flow in the cerebral circulation: Coupling 1-D, 0-D and cerebral auto-regulation models. Int J Numer Methods Fluids 56(8):1061–1067. Wiley Online Library
2. Köppl T, Schneider M, Pohl U, Wohlmuth B (2014) The influence of an unilateral carotid artery stenosis on brain oxygenation. Med Eng Phys 36(7):905–914. Elsevier

3. Köppl TT (2015) Multi-scale modeling of flow and transport processes in arterial networks and tissue. Technische Universität München
4. Alastruey JPKH, Parker KH, Peiró J, Byrd SM, Sherwin SJ (2007) Modelling the circle of Willis to assess the effects of anatomical variations and occlusions on cerebral flows. J Biomech 40(8):1794–1805. Elsevier
5. Alastruey JPKH, Parker KH, Peiró J, Byrd SM, Sherwin SJ (2007) Modelling the circle of Willis to assess the effects of anatomical variations and occlusions on cerebral flows. J Biomech 40(8):1794–1805
6. D'Angelo C (2007) Multiscale modelling of metabolism and transport phenomena in living tissues. ÉCOLE POLYTECHNIQUE FÉDÉRALE DE LAUSANNE
7. Deuflhard P, Bornemann F (2002) Scientific computing with ordinary differential equations, vol 42. Springer Science & Business Media
8. Thews G (1960) Ein verfahren zur bestimmung des O2-diffusionskoeffizienten, der O2-leitfähigkeit und des O2-löslichkeitskoeffizienten im gehirngewebe. Pflüger's Archiv für die gesamte Physiologie des Menschen und der Tiere, 271(2):227–244. Springer
9. Schmidt RF, Lang F, Heckmann M (1989) Human Physiology. Springer-Verlag, Berlin Heidelberg New York, Second completely Revised Edition
10. Grote J, Laue O, Eiring P, Wehler M (1996) Evaluation of brain tissue O2 supply based on results of PO2 measurements with needle and surface microelectrodes. J Autonom Nerv Syst 57(3):168–172. Elsevier

Chapter 6
Concluding Remarks

Abstract Considering the different chapters of this book, it can be seen that each chapter represents an important building block in Scientific Computing. In the following, we provide a short summary of them and give an outlook.

In Chap. 1, we have presented some important information on blood flow in vascular networks and vascular diseases from medical literature. The acquisition of knowledge from other disciplines is quite often the first step on the way to a numerical model, since many issues that are investigated by means of computer methods, arise in natural science and economic sciences. Thus, it is inevitable to become familiar with unknown fields such as physiology. Thereby, the goal is to extract relevant information for creating a mathematical model. In case of blood flow modeling, it is of great importance to study the characteristics of the different flows in a cardiovascular system. Additionally, it is crucial to learn the terminology and identify issues that can be studied more effectively by means of computer simulations.

At the end of the first chapter, we have identified the influence of unilateral stenoses on the supply of brain tissue and the impact of compensation mechanisms as an interesting issue. Investigating such a issue is difficult by classical means from medicine of physics. Moreover numerical simulations can produce a huge amount of data in a relatively short time. In context of medical tests quite often one is confronted with the problem that there is only a limited number of persons for a specific test. Due to that the significance of conventional studies is limited. Thus combining data from experiments with simulation data can enhance the significance of medical studies.

The content of Chap. 1 has been used in Chaps. 2 and 3 to derive a mathematical model for blood flow and transport processes in larger arteries. Since we are not interested in a spatial resolved flow profile, but only in averaged pressures and flow rates, 3D models have been reduced to 1D models. By this, the complexity of the model could be remarkably reduced, since for 1D simulations of networks a much smaller amount of data is required. For example, a detailed description of the vessel geometry can be omitted. Our considerations have resulted in a hyperbolic and non-linear first order PDE system. A hyperbolic and first order PDE system corresponds to the character of a convective and pulsating flow. This is due to the fact that first order hyperbolic equations do not exhibit second order derivatives that could describe

T. Köppl and R. Helmig, *Dimension Reduced Modeling of Blood Flow in Large Arteries*, Mathematical Engineering, https://doi.org/10.1007/978-3-031-33087-2_6

diffusion driven flows. An appropriate choice of model equations wrt. complexity and physical significance, is crucial for the validity of mathematical models. Another important aspect, which has been discussed in Chap. 3, is the mathematical analysis of the model equations. Before the choice of a numerical solution method is made, it is necessary to determine the regularity of the solutions. If the solutions have a high regularity, a method with a high approximation order can be chosen. Otherwise, the numerical effort expended for the higher approximation order does not yield any return. In addition, it is interesting to see whether the model exhibits conservation variables such as energy. However, it may happen that a even a higher order method does not account for conservation variables. The simulation of molecular and planetary motions e.g. is usually exerted by symplectic integrators instead of standard Runge-Kutta methods. Symplectic integrators are able to preserve the total energy of mechanical systems up to small approximation errors [1]. The analysis of the hyperbolic model equations shows that the solutions have a higher regularity if the boundary and initial conditions are smooth. According to the mathematical analysis that has been performed in Chap. 3, the place at which a shock occurs is located beyond the application-relevant scope.

This has been confirmed by a numerical test in the following Chap. 4. The test has been performed by means of the Lax-Wendroff method having consistency order 2 in both space and time. Since we can expect locally a higher regularity even for boundary conditions (such as (3.69)) that are only piecewise smooth, the usage of a second order method is justified. The method has been investigated wrt. consistency and stability both analytically and numerically. It has been shown that the Lax-Wendroff method has consistency order 2 in both place and time. In addition, we have seen that the Lax-Wendroff method is subject to a CFL-like time step size constraint. To solve the transport equation, we have used an implicit Euler method for the time and a finite volume discretization for the space discretization. This method has also been investigated wrt. its stability and consistency order. In contrast to the Lax-Wendroff method, it does not exhibit a CFL-like stepsize constraint. This allows a transport process to be simulated more flexibly, since the time stepsize can be selected independently of the grid width in space. Consequently, the transport process can be solved detached from the blood flow model. This can be attractive, if blood flow has a periodic behavior. Knowing the velocity field for a heartbeat, we can efficiently simulate a transport process for a long period of time. In the proof of consistency and convergence, we have restricted ourselves to positive velocity fields. However, it could be shown that the method for the linear transport model has consistency and convergence order 1 in both space and time. Despite of the fact that this method has a low order of approximation, numerical tests show that mass is globally conserved. On the other hand, local mass conservation can not be observed, since the method exhibits strong numerical dissipation. In this book, the Lax-Wendroff method and the implicit finite volume method are considered to give an introduction into the numerics of hyperbolic PDEs. Of course, there are more sophisticated methods in literature, which are listed at the end of Sect. 4.4.3. However, studying these methods creates an awareness of the problems arising in terms of numerical treatment of hyperbolic equations. The different features of these methods reveal that a solid knowledge of different numerical methods and their properties is essential for the production of meaningful simulation data. At the end of this chapter, we have used the numerical models to simulate blood flow and transport processes through an arterial network. We observe that the model can produce realistic blood pressure curves and blood volume distribution. Finally, the influence of the Windkessel effect on transport processes has been investigated. Thereby, it has been shown that the Windkessel effect is crucial for a fast transport of solutes in blood.

After studying in Chaps. 3 and 4 the mathematical model and the numerical methods, we apply the techniques from these chapters to a concrete issue from physiology. We reconsider an issue that has already been discussed in Chap. 1. The focus of this issue is studying the influence of a unilateral carotid stenosis, i.e. the narrowing of a carotid artery by deposits, on the supply of brain tissue. To compensate the reduced blood flow caused by the stenosis, two compensation mechanisms have been considered. These are on the one hand, the Circle of Willis, a ring-shaped vascular system located beneath the brain tissue, and metabolic autoregulation, a mechanism adjusting the flow resistance by means of the arterioles. For this purpose, the previous model had to be extended to include equations for the metabolic autoregulation. We have also considered arterial networks containing a complete and an incomplete Circle of Willis. Our simulations show that even severe narrowings of the carotid artery can essentially be compensated by the two mechanisms. If one of the two compensation systems is not working, it can be seen that a severe narrowing can no longer be compensated. This result is of high value for clinical applications, since it suggests that a patient with an incomplete Circle of Willis and a severe narrowing of the carotid artery should be treated first. Such an analysis, yielding new knowledge for the field of application, is usually the end of the work process in Scientific Computing. This means that in this book we have gone the long way from studying a foreign field to the investigation of an interesting problem.

It is obvious that in this book only a few aspects of dimensional reduced models for blood flows have been treated. At several places in the submitted book, we referred to further literature and research topics. In Chap. 2 several modeling approaches have been briefly discussed. In context of 1D-0D coupled blood flow models, we have distinguished between closed and open models. As one can see in Chap. 3, only open loop models have been considered i.e. the whole blood circulation has never been studied in this book. However, if one is also interested in a detailed description of the pulmonary circulation and the heart, a closed model such as in [2] could be introduced. Another important aspect that has been omitted in terms of modeling, is the inclusion of viscoelastic effects. In [3, 4] different ways have been described how the existing differential equation system can be extended such that viscoelastic effects can be considered.

Wrt. possible numerical solution methods we have designed and analyzed two simple methods. Of course, there is a variety of other numerical methods. For example, one could use Discontinuous Galerkin methods for the space discretization and suitable time integrators as in [5, 6] to obtain a higher approximation accuracy. However, there is the risk of numerical artifacts such as oscillations in the vicinity of non-differentiable points. This can be prevented by means of stabilization techniques like slope limiters [7–10]. Thereby, one has to ensure that smooth extrema are not damped by stabilization techniques. However, this can be avoided by an adaptive damping strategy. Another interesting class of methods that have been used for the numerical solution of 1D models are machine learning techniques. Since the effort to produce training data by 1D models is low, a large amount of training data can be generated in short time. Two machine learning methods that have already been applied to 1D blood flow models are Physics Informed Neural Networks (PINNs) [11] or Kernel methods [12]. Physics Informed Neural Networks integrate the residuals of the model equations into the loss function. By this they directly incorporate the features of the mathematical model. Kernel methods represent the solution as a linear combination of kernel functions such as the Gaussian kernels. Unlike the weights and hyperparameters in a neural network, kernel methods optimize the coefficients of the linear combination and the shape parameters of the kernel functions.

Other aspects not covered in this book concern an efficient implementation of the solution methods and the use of effective parallelization techniques. Apart from the estimation of the Windkessel parameters, we have not discussed the calibration of model parameters. However, this is essential for the use of the numerical model in context of patient-specific simulations. A variety of calibration algorithms can be studied at this point. Besides classical solution methods [13], some quantum algorithms are also of great interest since their complexity is superior to classical methods. If the performance of quantum computers is further improved they could be applied to some interesting issues in future [14–17].

References

1. Hairer E, Wanner G, Lubich C (2006) Symplectic integration of Hamiltonian systems. In: Geometric numerical integration. Springer, pp 179–236
2. Acosta S, Puelz C, Rivière B, Penny D, Brady K, Rusin C (2017) Cardiovascular mechanics in the early stages of pulmonary hypertension: a computational study. Biomech Model Mechanobiol 16(6):2093–2112. Springer
3. Chen P, Quarteroni A, Rozza G (2013) Simulation-based uncertainty quantification of human arterial network hemodynamics. Int J Numer Methods Biomed Eng 29(6):698–721. Wiley Online Library
4. Čanić S, Tambača J, Guidoboni G, Mikelić A, Hartley C, Rosenstrauch D (2006) Modeling viscoelastic behavior of arterial walls and their interaction with pulsatile blood flow. SIAM J Appl Math 67(1):164–193. SIAM
5. Luo H, Baum JD, Löhner R (2008) A discontinuous Galerkin method based on a Taylor basis for the compressible flows on arbitrary grids. J Comput Phys 227:8875–8893
6. Puelz C, Čanić S, Rivière B, Rusin C (2017) Comparison of reduced models for blood flow using Runge–Kutta discontinuous Galerkin methods. Appl Numer Math 11:114–141. Elsevier
7. Köppl T, Wohlmuth B, Helmig R (2013) Reduced one-dimensional modelling and numerical simulation for mass transport in fluids. Int J Numer Methods Fluids 72(2):135–156. Wiley Online Library
8. Krivodonova L (2007) Limiters for high-order discontinuous Galerkin methods. J Comput Phys 226(1):879–896. Elsevier
9. Kuzmin D (2011) Slope limiting for discontinuous Galerkin approximations with possibly non-orthogonal Taylor basis. Int J Numer Methods Fluids 1:1–12

10. Kuzmin D (2010) A vertex-based hierarchical slope limiter for p-adaptive discontinuous Galerkin methods. J Comput Appl Math 233:3077–3085
11. Kissas G, Yang Y, Hwuang E, Witschey W, Detre J, Perdikaris P (2020) Machine learning in cardiovascular flows modeling: predicting arterial blood pressure from non-invasive 4D flow MRI data using physics-informed neural networks. Comput Methods Appl Mech Eng 358:112623. Elsevier
12. Koeppl T, Santin G, Haasdonk B, Helmig R (2018) Numerical modelling of a peripheral arterial stenosis using dimensionally reduced models and kernel methods. Int J Numer Methods Biomed Eng 34(8):e3095. Wiley Online Library
13. Ismail M, Wall W, Gee M (2013) Adjoint-based inverse analysis of windkessel parameters for patient-specific vascular models. J Comput Phys 244:113–130. Elsevier
14. Alberts GJN, Rol MA, Last T, Broer BW, Bultink CC, Rijlaarsdam MSC, Van Hauwermeiren AE (2021) Accelerating quantum computer developments. EPJ Quantum Technol 8(1):18. Springer, Berlin, Heidelberg
15. Moguel E, Berrocal J, García-Alonso J, Murillo JM (2020) A roadmap for quantum software engineering: applying the lessons learned from the classics. Q-SET@ QCE, pp 5–13
16. Quantum computing: progress and prospects. National Academies of Sciences, Engineering, and Medicine and others (2019). National Academies Press
17. Seidel R, Tcholtchev N, Bock S, Becker CK-U, Hauswirth M (2022) Efficient floating point arithmetic for quantum computers. IEEE Access 10:72400–72415. IEEE

Chapter 7
Data and Mathematical Tools

7.1 Leibniz Rule

For continuous differentiable functions χ, φ and f it holds:

$$\frac{d}{d\omega}\int_{\chi(\omega)}^{\varphi(\omega)} f(z,\omega)\, dz = \int_{\chi(\omega)}^{\varphi(\omega)} \frac{\partial}{\partial\omega} f(z,\omega)\, dz$$
$$+ f(\varphi(\omega),\omega)\frac{d}{d\omega}\varphi(\omega) - f(\chi(\omega),\omega)\frac{d}{d\omega}\chi(\omega).$$

7.2 Simplification of the Navier-Stokes Equations

We consider the incompressible Navier-Stokes equations modeling blood flow in a single cylindrical vessel and represent them by means of cylindrical coordinates (z, r, θ). The velocity field is denoted by $\mathbf{u} = (u_z, u_r, u_\theta)$. Moreover, we assume that flow is rotationally symmetric and the z-coordinate is aligned with the main axis of the vessel. From this it follows that $u_\theta = 0$. For this reason, we only need to consider the momentum equations for the r and z components:

$$\frac{\partial u_z}{\partial t} + u_r\frac{\partial u_z}{\partial r} + u_z\frac{\partial u_z}{\partial z} + \frac{1}{\rho}\frac{\partial p}{\partial z} = \mu\left[\frac{\partial^2 u_z}{\partial r^2} + \frac{1}{r}\frac{\partial u_z}{\partial r} + \frac{\partial^2 u_z}{\partial z^2}\right]$$
$$\frac{\partial u_r}{\partial t} + u_r\frac{\partial u_r}{\partial r} + u_z\frac{\partial u_r}{\partial z} + \frac{1}{\rho}\frac{\partial p}{\partial r} = \mu\left[\frac{\partial^2 u_r}{\partial r^2} + \frac{1}{r}\frac{\partial u_r}{\partial r} - \frac{u_r}{r^2} + \frac{\partial^2 u_r}{\partial z^2}\right].$$

μ is the viscosity parameter. The goal of this section is to show that the pressure p is constant with respect to the radial coordinate. In a first step we introduce the following characteristic quantities:

T. Köppl and R. Helmig, *Dimension Reduced Modeling of Blood Flow in Large Arteries*, Mathematical Engineering, https://doi.org/10.1007/978-3-031-33087-2_7

- U_r and U_z are the characteristic radial and axial velocities
- L is the characteristic length, R_0 is the characteristic vessel radius

The non-dimensional variables read as follows:

$$r = R_0\tilde{r},\ z = K\tilde{z},\ t = \frac{L}{U_z}\tilde{t},\ u_z = U_z\tilde{U}_r,\ p = \rho U_z^2\tilde{p}.$$

ρ is the density of blood. In the following, we assume that the parameter ε given by

$$\varepsilon = \frac{R_0}{L} = \frac{U_r}{U_z}$$

is small. This is a typical feature of larger arteries. For parts of the aorta is holds e.g.: $\varepsilon \in \mathcal{O}\left(10^{-2}\right)$. Moreover, we assume that it holds [1, Sect. 2.1]:

$$\frac{\mu}{R_0 U_z} \in \mathcal{O}(\varepsilon).$$

The r-component of the momentum equation in non-dimensional form reads as follows:

$$\begin{aligned}\frac{U_z}{L}\frac{\partial}{\partial\tilde{t}}(U_r\tilde{u}_r) + \frac{U_r\tilde{u}_r}{R_0}\frac{\partial}{\partial\tilde{r}}(U_r\tilde{u}_r) + \frac{U_z\tilde{u}_z}{L}\frac{\partial(U_r\tilde{u}_r)}{\partial\tilde{z}} + \frac{1}{\rho}\frac{U_z^2\rho}{R_0}\frac{\partial\tilde{p}}{\partial\tilde{r}} \\ = \mu\left[\frac{U_r}{R_0^2}\frac{\partial^2\tilde{u}_r}{\partial\tilde{r}^2} + \frac{U_r}{R_0^2\tilde{r}}\frac{\partial\tilde{u}_r}{\partial\tilde{r}} - \frac{U_r\tilde{u}_r}{R_0^2\tilde{r}^2} + \frac{U_r}{L^2}\frac{\partial^2\tilde{u}_r}{\partial\tilde{z}^2}\right].\end{aligned}$$

Division by U_z^2 and multiplying by R_0 yields:

$$\begin{aligned}\frac{U_r R_0}{U_z L}\frac{\partial\tilde{u}_r}{\partial\tilde{t}} + \frac{U_r^2\tilde{u}_r}{U_z^2}\frac{\partial\tilde{u}_r}{\partial\tilde{r}} + \frac{U_r R_0\tilde{u}_z}{U_z L}\frac{\partial\tilde{u}_r}{\partial\tilde{z}} + \frac{\partial\tilde{p}}{\partial\tilde{r}} \\ = \mu\left[\frac{U_r}{R_0 U_z^2}\frac{\partial^2\tilde{u}_r}{\partial\tilde{r}^2} + \frac{U_r}{R_0 U_z^2\tilde{r}}\frac{\partial\tilde{u}_r}{\partial\tilde{r}} - \frac{U_r\tilde{u}_r}{U_z^2 R_0^2\tilde{r}^2} + \frac{R_0 U_r}{L^2 U_z^2}\frac{\partial^2\tilde{u}_r}{\partial\tilde{z}^2}\right].\end{aligned}$$

Due to $\varepsilon = R_0/L = U_r/U_z$ and $\frac{\mu}{R_0 U_z} \in \mathcal{O}(\varepsilon)$ each term in the equation from above has a prefactor ε^2 apart from the term $\partial u/\partial r$. Thus it holds:

$$\frac{\partial\tilde{p}}{\partial\tilde{r}} = 0.$$

Therefore, the pressure is constant across the cross-sectional area.

7.3 Lemma 1

Let $g : [a, b] \to \mathbb{R}$ be a continuous function. Then there are for each $\varepsilon > 0$ step functions $\varphi,\ \Psi : [a, b] \to \mathbb{R}$ having the following properties [2]:

(a) $\varphi(z) \leq g(z) \leq \Psi(z),\ \forall z \in [a, b]$
(b) $|\varphi(z) - \Psi(z)| < \varepsilon,\ \forall z \in [a, b]$

A function $\varphi : [a, b] \to \mathbb{R}$ or Ψ is referred to as step function, if there is a decomposition

$$a = z_0 < z_1 < \cdots < z_{n-1} < z_n = b$$

of the interval $[a, b]$ and constants $c_1, \ldots, c_n \in \mathbb{R}$, such that:

$$\varphi(z) = c_k,\ \forall z \in [z_{k-1}, z_k),\ \varphi(b) = c_n,\ 1 \leq k \leq n.$$

7.4 Fundamental Lemma of Calculus of Variations

Let $I = [a, b] \subset \mathbb{R}$ be an interval and let $f : I \to \mathbb{R}$ be a continuous function. For each continuous differentiable function $g : I \to \mathbb{R}$ with $g(a) = g(b) = 0$ it holds:

$$\int_a^b f(z)g(z)\, dz = 0.$$

Then f is equal to the zero function [3].

7.5 Lemma 2

Let $I = [a, b] \subset \mathbb{R}$ be an interval and let $f : I \to \mathbb{R}$ be a continuous function. Furthermore, let $\mathcal{I}$ be the set of all intervals in I. If it holds

$$\int_{I_i} f(z)\, dz = 0, \quad \forall I_i = \left[z_1^{(i)}, z_2^{(i)}\right] \in \mathcal{I}, \tag{7.1}$$

then f is equal the zero function.

Proof Let $g \in C^1[a, b]$ be an arbitrary function with $g(a) = g(b) = 0$. We consider the following integral:

$$\int_a^b f(z)g(z)\, dz.$$

According to Lemma 1 there are two step functions φ and Ψ for g such that:

$$\varphi(z) \leq g(z) \leq \Psi(z), \ \forall z \in [a, b] .$$

We represent the step functions φ and Ψ for a decomposition

$$a = z_0 < z_1 < \cdots < z_{n-1} < z_n = b$$

as follows:

$$\varphi(z) = c_k, \ \forall z \in \left[z_{k-1}, z_k\right), \ \varphi(b) = c_n, \ 1 \leq k \leq n.$$

$$\Psi(z) = d_k, \ \forall z \in \left[z_{k-1}, z_k\right), \ \Psi(b) = d_n, \ 1 \leq k \leq n.$$

Then it holds:

$$\sum_{k=0}^{n} \int_{z_{k-1}}^{z_k} f(z) c_k \, dz \leq \int_a^b f(z) g(z) \, dz \leq \sum_{k=0}^{n} \int_{z_{k-1}}^{z_k} f(z) d_k \, dz$$

From Eq. (7.1) it follows:

$$\int_{z_{k-1}}^{z_k} f(z) d_k \, dz = d_k \int_{z_{k-1}}^{z_k} f(z) \, dz = 0, \quad \int_{z_{k-1}}^{z_k} f(z) c_k \, dz = c_k \int_{z_{k-1}}^{z_k} f(z) \, dz = 0$$

This means that:

$$\int_a^b f(z) g(z) \, dz = 0$$

and thus f is a zero function according to the **fundamental lemma of calculus of variations**. □

7.6 Derivation of Reynold's Transport Theorem

Let $V_t \subset \mathbb{R}^3$ be a volume that can change its shape in time and $f : \mathbb{R}^3 \to \mathbb{R}$ a continuous differentiable function. $\mathbf{u}$ denotes a velocity field, deforming the boundary of V_t and n is the outer normal. Then for the time derivative of

$$\Phi = \int_{V_t} f \, dV$$

it holds:

$$\frac{d\Phi}{dt} = \int_{V_t} \frac{\partial f}{\partial t} \, dV + \int_{\partial V_t} f \mathbf{u} \cdot n \, dS.$$

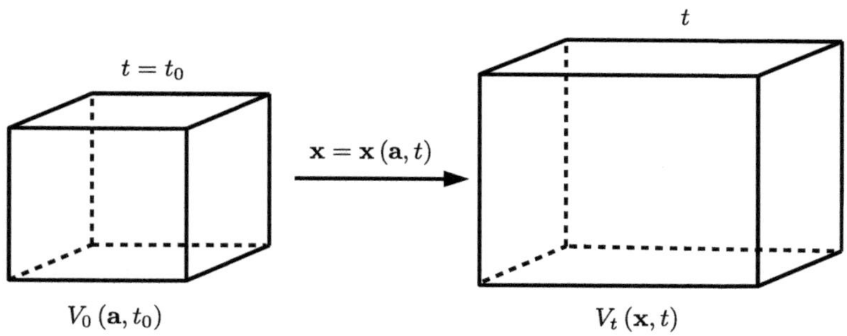

Fig. 7.1 Deformation of the volume V_0

Proof Let us consider a volume V_0 that is fixed in space for $t = 0$. The points in $V_t \subset \mathbb{R}^3$ are described by $\mathbf{x}$, while $\mathbf{a}$ describes the points in V_0. The determinant of the jacobian and the mapping assigning the vector $\mathbf{a}$ to the vector $\mathbf{x}$, govern the transformation of an infinitesimal volume dV_0 to an infinitesimal volume dV at time t (see Fig. 7.1).

$$dV = J \cdot dV_0.$$

It depends on the derivatives $\partial x_i / \partial a_j,\ x_i \in \{x, y, z\}$:

$$J = \left| \frac{\partial (x, y, z)}{\partial (a_x, a_y, a_z)} \right| = \begin{vmatrix} \frac{\partial x}{\partial a_x} & \frac{\partial x}{\partial a_y} & \frac{\partial x}{\partial a_z} \\ \frac{\partial y}{\partial a_x} & \frac{\partial y}{\partial a_y} & \frac{\partial y}{\partial a_z} \\ \frac{\partial z}{\partial a_x} & \frac{\partial z}{\partial a_y} & \frac{\partial z}{\partial a_z} \end{vmatrix} = \sum_{i,j,k} \epsilon_{ijk} \frac{\partial x}{\partial a_i} \frac{\partial y}{\partial a_j} \frac{\partial z}{\partial a_k}, \tag{7.2}$$

where ϵ_{ijk} is the permutation symbol (Levi-Civitá-Symbol). This symbol is a tensor of third order:

$$\epsilon_{ijk} = -\epsilon_{ikj} = -\epsilon_{jik} \text{ and } \epsilon_{123} = 1.$$

If two indices are equal, ϵ_{ijk} is equal to zero. Using this notation, we compute:

$$\begin{aligned} \frac{d\Phi}{dt} &= \frac{d}{dt} \int_{V_t} f\, dV = \int_{V_0} \frac{d}{dt} \left[f(\mathbf{x}(\mathbf{a}, t), t) \cdot J \right] dV_0 \\ &= \int_{V_0} \left[J \left(\left. \frac{\partial f}{\partial t} \right|_{\mathbf{x}} + \sum_i \frac{dx_i}{dt} \left. \frac{\partial f}{\partial x_i} \right|_t \right) + f \frac{dJ}{dt} \right] dV_0 \\ &= \int_{V_0} \left[J \left(\frac{\partial f}{\partial t} + (\mathbf{u} \cdot \nabla) f \right) + f \frac{dJ}{dt} \right] dV_0, \end{aligned}$$

where $\mathbf{u} = (u_1, u_2, u_3)^T$ denotes the velocity field deforming V_0:

$$u_i = \frac{dx_i}{dt}, \ x_i \in \{x, y, z\}.$$

Next, we compute the time derivative of J:

$$\begin{aligned}
\frac{dJ}{dt} &= \frac{d}{dt} \sum_{i,j,k} \epsilon_{ijk} \frac{\partial x}{\partial a_i} \frac{\partial y}{\partial a_j} \frac{\partial z}{\partial a_k} \\
&= \sum_{i,j,k} \epsilon_{ijk} \left(\frac{\partial u_1}{\partial a_i} \frac{\partial y}{\partial a_j} \frac{\partial z}{\partial a_k} + \frac{\partial x}{\partial a_i} \frac{\partial u_2}{\partial a_j} \frac{\partial z}{\partial a_k} + \frac{\partial x}{\partial a_i} \frac{\partial y}{\partial a_j} \frac{\partial u_3}{\partial a_k} \right).
\end{aligned}$$

Using

$$\frac{\partial u_k}{\partial a_i} = \sum_l \frac{\partial u_k}{\partial x_l} \frac{\partial x_l}{\partial a_i}, \ k \in \{1, 2, 3\} \text{ it follows:}$$

$$\frac{dJ}{dt} = \sum_{i,j,k,l} \epsilon_{ijk} \left(\frac{\partial u_1}{\partial x_l} \frac{\partial x_l}{\partial a_i} \frac{\partial y}{\partial a_j} \frac{\partial z}{\partial a_k} + \frac{\partial x}{\partial a_i} \frac{\partial u_2}{\partial x_l} \frac{\partial x_l}{\partial a_j} \frac{\partial z}{\partial a_k} + \frac{\partial x}{\partial a_i} \frac{\partial y}{\partial a_j} \frac{\partial u_3}{\partial x_l} \frac{\partial x_l}{\partial a_k} \right)$$

For the first term within the brackets we obtain:

$$\begin{aligned}
l = 1 &: \ \frac{\partial u_1}{\partial x} \frac{\partial x}{\partial a_i} \frac{\partial y}{\partial a_j} \frac{\partial z}{\partial a_k} \\
l = 2 &: \ \frac{\partial u_1}{\partial y} \frac{\partial y}{\partial a_i} \frac{\partial y}{\partial a_j} \frac{\partial z}{\partial a_k} \ \text{yields multiplied with } \epsilon_{ijk} \text{ and summation } 0 \\
l = 3 &: \ \frac{\partial u_1}{\partial z} \frac{\partial z}{\partial a_i} \frac{\partial y}{\partial a_j} \frac{\partial z}{\partial a_k} \ \text{yields multiplied with } \epsilon_{ijk} \text{ and summation } 0.
\end{aligned}$$

In case of the second term there is only one non-vanishing summand for $l = 2$. Similarly, for the third term, there is only a single non-vanishing summand for $l = 3$. The overall result is:

$$\frac{dJ}{dt} = \sum_{i,j,k} \epsilon_{ijk} \left[\frac{\partial u_1}{\partial x} + \frac{\partial u_2}{\partial y} + \frac{\partial u_3}{\partial z} \right] \frac{\partial x}{\partial a_i} \frac{\partial y}{\partial a_j} \frac{\partial z}{\partial a_k} = (\nabla \cdot \mathbf{u}) \, J.$$

Thus from Gaussian theorem the Reynold's transport theorem results:

$$\begin{aligned}
\frac{d\Phi}{dt} &= \int_{V_0} \left[\frac{\partial f}{\partial t} + (\mathbf{u} \cdot \nabla) \, f + f \, (\nabla \cdot \mathbf{u}) \right] J \, dV_0 \\
&= \int_{V_t} \left[\frac{\partial f}{\partial t} + \nabla \cdot (f\mathbf{u}) \right] dV = \int_{V_t} \frac{\partial f}{\partial t} \, dV + \int_{\partial V_t} f\mathbf{u} \cdot n \, dS.
\end{aligned}$$

7.7 Fourier Series

Each 2π-periodic function $f \in L^2(\mathbb{R})$ can be represented as a series

$$f(z) = \sum_{k \in \mathbb{Z}} c_k e^{ikz}, \text{ mit } c_k \in \mathbb{C},$$

where the coefficients c_k are defined as follows:

$$c_k = \frac{1}{2\pi} \int_{-\pi}^{\pi} f(z) \cdot e^{-ikz}\, dz, \ \forall k \in \mathbb{Z}.$$

A finite series with N summands has the following form:

$$f(z) = \sum_{k=-\frac{N}{2}}^{\frac{N}{2}-1} c_k e^{ikz}.$$

7.8 Weak Solutions of a Hyperbolic PDE

Let us consider a scalar hyperbolic problem:

$$\begin{aligned} \frac{\partial u}{\partial t} + \frac{\partial}{\partial z} f(u) = 0,\ z \in \mathbb{R},\ t > 0 \\ u(z, 0) = u_0(z),\ z \in \mathbb{R},\ t = 0, \end{aligned} \tag{7.3}$$

with: $u_0 \in L^\infty(\mathbb{R})$. The flux function is given by a function f

$$f : \mathbb{R} \to \mathbb{R},\ u \mapsto f(u).$$

Considering non-differentiable initial conditions, we require a weaker solution concept that is not base on $u_0 \in C^1(\mathbb{R})$. Thus in the following, it is assumed that: $u_0 \notin C^1(\mathbb{R})$.

Definition 7.1 A function $u \in L^\infty(\mathbb{R} \times [0, \infty))$ is weak solution of (7.3), if the integral equation

$$\int_{\mathbb{R}} \int_0^\infty u\Phi_t + f(u)\Phi_z\, dtdz + \int_{\mathbb{R}} u_0(z)\Phi(z, 0)\, dz = 0 \tag{7.4}$$

holds for all $\Phi \in C_0^\infty(\mathbb{R} \times [0, \infty))$ [4]. The functions Φ are called test functions and $C_0^\infty(\mathbb{R} \times [0, \infty))$ is the space of test functions.

Under certain conditions, it can be shown that (7.3) has a unique weak solution:

Definition 7.2 A weak solution u is called entropy solution, if the inequality holds

$$u(z+x,t) - u(z,t) \le C\left(1+\frac{1}{t}\right)z,$$

for a positive constant C and almost all $x, z \in \mathbb{R},\ t > 0, z > 0$.

Theorem 7.1 *If $f \in C^2(\mathbb{R}, \mathbb{R})$ and f is convex, then there exists a unique entropy solution u of* (7.3) *(up to a set of zero measures).*

A proof of this statement can be found e.g. in [4, Chap. 3.4, Theorem 3]. To better understand this solution concept, we show that the linear scalar transport equation with a discontinuous initial condition has a weak solution which is unique according to Theorem 7.1:

$$\frac{\partial u}{\partial t} + a \cdot \frac{\partial u}{\partial z} = 0,\ z \in \mathbb{R},\ t > 0,\ a > 0, \tag{7.5}$$

$$u(z,0) = u_0(z) = \begin{cases} 1 & \text{for } z \le 0, \\ 0 & \text{for } z > 0. \end{cases}$$

A hyperbolic problem (7.5) with an initial condition u_0 that is discontinuous at some places and piecewise constant, is called a Riemann problem. Now we show that the function $u(z,t) = u_0(z - a \cdot t)$ is a weak solution of (7.5). Obviously, it exhibits a discontinuity along the curve

$$I(z) = \begin{pmatrix} z \\ \frac{1}{a} \cdot z \end{pmatrix},\ z > 0.$$

I divides the space time domain $\mathbb{R} \times [0, \infty)$ in a domain Ω_l and a domain Ω_r. In Ω_l the function attains the value 1 and in the domain Ω_r the value 0. The function value 1 is propagated along the characteristic curves

$$t(z) = \frac{z - z_0}{a},\ z_0 \le 0$$

in the space time domain (see Figs. 7.2 and 7.3) from left to right. We decompose the first integral in (7.4) into an integral over Ω_l and Ω_r:

$$\int_{\mathbb{R}}\int_0^{\infty} u\Phi_t + f(u)\Phi_z\, dtdz = \int_{\Omega_l} u \cdot \Phi_t + a \cdot u \cdot \Phi_z\, dtdz + \underbrace{\int_{\Omega_r} u \cdot \Phi_t + a \cdot u \cdot \Phi_z\, dtdz}_{=0}.$$

The integral over Ω_l is now further transformed using partial integration in several space dimensions. For this purpose we set:

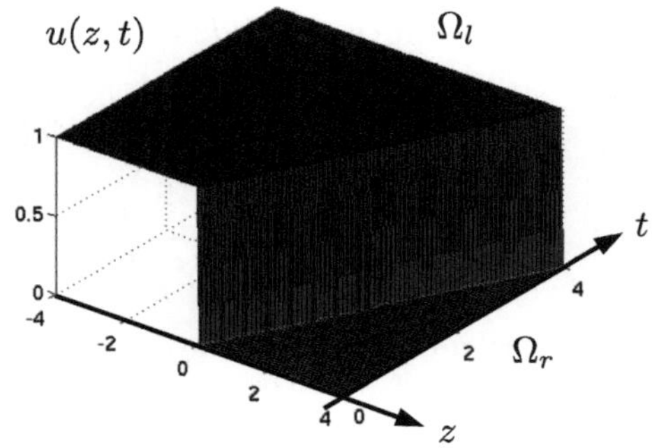

Fig. 7.2 Decomposition of the space time domain into Ω_l and Ω_r. In Ω_l the solution of Problem (7.5) attains the value 1, on Ω_r there is the value 0

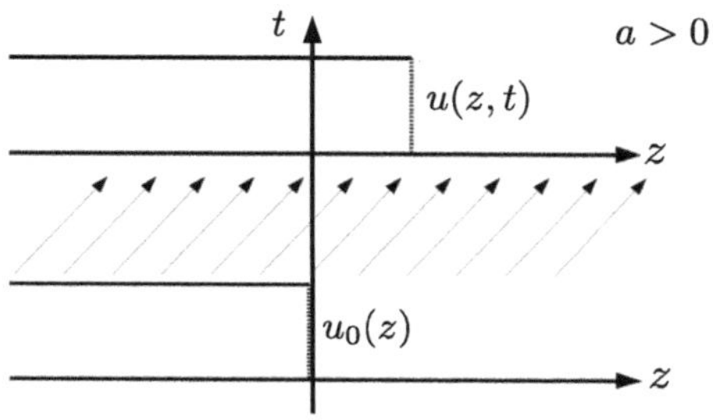

Fig. 7.3 Characteristic curves and movement of a function u_0 for $a > 0$ in the space time domain

$$\tilde{\mathbf{F}} = \begin{pmatrix} a \cdot u \\ u \end{pmatrix} \text{ und } \varphi = \Phi.$$

To compute the boundary integral wrt. Ω_l we have to consider the negative real axis $\mathbb{R}_-$ and the curve I, since the test function Φ has compact support (see Fig. 7.4). All in all, we obtain:

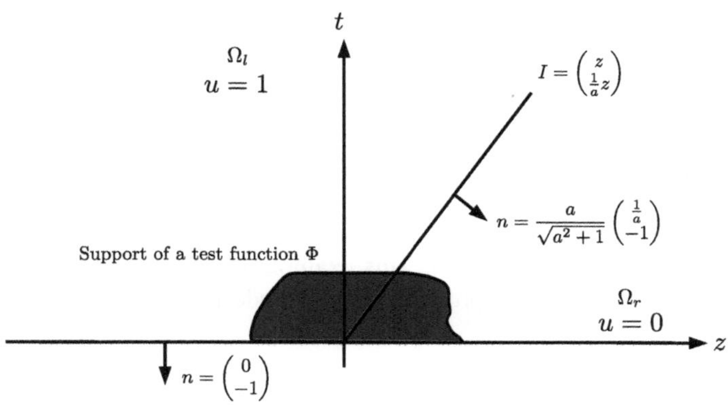

Fig. 7.4 Normals at the boundaries of Ω_l

$$\int_{\Omega_l} u \cdot \Phi_t + a \cdot u \cdot \Phi_z \, dtdz = \int_{\Omega_l} \begin{pmatrix} a \cdot u \\ u \end{pmatrix} \cdot \nabla \Phi \, dtdz \tag{7.6}$$
$$= \int_{\mathbb{R}_-} \Phi \begin{pmatrix} a \cdot u \\ u \end{pmatrix} \cdot n \, dS + \int_I \Phi \begin{pmatrix} a \cdot u \\ u \end{pmatrix} \cdot n \, dS$$
$$- \int_{\Omega_l} \Phi \operatorname{div} \begin{pmatrix} a \cdot u \\ u \end{pmatrix} dtdz.$$

The normal n on $\mathbb{R}_-$ is given by:

$$n = \begin{pmatrix} 0 \\ -1 \end{pmatrix},$$

while the normal on I is given by

$$n = \frac{a}{\sqrt{1+a^2}} \begin{pmatrix} \frac{1}{a} \\ -1 \end{pmatrix}.$$

From this, we obtain:

$$\int_{\mathbb{R}_-} \Phi \underbrace{\begin{pmatrix} a \cdot u \\ u \end{pmatrix}}_{u=1} \cdot \begin{pmatrix} 0 \\ -1 \end{pmatrix} dS = -\int_{\mathbb{R}_-} \Phi \, dS = -\int_{\mathbb{R}_-} \Phi \, u_0 \, dS = -\int_{\mathbb{R}} \Phi \, u_0 \, dS \tag{7.7}$$

$$\int_I \Phi \begin{pmatrix} a \cdot u \\ u \end{pmatrix} \cdot \frac{a}{\sqrt{1+a^2}} \begin{pmatrix} \frac{1}{a} \\ -1 \end{pmatrix} dS = \int_I \Phi \frac{a}{\sqrt{1+a^2}} (u - u) \, dS = 0. \tag{7.8}$$

Moreover, it holds due to $u \equiv 1$ on Ω_l:

$$\int_{\Omega_l} \Phi \operatorname{div} \begin{pmatrix} a \cdot u \\ u \end{pmatrix} dtdz = 0. \tag{7.9}$$

Combining (7.7)–(7.9) with (7.6), one obtains:

$$\int_{\mathbb{R}} \int_0^\infty u\Phi_t + f(u)\Phi_z \, dtdz = -\int_{\mathbb{R}} \Phi \, u_0 \, dz.$$

Thus u is a weak solution of (7.5). This example shows that even discontinuous functions can solve a Cauchy problem. To conclude this section, we make a few remarks:

- The weakening of the solution concept allows the inclusion of non-differentiable initial conditions and it facilitates the solution of problems with intersecting characteristic curves.

As an example one can consider Burgers equation:

$$\frac{\partial u}{\partial t} + \frac{\partial}{\partial z}\left(\frac{u^2}{2}\right) = 0$$

combined with the initial condition

$$u_0(z, 0) = \begin{cases} 1, & z \leq 0, \\ 1 - z, & 0 < z < 1, \\ 0, & z \geq 1. \end{cases} \tag{7.10}$$

For this problem the characteristic curves read as follows:

$$z(t) = \begin{cases} t + z_0, & z_0 \leq 0, \\ (1 - z_0)\,t + z_0, & 0 < z_0 < 1, \\ z_0, & z_0 \geq 1. \end{cases}$$

We observe that the characteristic curves are not intersecting for $t < 1$. For $t \geq 1$ the characteristic curves can intersect and thus discontinuities can occur. This means that the solution can not be considered as a classical solution anymore. Uniqueness of a physical meaningful solution can be obtained from a jump condition (Rankine-Hugoniot condition) and a entropy condition [4] (Fig. 7.5).

- The weak solution concept forms the basis for the derivation of numerical methods (DG methods and Finite Volume methods), which are not discretizing the PDE but an integral equation such as (7.4).
- Even simple examples show that characteristic curves are in general not given by straight lines. In case of non-linear problems the characteristic curves can be curved. Setting $a(z, t, u) = t \cdot z$ and $f \equiv 0$ in (4.45), then holds:

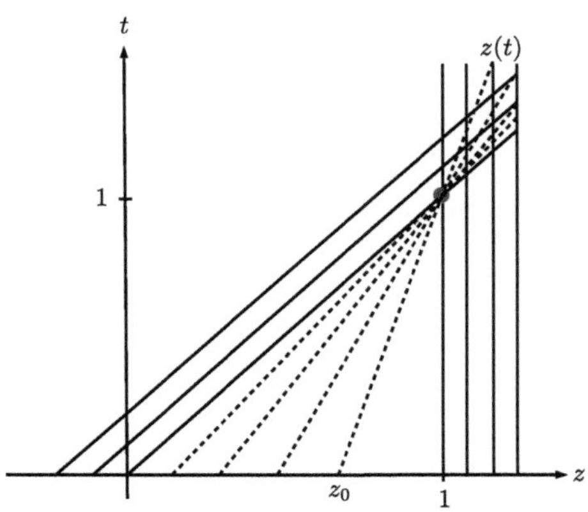

Fig. 7.5 Characteristic curves $z(t)$ for the Burgers equation combined with the initial condition (7.10)

$$z(t) = z_0 \cdot e^{\frac{t^2}{2}}.$$

- Since the system of Eq. (3.29a) exhibits non-linear terms, it is possible, as in case of the Burgers equation that shocks can occur. This has been analyzed in Sect. 4.4. It turns out in case of the blood flow model, a shock occurs after several meters. However, since we usually simulate shorter flow paths in arterial networks, we can assume that we can exclude the formation of shocks in terms of the blood flow problem.

7.9 Gaussian Theorem

Let $\Omega \subset \mathbb{R}^n$ be a compact set with a piecewise smooth boundary $\partial\Omega$. The outer normals on $\partial\Omega$ are given by n. In addition, let $\tilde{\mathbf{F}}$ be a continuous differentiable vector field that is defined on an open set V with $\Omega \subset V \subseteq \mathbb{R}^n$. Then the following relationship holds:

$$\int_\Omega \operatorname{div} \tilde{\mathbf{F}}\, dV = \int_{\partial\Omega} \tilde{\mathbf{F}} \cdot n\, dS.$$

7.10 Partial Integration in $\mathbb{R}^n$

Let $\Omega \subset \mathbb{R}^n$ be a compact set with a piecewise smooth boundary $\partial\Omega$. The outer normals on $\partial\Omega$ are given by n. In addition, let $\tilde{\mathbf{F}}$ be a continuous differentiable vector field that is defined on an open set V with $\Omega \subset V \subseteq \mathbb{R}^n$. φ is a continuous differentiable function on Ω. Then it holds:

$$\int_\Omega \varphi \operatorname{div} \tilde{\mathbf{F}}\, dV = \int_{\partial\Omega} \varphi \left(\tilde{\mathbf{F}} \cdot n\right) dS - \int_\Omega \tilde{\mathbf{F}} \cdot \nabla\varphi\, dV.$$

7.11 Taylor Expansion of a Vector Valued Function

Let us consider a function

$$f : \mathbb{R}^n \to \mathbb{R}^m,\ x = \begin{pmatrix} x_1 \\ \vdots \\ x_n \end{pmatrix} \mapsto \begin{pmatrix} f_1(x_1, \ldots, x_n) \\ \vdots \\ f_m(x_1, \ldots, x_n) \end{pmatrix}$$

as well as two vectors $x_0 \in \mathbb{R}^n$ and $x_0 + \Delta x \in \mathbb{R}^n$ such that

$$x_0 + \Delta x = \begin{pmatrix} x_{0,1} + \Delta x_1 \\ \vdots \\ x_{0,n} + \Delta x_n \end{pmatrix}.$$

Then it holds for a taylor expansion around x_0:

$$\begin{aligned} f(x_0 + \Delta x) &= f(x_0) + Df(x_0)\,\Delta x + \frac{1}{2} H_f(x_0)\,\Delta x^2 + \frac{1}{6} T_f(\xi)\,\Delta x^3 \\ &= f(x_0) + Df(x_0)\,\Delta x + \frac{1}{2} H_f(x_0)\,\Delta x^2 + \mathcal{O}\left(\Delta x^3\right). \end{aligned}$$

$\xi \in \mathbb{R}^n$ is a vector for which it holds:

$$x_{0,i} \leq \xi_i \leq x_{0,i} + \Delta x_i, \; i \in \{1, \ldots, n\}.$$

The different components are defined as follows:

$$\begin{aligned} \left(Df(x_0)\,\Delta x\right)_j &= \sum_{i=1}^{n} \partial_{x_i} f_j(x_0)\,\Delta x_i, \\ \left(H_f(x_0)\,\Delta x^2\right)_j &= \sum_{k=1}^{n} \sum_{i=1}^{n} \Delta x_k \partial^2_{x_k, x_i} f_j(x_0)\,\Delta x_i, \\ \left(T_f(\xi)\,\Delta x^3\right)_j &= \sum_{k=1}^{n} \sum_{i=1}^{n} \sum_{l=1}^{n} \Delta x_k \partial^3_{x_k, x_i, x_l} f_j(\xi)\,\Delta x_i \Delta x_l, \end{aligned}$$

with $j \in \{1, \ldots, m\}$.

7.12 Discrete Orthogonality

Let $\Delta z = \frac{2\pi}{N}$. Then it holds for $r - m \neq N$:

$$\Delta z \sum_{l=0}^{N-1} e^{i(r-m)l\Delta z} = 2\pi\,\delta_{rm}.$$

Proof For $r = m$ it follows directly:

$$\Delta z \sum_{l=0}^{N-1} 1 = 2\pi.$$

If $r \neq m$ it holds, one can show by means of the geometric series:

$$\sum_{l=0}^{N-1} e^{i(r-m)l\Delta z} = \frac{1-e^{-i(r-m)N\Delta z}}{1-e^{-i(r-m)\Delta z}} = \frac{1-e^{-i(r-m)2\pi}}{1-e^{-i(r-m)\Delta z}} = \frac{1-1}{1-e^{-i(r-m)\Delta z}} = 0.$$

The denominator in the geometric formula is different from zero, if it holds that $(r-m)\,\delta z = 2\pi$ or $r-m = N$. However, this condition is excluded by our assumptions. Thus the statement is proved.

7.13 Network Containing the Circle of Willis

Table 7.1 Data for the network in Fig. 5.1 [1]. The units of the elasticity parameters, peripheral resistances and capacities read as follows: $\left[10^6 \text{ Pa}\right]$, $\left[10^2 \text{ Pa s cm}^{-3}\right]$ and $\left[10^{-2}\text{cm}^3 \text{ Pa}^{-1}\right]$

ID	Length l [cm]	Radius r [cm]	Wall thickness h [cm]	E-Modul E	Periph. Res. R	Periph. Cap. C	Number grid nodes
1	4.0	1.20	0.163	0.4	–	–	11
2	2.0	1.12	0.126	0.4	–	–	7
3	3.4	0.62	0.080	0.4	–	–	11
4	3.9	1.07	0.115	0.4	–	–	11
5	20.8	0.25	0.063	0.4	–	–	51
6	8.35	0.25	0.063	0.4	–	–	23
7	3.4	0.423	0.067	0.4	–	–	21
8	15.6	0.999	0.11	0.4	1.8	0.387	46
9	3.4	0.423	0.067	0.4	–	–	11
10	17.7	0.15	0.038	0.8	54.3	0.0127	46
11	17.7	0.20	0.050	0.8	–	–	46
12	17.7	0.20	0.050	0.8	–	–	46
13	17.7	0.15	0.038	0.8	54.3	0.0127	46
14	14.8	0.136	0.034	0.8	–	–	33
15	42.2	0.403	0.067	0.4	26.8	0.0258	51
16	42.2	0.403	0.067	0.4	26.8	0.0258	51
17	14.8	0.136	0.034	0.8	–	–	33
18	0.5	0.2	0.05	1.6	–	–	5

(continued)

Table 7.1 (continued)

ID	Length l [cm]	Radius r [cm]	Wall thickness h [cm]	E-Modul E	Periph. Res. R	Periph. Cap. C	Number grid nodes
19	1.5	0.073	0.018	1.6	–	–	7
20	1.5	0.073	0.018	1.6	–	–	7
21	0.5	0.2	0.05	1.6	–	–	5
22	2.9	0.162	0.04	1.6	–	–	5
23	11.9	0.143	0.036	1.6	59.7	0.0116	26
24	11.9	0.143	0.036	1.6	59.7	0.0116	26
25	1.2	0.117	0.029	1.6	–	–	6
26	1.2	0.117	0.029	1.6	–	–	6
27	0.5	0.107	0.027	1.6	–	–	5
28	0.5	0.107	0.027	1.6	–	–	5
29	10.3	0.12	0.03	1.6	84.8	0.0082	21
30	10.3	0.12	0.03	1.6	84.8	0.0082	21
31	0.5	0.074	0.019	1.6	–	–	5
32	8.6	0.105	0.026	1.6	110.8	0.0062	21
33	8.6	0.105	0.026	1.6	110.8	0.0062	21
34	8.35	0.25	0.063	0.4	–	–	23

7.14 Data Set for Network 43

Table 7.2 Data for the network in Fig. 4.8 [1, 2]. The units of the elasticity parameters, peripheral resistances and capacities read as follows: $\left[10^6 \text{ Pa}\right]$, $\left[10^2 \text{ Pa s cm}^{-3}\right]$ and $\left[10^{-2}\text{cm}^3 \text{ Pa}^{-1}\right]$

Artery	Length l [cm]	Radius r [cm]	Wall thickness h [cm]	E-modulus E	Periph. Res. R	Periph. Cap. C
1	4.0	1.470	0.163	0.4	–	–
2	2.0	1.120	0.130	0.4	–	–
3	3.4	0.620	0.080	0.4	–	–
4	3.4	0.423	0.063	0.4	–	–
5	17.7	0.370	0.045	0.4	–	–
6	17.6	0.380	0.045	0.8	35.7868	$6.322 \cdot 10^{-3}$
7	17.7	0.380	0.045	0.8	35.7868	$6.322 \cdot 10^{-3}$
8	3.9	1.200	0.120	0.4	–	–
9	20.8	0.370	0.063	0.4	–	–
10	17.0	0.380	0.045	0.8	35.7868	$6.322 \cdot 10^{-3}$

(continued)

Table 7.2 (continued)

Artery	Length l [cm]	Radius r [cm]	Wall thickness h [cm]	E-modulus E	Periph. Res. R	Periph. Cap. C
11	17.7	0.380	0.045	0.8	35.7868	$6.322 \cdot 10^{-3}$
12	5.2	1.120	0.110	0.4	–	–
13	3.4	0.470	0.066	0.4	–	–
14	14.8	0.185	0.045	0.8	26.8401	0.109154
15	42.2	0.310	0.067	0.4	60.1	$3.0955 \cdot 10^{-3}$
16	14.8	0.185	0.045	0.8	26.8401	0.109154
17	42.2	0.310	0.067	0.4	60.1	$3.0955 \cdot 10^{-3}$
18	8.00	0.175	0.049	0.4	13.9	0.013384
19	10.4	1.017	0.100	0.4	–	–
20	5.3	0.920	0.090	0.4	–	–
21	2.0	0.588	0.064	0.4	13.9	0.013384
22	5.9	0.499	0.069	0.4	9.93563	0.020005
23	1.0	0.834	0.080	0.4	–	–
24	3.2	0.350	0.053	0.4	11.3	0.016464
25	1.0	0.794	0.080	0.4	–	–
26	3.2	0.350	0.053	0.4	11.3	0.016464
27	10.6	0.665	0.075	0.4	–	–
28	5.0	0.194	0.043	0.4	68.8	$2.7041 \cdot 10^{-3}$
29	1.0	0.631	0.065	0.4	–	–
30	5.9	0.470	0.060	0.4	–	–
31	5.9	0.470	0.060	0.4	–	–
32	14.5	0.320	0.053	0.4	–	–
33	5.0	0.200	0.040	1.6	79.36	2.33443
34	14.5	0.320	0.053	0.4	–	–
35	5.0	0.200	0.040	1.6	79.36	2.33443
36	44.4	0.235	0.050	0.8	–	–
37	12.7	0.260	0.047	0.8	47.7	$3.9003 \cdot 10^{-3}$
38	44.4	0.235	0.050	0.8	–	–
39	12.7	0.260	0.047	0.8	47.7	$3.9003 \cdot 10^{-3}$
40	32.2	0.190	0.045	1.6	47.7	$3.9003 \cdot 10^{-3}$
41	34.4	0.170	0.039	1.6	55.9	$3.3281 \cdot 10^{-3}$
42	32.2	0.190	0.045	1.6	47.7	$3.9003 \cdot 10^{-3}$
43	34.4	0.170	0.039	1.6	55.9	$3.3281 \cdot 10^{-3}$

References

1. Čanić S, Kim E (2003) Mathematical analysis of the quasilinear effects in a hyperbolic model blood flow through compliant axi-symmetric vessels. Math Methods Appl Sci 26(14):1161–1186. Wiley Online Library
2. Forster O (1978) Analysis. Springer
3. Leitmann G (2013) The calculus of variations and optimal control: an introduction, vol 24. Springer Science & Business Media
4. Evans L (1998) Partial differential equations. American Mathematical Society, Providence, Rhode Island

Glossary

Aneurysm: Local extension of the vessel wall by more than 50% of the vessel diameter. A rupture of an aneurysm can cause a severe internal bleeding and rapid death.

anterior: The front surface of an object.

Atheromatosis: Injury of an arterial vessel wall.

Arteriosclerosis: Accumulation of cholesterol, calcium phosphate, blood clots and connective tissue at or in a vessel wall, which can lead to narrowing or even occlusion of the affected blood vessel.

Arteriovenous malformation: Malformation of blood vessels, arising if arteries are directly connected to veins without a capillary bed.

Diastole: Relaxation phase in the cardiac cycle in which the ventricles are filled with blood. Besides the main ventricular diastole, there is also an atrial diastole. Diastole of the atria occurs during the systole of the ventricles.

distal: Distant from the heart.

Endothelium: Cell layer at the inside of a vessel. It acts as a mediator between blood flow and the rest of the vessel wall.

Erythrocytes: Also known as "red blood cells". This is the most frequent type of blood cells. Their main function is to transport oxygen from the lungs to the organs.

Hemoglobin: Iron-containing protein complex found in red blood cells, to which oxygen molecules can be attached.

Hematocrit value: Volume fraction of red blood cells with respect to the total blood volume. It determines the viscosity of blood.

Lesion: Injury or disruption of an anatomical object or body function.

T. Köppl and R. Helmig, *Dimension Reduced Modeling of Blood Flow in Large Arteries*, Mathematical Engineering, https://doi.org/10.1007/978-3-031-33087-2

Leukocytes: Also known as: "White blood cells". Type of blood cell that is produced in the bone marrow. It supports the immune system.

Metabolism: Chemical processes taking place in a cell. Thereby nutrients and oxygen are turned into energy and waste products.

Metabolites: Waste products of the metabolic process.

posterior: The backside of an object.

proximal: Close to the heart.

Stenosis: Narrowing of a blood vessel.

Stent: Short narrow tube consisting of a metal mesh. Stents are used to dilate narrowed vessels.

Systole: Phase in the cardiac cycle in which the ventricles contract. As a result blood is pushed into the pulmonary and systemic circulations.

Thrombus: Blood clot in a vessel.

Platelets: Blood cells causing blood clotting.

Vasoconstriction: Mechanism decreasing the diameter of a vessel. This is achieved by small muscles located in the vessel walls or wrapped around the vessel walls.

Vasodilatation: Mechanism that increases the diameter of a vessel (Vasodilatation is the counterpart of vasoconstriction).

Windkessel effect: Mechanism that mitigates the large difference between systolic and diastolic pressure. This is accomplished by the elasticity of the arterial walls near the heart (especially the aorta). The name is deduced from the early fire brigades, containing large chambers filled with air (Windkessel). The elasticity of the arterial walls makes it possible to store blood in the arteries for a certain period of time. Thereby a continuous supply of blood is ensured.

GPSR Compliance

The European Union's (EU) General Product Safety Regulation (GPSR) is a set of rules that requires consumer products to be safe and our obligations to ensure this.

If you have any concerns about our products, you can contact us on ProductSafety@springernature.com

In case Publisher is established outside the EU, the EU authorized representative is:

Springer Nature Customer Service Center GmbH
Europaplatz 3
69115 Heidelberg, Germany

Batch number: 09151316

Printed by Printforce, the Netherlands